Quantum Mechanics, Quantum Field Theory
geometry, language, logic

by R. Mirman

Group Theory: An Intuitive Approach
(Singapore: World Scientific Publishing Co., 1995)

Group Theoretical Foundations of Quantum Mechanics
(Commack, NY: Nova Science Publishers, Inc., 1995)

Massless Representations of the Poincaré Group
electromagnetism, gravitation, quantum mechanics, geometry
(Commack, NY: Nova Science Publishers, Inc., 1995)

Point Groups, Space Groups, Crystals, Molecules
(Singapore: World Scientific Publishing Co., 1999)

Quantum Mechanics, Quantum Field Theory
geometry, language, logic
(Huntington, NY: Nova Science Publishers, Inc., 2001)

Quantum Field Theory, Conformal Group Theory, Conformal Field Theory: Mathematical and conceptual foundations, physical and geometrical applications (Huntington, NY: Nova Science Publishers, Inc., 2001)

Wavefunction illustrations (with imaginary parts indicated by the color) on the covers are drawn with MAPLE, and enhanced, using instructions

on the front cover:

```
ys := exp(-((y+3.5)**2/3))-1.4*exp(-((y-3)**2/3));
ym := 1.2*exp(-((y+4)**2/11))+exp(-((y-4)**2/15));
wu := 15*exp(-(x+2)**2/0.75)*ym + 12*exp(-((x-2)**2)/1.02)*ys;
plot3d(wu,x=-4.5..4.5,y=-7.5..7.5,color=sin(x*y),style=patch,grid=[80,80],orientation=[45,60]);

ys2 := exp(-((y+3.5)**2/3))-1.4*exp(-((y-3)**2/3));
ym2 := 1.2*exp(-((y+4)**2/11))+exp(-((y-4)**2/15));
wu := 15*exp(-(x+2)**2/0.75)*ym2 + 12*exp(-((x-2)**2)/1.02)*ys2;
plot3d(wu*cos((y+2)/0.05),x=-5..3,y=-10..10,color=cos(x*y),style=patch,
grid=[80,80],axes=framed,orientation=[45,60]);

w1 := exp(-(x+3)**2/2)*exp(-((y+4)**2/5));
plot3d(w1*cos((y+2)/0.01),x=-4..0,y=-5..2,color=cos(x*y),style=patch, grid=[80,80],orientation=[45,60]);

wpl1 := plot3d(ws1,x=-15..15,y=-(0.001)..(0.001), grid=[80,80],orientation=[90,45]):
wpl3 := plot3d(ws3,x=-15..15,y=-(0.001)..(0.001), grid=[80,80],orientation=[90,45]):
wpl5 := plot3d(ws5,x=-15..15,y=-(0.001)..(0.001), grid=[80,80],orientation=[90,45]):
display({wpl1,wpl3,wpl5},color=maroon);
```

on the spine:

```
vl := 2; vb := 4;
zr := plot3d(vl-0.5,x=-1.02..0.85,y=-0.001..0.001,color=green,axes=none,thickness=3):
zrp := plot3d(vl-0.45,x=-1.02..0.85,y=-0.001..0.001,color=green,axes=none,thickness=3):
zrl := plot3d(vl-0.53,x=-1.02..0.85,y=-0.001..0.001,color=green,axes=none,thickness=3):
zrpp := plot3d(vl-0.43,x=-1.02..0.85,y=-0.001..0.001,color=green,axes=none,thickness=3):
br := plot3d(vb+0.548,x=0.85..3,y=-0.001..0.001,color=red,axes=none,thickness=3):
brp := plot3d(vb+0.528,x=0.85..3,y=-0.001..0.001,color=red,axes=none,thickness=3):
stf := plot3d(-cos(12*(x+1))+3,x=-1..0.84,y=-0.1..0.1,color=cos(x)*sin(x),axes=none,thickness=3):
stb := plot3d(4*exp(-x)+2.26,x=0.84..3,y=-0.1..0.1,color=sin(x),axes=none,thickness=3):
display(zrl,zr,zrp,zrpp,br,brp,stf,stb,orientation=[270,110]);
```

and on the back cover:

```
plot3d(exp(-x)*exp(-abs(y)**(1/4))*cos(x/0.05),x=-4.5..-1,y=-
2.3*abs(x**2.5)..2.3*abs(x**2.5),style=patch,color=4*cos(((x+1))**2),orientation=[255,70]);
```

Quantum Mechanics, Quantum Field Theory
geometry, language, logic

R. Mirman

AN AUTHORS GUILD BACKINPRINT.COM EDITION

Quantum Mechanics, Quantum Field Theory
geometry, language, logic

AN AUTHORS GUILD BACKINPRINT.COM EDITION

Published by iUniverse, Inc.

For information address:
iUniverse, Inc.
2021 Pine Lake Road, Suite 100
Lincoln, NE 68512
www.iuniverse.com

Originally published by Nova Science Publishers, Inc.

ISBN-13: 978-0-595-33690-6
ISBN-10: 0-595-33690-6

Printed in the United States of America

Preface

This book is likely to distress, even infuriate, many readers — it should. A purpose of this preface is to attempt to explain its objectives — not to mollify readers but, along with the distress and exasperation, to stimulate them. There is much in these subjects worth exploring, likely much more than generally realized. Such exploration is thus a task of potentially great value. But there are more important reasons.

P.1 Physicists are encouraging irrationality, and this must be fought

There is great interest in physics in the foundations of quantum mechanics and quantum field theory, the topics of most of the book (one aim is to erase the artificial distinction between them). But there is more motivation than simple presentation. If there are places at which the book seems polemical, than perhaps the number of these is insufficient. We all have a stake in the reputation, credibility and standards of science. And these are being undermined — by physicists. This is something that must be strongly opposed.

Scientists have severely negative opinions about crackpot subjects — astrology, flying saucers, alien abduction, creationism and so on — that are so popular. Indeed there has been much support for irrationalism in academic disciplines also, like literature, history, anthropology, sociology. There seems to be a postmodern attitude that science is a product of the power structure and that scientific laws are dependent on society. Almost all scientists, physicists, would agree that these beliefs must be vigorously fought. But many physicists rather than fighting such beliefs are working hard to encourage them.

Discussions of topics such as relativity, quantum mechanics, chaos, instead of trying to show their sensibility, often reinforce the belief that common sense cannot be relied on — anything is possible. When physicists loudly proclaim how weird quantum mechanics is, how bizarre is the universe, those who take crackpot subjects seriously, who want to believe that the universe really is really weird, are greatly encouraged, inspired, stimulated — producing far more nonsense, thus much harm, including to science.

Scientists have often been criticized by other scientists for failing to speak out in support of science, and against irrationality. But that is only part of the problem, worse is that scientists are speaking out — in support of irrationality, of the goofy, and are deceiving society. That is a fundamental cause of many of the problems in trying to get people to think and act rationally.

And there is much discussion about the need to lobby Congress, and so on. Yet ultimately influence on policy depends on our credibility; since most people do not understand physics, this is really all we have. If we lie to people about physics, why should they believe us? The attitude shown by the discussion of things like the Schrodinger's cat experiment (sec. II.2.c.ii, p. 69; sec. II.2.f, p. 73), is not only unethical, but in the end it could be disastrous. It must be exposed and fought.

Scientists often pander to the public's appetite for the weird in part because of the misguided belief that it stimulates public interest and support — but worse there are a lot of scientists who are themselves beguiled by the mysterious, the absurd. Instead of fighting nonsense, they are agreeing with and encouraging it. What should the public then think about science? If that is what we do then why complain about how it is misunderstood? Why complain about crackpots? Too many scientists, and those involved with science, encourage crackpots by saying things like the universe is bizarre, quantum mechanics is weird and so on.

Yet what is really weird is that so many physicists have a compulsion to flaunt their incompetence, their confusion, by proclaming how outlandish the universe is. This is something worth much study by psychologists, by psychiatrists, and would also make an interesting subject of study for sociologists of science. No competent physicist could ever possibly think that the universe is in any way strange.

Such behavior hurts all physics, all science. Outsiders do not distinguish different groups of physicists. Damage done by anyone hurts all, indiscriminately. No matter how unfair it is, what each does affects everyone else. All the nonsense about quantum mechanics may come back to haunt us when it is realized how nonsensical, how untruthful, it very much is.

We physicists are in no position to criticize improper or irrational behavior if ours is worse. We cannot complain about how little the public knows about science, if we misinform it — as we too often do, and not only in proclaimations about quantum mechanics. These discussions are truly a disservice to science [van Kampen (1991)].

Indeed many people object to scientists "playing God", something that seems ridiculous to real scientists. But when leaders of science, especially physicists, say that doing physics is trying to know God's thinking, doesn't that encourage people to believe that scientists are "playing God", and try to stop it [Crease (2000)]?

What is anyone doing to preserve the essential credibility of the scientific community? The silence is total. That is why it is so important to speak out so strongly. The anti-rationality being put forth by far too many physicists should be severely criticized. It is of paramount importance to combat such outlooks, and this must start with physicists, and the misconceptions about nature that they encourage; all such proclamations of the spookiness of nature — and by physicists — must be aggressively contested. The universe is certainly not weird. Fighting this nonsense is a main purpose of the book.

P.1.a Quantum mechanics is sensible; weirdness comes from carelessness

Standard, accepted, quantum mechanics is completely correct, completely reasonable, totally necessary [Mirman (1995b)], with no paradoxes, no weirdness, no real problems (even though it is nonintuitive). Although no one may understand quantum mechanics perfectly, it is really not difficult to grasp even though it is nonintuitive. All mysteries, strangeness, come only from carelessness and confusion. One goal of the book is to show this, point out errors, misuse of language, misunderstanding. It is hoped that this will clarify quantum mechanics and help readers, and through them others, to better understand it — or at least force people to be more careful, to think (and not only) before they say anything. And that leads to understanding. The medicine may be unpleasant, and many readers will resent it, but perhaps it will lead to improvements.

Quantum mechanics seems weird, not because it is, but because it isn't, because conclusions frequently drawn from it are just the opposite of what it gives, and in fact often violate its fundamental rules (sec. II.3.b, p. 76; sec. IV.2.e, p. 157; sec. IV.2.i, p. 161; sec. V.2, p. 201; sec. V.2.d, p. 205; sec. V.4.d.i, p. 229; sec. V.7.d.i, p. 257; sec. V.7.f.ii, p. 260; sec. V.7.f.iii, p. 261; sec. V.7.g, p. 262), seen perhaps most spectacularly with the EPR "paradox" (sec. V.3.e, p. 222). Results often mistakenly drawn from quantum mechanics can contradict its basic principles, and there are even conclusions drawn about physics that are in conflict with elementary mathematics, elementary algebra (sec. V.4.i, p. 238).

It is unlikely that anyone studying quantum mechanics really believes that Schrodinger's cat can be treated quantum mechanically, or disagrees with the discussions here (sec. II.2.c.ii, p. 69; sec. II.2.f, p. 73) — although most people will disagree with the way these are stated. However this actual view is not the impression given to others, who are lead to believe that the superposition of dead and alive states is really meaningful. There was a picture in a science magazine a few years ago of a cat alive and a cat dead, with the statement that a new view of quantum mechanics (which is wrong) frees the cat from this terrible situation. The people who talk about this experiment as if it should be taken seriously, and even those who just use such terms as "Schrödinger's cat state" for something like a SQUID, are not merely misleading others, but lying to them, and they know that they are doing so. We constantly complain about the public's lack of knowledge of science. But if we mislead them, lie to them, what do we expect?

There are many other cases: for example attempts to quantize gravity, are supported by governments — taxpayers — and have been reported many times in the media, and will be in the future. All specialists in gravitation know that general relativity is incompatible with quantum mechanics. How many can explain why they believe that or what the incompatibility is, or have references to discussions of these? Is there even a single one? Are there papers on the subject explaining this? Do people have even the slightest hint of why this should be true, or of what they want to do when they quantize gravity? How will they know when they have succeeded in what they are trying to do? How many know what they themselves are trying to do? Suppose that it turns out, as is very likely (sec. V.4.j.iv, p. 242), that this is all nonsense, how will that look? How much credibility will physicists, not merely those working in this field, still have? What will the reaction be to those who spend so much money to solve a nonexistent problem?

P.1.b The centrality of language

Quite related to the exposition and elimination of errors and misunderstandings in the discussions (but really not the interpretation) of quantum mechanics is the need to clarify the role of language. Misuse of language, or of what is presented as language, is a major cause of so much confusion (and of course, not only in quantum mechanics, not only in physics, not only in science, but in all of life).

Choice of names can often be harmful; many [Zurek (1997)] refer to quantum weirdness so often that it appears the name quantum mechanics has been changed. Soon there will be schools offering courses in quantum weirdness.

Stressed here very strongly is the proper, careful use of language. Precision of language and definitions is much emphasized. The book attempts to be very precise in definitions of terms, and of what is said, stating experimental meanings of concepts, even if these are very indirect. Perhaps this will, at least slightly, illustrate the proper use of language in physics, and be widely copied. Also, in a sense the book might serve as a rough sort of dictionary to provide definitions of some terms and concepts readers use (or at least to remind them that these terms can, and should, have definitions).

The standards of the book in the use of language are very high, and even though they have not been met (perhaps cannot be met) at least what has failed to be accomplished may motivate others to do better so raise the standards of all. This emphasis should spur readers to notice failings (in themselves and others) so sensitize them to be more careful with their own language. Sensitizing readers to the use and misuse of words should encourage them do better work, and is also valuable for students (especially if their teachers are sensitized), helping train them to do physics (and other work) carefully — and more productively. Physics is a liberal art, for one reason because it provides training in how to think — training that physicists too often do not use — and this should include how to think about and use language. Physics is a fine way of learning about language, and this increases its value as a liberal art. One aim of the book is to be a useful supplementary text, especially, but not only, in quantum mechanics (some material could be interesting to people concerned with language, like linguists and English professors).

Questions of language have been discussed before [Mirman (1995a), sec. P.4, p. xiii], and some points here are based on that.

The book starts with a discussion of the foundations of quantum mechanics and quantum field theory. In order to understand quantum mechanics it is necessary to briefly consider what might be called the philosophy of science, which is done at the beginning, emphasizing the — experimental, operational — meaning of terminology, that is how and whether language has content. The philosophy of science is too often treated in some abstract way, describing the views of a particular person writing about it (perhaps of idealizations of what that person believes it should be), but with no clear connection (if any) to actual science. It ought instead be regarded as an observational science, like astronomy, the approach here. We try to understand what science is — as it is practiced — and how that affects what we do in science, and how its implementation helps us understand, and causes us to misunderstand, the universe that we are trying to study.

Philosophers should find at least part of the book relevant, and provoking them might lead to clarification of the content of physical laws.

P.2 A fundamental view of the proper design of theories

There is one aspect of the philosophy that underlies these discussions that should be mentioned. The approach attempts to be very conservative, being based on what is well-known and strongly supported, and if leaving that going as small a distance as possible. In developing theories and systems whether to look at more speculative possibilities is a matter of judgment and taste. But these are not attractive ways of attacking physical problems, especially if there is no indication that they are necessary, and are highly conjectural. The safest approach is to start with what is known and build on that. Otherwise the number of possibilities becomes very large, and it is unlikely that ones chosen are correct or relevant, so the chances are great of wasting much time. It is best as seen in all these books [Mirman (1995a,b,c,1999)] to start with what is already strongly validated. Of course going beyond that, at times necessary, requires risks. But since everything is close to what is known to be correct, theoretically and experimentally, these risks are much less.

There is another reason for disliking speculative work. Physics requires more than, say, a set of equations for a particular phenomenon. It may be possible to find ones that seem reasonable, but cannot possibly be correct because underlying assumptions (like the dimension) are not possible, because resultant theories are incomplete, and incompletable [Mirman (1995b,c)]. Great care in needed with change — everything can unravel.

The history of physics (and really all science) shows that advances are made not by daydreams, not by wishful thinking, but by small, often accidental, steps, with strong rationales, making careful changes in existing theories, just enough to get agreement with experiment. And then many times these lead to the most unexpected, startling, consequences.

Thus Planck made a minor mathematical modification in the derivation of the blackbody radiation formula, merely changing an integral to a sum, and to his regret started (but only started) physics on the road to theories then inconceivable. Daydreamers would never have thought of these.

Bohr took the classical model of the atom and modified it with arbitrary assumptions guessed at to give the correct experimental results, and developed a model (inconsistent) of the hydrogen atom (only). And that, by a series of steps, lead to quantum mechanics, a theory unimaginable in Bohr's time.

Special relativity was developed, not from wishful thinking, but from necessary steps. The mathematics was found by trying to use classical theories, just adding some assumptions needed to get correct formulas — the Lorentz transformations. Einstein then realized the significance of these.

Dirac tried to find a mathematical solution to a well-defined — and unavoidable — problem: what is the correct generalization of Schrödinger's equation compatible with relativity, linear in the time, and whose solutions obey the Klein-Gordon equation? This had to be done, it was just a matter of finding the right mathematics. And, among other consequences, it predicted anti-matter. (Actually this was not the generalization, but a generalization. The most general equations are given by the Poincaré invariants and the labeling operators (sec. III.3.d, p. 127).)

What is striking about these, and so many other, discoveries, is that they could not have been found much earlier, perhaps five years, (very rarely) ten years at the most. The information need for them was not available. Important discoveries are made, not because brilliant people suddenly appear, but because they become ripe for realization, because people are available to make them, but people are always available. When discoveries are ready to be made, people are there to do so.

Genius consists, to a large extent, of being in the right place at the right time, and having a right place to be in at just the right time — and the possession of the specific abilities needed at that time and place.

The history of science strongly emphasizes this.

Physics is not done by having everyone get together and wishing as hard as possible. Too often (what is claimed to be) physics is driven by mob psychology. But no matter what drives physicists, mob psychology does not drive nature.

Much too frequently physicists are interested in only trying (and then getting very excited about) the wildest, most implausible systems, the ones furthest from what is well-supported, furthest from experiment, least likely to be correct, rather than the best supported, most credible ones, those most likely to succeed.

While the history of physics shows that the soundest, most productive criteria are experiment, logic and well-supported concepts and theories, physicists often seem to be searching for any criteria but these. One currently fashionable word is "beauty", with the belief that theories are correct if they are "beautiful". Of course beauty is in the eyes of the beholder (and in physics usually the holder of grants). What in our culture might be regard as very attractive, in the culture of nature can be very unattractive. Nature need not agree with our aesthetics (no matter how much we belief it should), and what nature finds beautiful is correct physics. Although these ideas are widespread, too widespread, there are wiser people [Gingerich, (2000)] who realize that aesthetics can be dangerously seductive, possibly sheer quicksand for the unwary.

Indeed it is useful to ask why these strange concepts have become so elevated in discussions of physics [Drummond (2000)]. Unfortunately the question of "Is it true?" has been replaced by ones like "Is it elegant?". Such correct questions play ever decreasing roles in modern physics, and misleading ones ever more. Let us hope this does not go on for evermore, otherwise physics will not.

Here then we try to eliminate speculation, unfounded, unlikely notions, and present arguments that seem logically unavoidable. The reader may find it useful to try to discover holes in them. That will provide opportunities to improve the logic and arguments, so to help create a stronger, perhaps unavoidable, foundation for physics.

The purpose of reading this book should not be to absorb, but to disagree and to think — and then go beyond.

P.3 The underlying theme of the book

The subjects of these books may seem somewhat disparate, but all are related to the underlying aim: greater understanding of laws of physics, and reasons for them. What they provide is of course not complete. But part of what they impart are hints, suggestions, and some deciphering of these. Perhaps these will aid in gathering further understanding, in finding ways of going further.

A fundamental view of this book is that physics is determined (completely?) by the geometry in which it takes place, and that this geometry is determined by physics being able to exist in it, these related and expressed by group theory [Mirman (1995b,c)], the theory of the transformation group of our space, the Poincaré group, and as we see the conformal group, and their representations [Mirman (2001)].

These we wish to probe, both mathematically and physically. But physics is more than a set of experiments, it involves relating our thinking and our experiences. Thus to gain insight into physics we need insight into how we describe our experiences, in particular our interactions with what we believe is the external world. Our descriptions, sometimes accurate but too often misleading, are really what we think are physical laws. So to infer laws we must know why we describe them as we do, and why these descriptions are too frequently meaningless and confusing. These then form a large part of our study.

Much of this, the Poincaré group, quantum mechanics, quantum field theory, is well-known, even if incorrectly known. But we can go further and try to explore a subject for whose relevance there are compelling arguments, the conformal group — this being much less well-known, gives much to explore. And unfortunately there is so much that we can only point out why it is relevant, and suggest some of the questions whose consideration likely will lead to deeper insight into group theory, mathematics, and, especially, physics [Mirman (2001)].

Essential purposes of this whole discussion are to show how all these topics are related, and how all shed light on what physics is, and why it must be that way, and, perhaps most, to indicate how to go further and deeper, and to encourage and stimulate thoughtful and adventurous readers.

P.4 ACKNOWLEDGEMENTS

Comments, communications, criticisms, disagreements, papers, suggestions and other information, even encouragement, over the last few years from various people have stimulated thinking that lead to some of the present commentaries. These include Dharam V. Ahluwalia, D. M. Appleby, Dan Baleanu, John Barrow, Carl H. Brans, Ariel Caticha, George Chapline, Richard Cleve, Scott Dodelson, S. Goldstein, Mark J. Hadley, Robert C. Hilborn, Werner Hofer, Bernard Jancewicz, Vladimir Kisil, Bernd Kuckert, John Leslie, Pertti Lounesto, Elihu Lubkin, Gloria Lubkin, Malcolm MacCallum, Philip Mannheim, John Norton, Robert L. Park, David I. Santiago, Phillip F. Schewe, Anwar Y. Shiekh, Dan Solomon, Andrew Steane, Victor J. Stenger, Leo Stodolsky, Mark Stuckey, James T. Wheeler, Raj Vatsya, H. Dieter Zeh, and particularly Ruth E. Kastner and N. David Mermin, whose influence is seen at several places. Many of these to whom gratitude is due will themselves be grateful to know that their contribution has been disagreement. But disagreements can often be quite stimulating.

And for much help, this is to thank the Department of Natural Sciences of Baruch College (City University of New York).

Table of Contents

I Preface v

P.1 Physicists are encouraging irrationality, and this must be fought . . . v

P.1.a Quantum mechanics is sensible; weirdness comes from carelessness . . . vii

P.1.b The centrality of language . . . viii

P.2 A fundamental view of the proper design of theories . . . ix

P.3 The underlying theme of the book . . . xii

P.4 ACKNOWLEDGEMENTS . . . xii

I Language, concepts and physics 1

I.1 Geometry, field theory, and the description of nature . . . 1

I.1.a Necessity for examining foundations . . . 2

I.1.b What is reality, and is it only classical? . . . 2

I.1.b.i *What criteria should we use for reality?* . . . 3

I.1.b.ii *Is there really a difference between quantum reality and classical reality?* . . . 4

I.1.b.iii *Does the universe exist if we do not observe it?* . . . 4

I.2 Why some words have meaning . . . 5

I.2.a Constructs are needed for coherence . . . 5

I.2.b Space and coordinates are constructs . . . 6

I.2.c Language also varies from precise to nonsense . . . 7

I.2.d Meaning is pragmatic . . . 7

I.2.e How electrons differ from virtual particles . . . 7

I.2.f Meaning comes from being a (necessary) part of a web . . . 8

I.2.f.i *Measuring instruments* . . . 8

I.2.f.ii *Is a concept needed?* . . . 9

I.2.f.iii *Quantities defined indirectly* . . . 10

I.2.f.iv *What does the phrase "very small distances" say?* . . . 10

I.2.f.v *Hollow rigorous formalisms* . . . 11

I.2.g Physics does have meaning . . . 12

I.3 Quantum field theory and what we need to study it . . . 12

I.3.a The need for interpretation . . . 12

I.3.b The relevance of the conformal group as a transformation group . . . 13

I.3.c Nonintuitive, thus confusing . . . 13

I.3.d Why do we disagree? . . . 14

I.4 Frameworks and theories . . . 14

I.4.a In science there are many frameworks, and far more theories . . . 15

I.4.a.i *Classical physics and quantum mechanics as frameworks* 15
I.4.a.ii *Ptolemic and Copernician views and evolution are frameworks* 15
I.4.a.iii *Quantum field theory is a framework* 16
I.4.b Theories are formed within a framework 16
I.4.c Frameworks are not true or false, but useful or not 17
I.4.c.i *Criteria differ for theories and frameworks* 18
I.4.c.ii *Usefulness and correspondence to reality* . 18
I.4.c.iii *Tools, not truths* 19
I.4.c.iv *Decisions about value are subjective* 19
I.4.c.v *Theories, physical and phenomenological* . 20
I.5 Underlying assumptions 20
I.5.a Are there assumptions if there can be no other universe? 21
I.5.b The necessity for regularity, and the need for inertia 22
I.5.c Assumptions, formalism and language 23
I.5.c.i *Assumptions come from language* 23
I.5.c.ii *Where are the assumptions and what can we do about them?* 24
I.5.c.iii *What can be postulated about quantum mechanics?* 24
I.5.d Is physics based on our biology? 26
I.5.e Why must the world be what it is? 27
I.5.f The wonder of simplicity 27
I.6 Fundamental definitions and their rationales 28
I.6.a What is a geometry? 28
I.6.b Why space is flat 30
I.6.c Space, time, now, then and causality 31
I.6.d What is a field theory? 32
I.6.e What do we mean by coupling constants? 33
I.7 Transformations and groups, their meaning and use . . . 34
I.7.a Reference systems, geometry and physics 35
I.7.a.i *Coordinate systems, and how they are given* 35
I.7.a.ii *Physical coordinate system are needed* . . . 35
I.7.a.iii *Defining geometry, mathematically and physically* 36
I.7.a.iv *Physical theories must make physical sense, not be rigorous mathematically* 37
I.7.a.v *Starting with mathematics, then going to physics* 37
I.7.b What is a transformation group? 37
I.7.b.i *Transformations define the points of the geometry* 38
I.7.b.ii *Defining the Poincaré group* 38

I.7.b.iii *Inhomogeneous groups and solvable groups* 39
I.7.b.iv *The relevance and implications of transformation groups* 39
I.7.b.v *Invariance and transformation groups* . . . 40
I.7.b.vi *Invariance, covariance and gravitation* . . . 41
I.7.b.vii *What is the effect of translations with interactions?* 41
I.7.b.viii *Why transformations determine physics* . . 42
I.7.b.ix *Consistency and group theory* 43

I.8 Why states belong to representations, even without invariance 44
I.8.a Statefunctions are basis states 44
I.8.a.i *Statefunctions are basis states even without invariance* 45
I.8.a.ii *Statefunctions are physically required to be basis vectors* 46
I.8.a.iii *Why angular momentum values are limited* 47
I.8.b Why fundamental objects are given by irreducible representations 48
I.8.b.i *Representations of what groups are required to be irreducible?* 48
I.8.b.ii *How reducible and irreducible representations describe objects differently* 49
I.8.b.iii *The representation is irreducible* 50
I.8.b.iv *Differences between elementary and composite objects* 50
I.8.b.v *Interactions lead to irreducible representations* 51
I.8.b.vi *Translation groups and how they differ* . . 52
I.8.b.vii *Why the deuteron is composite* 53

II Foundations of quantum theory **54**
II.1 Comprehensibility of quantum theory 54
II.1.a The reasons for physical laws are based on logic and experiment, not history or the occult 55
II.1.b Complex numbers 55
II.1.b.i *Complex numbers from group generators* . 56
II.1.b.ii *Implausibility of other number systems* . . . 56
II.1.b.iii *Hermiticity and unitarity* 58
II.1.b.iv *Why probability is relevant to hermiticity* . 58
II.1.c Probability and the absolute square of statefunctions 60
II.1.c.i *The phase cannot appear* 60
II.1.c.ii *How probability comes from the square* . . 61
II.1.c.iii *Probability is proportional to energy* 61
II.1.c.iv *How the formalism gives energy, so probability* 62

II.1.d Gleason's theorem 63
II.2 Wavefunctions don't collapse, oversimplifications do . . . 64
II.2.a Processes are always continuous 65
II.2.a.i *How we make it seem as if wavefunctions collapse* 65
II.2.a.ii *Measurement and collapse* 66
II.2.a.iii *What is a measurement, and why they occur continuously* 67
II.2.b Wavefunctions must be of complete systems, including brains 67
II.2.c Macroscopic states do not distinguish different microscopic wavefunctions 68
II.2.c.i *Statefunctions for microscopic and classical systems are fundamentally equivalent* . . . 69
II.2.c.ii *Macroscopic objects prevent interference* . 69
II.2.c.iii *Does observation force a quantum mechanical system into a state?* 70
II.2.c.iv *When does a recording occur?* 71
II.2.c.v *How a measurement takes place* 71
II.2.d Does consideration of brains lead to an infinite regression? 72
II.2.e Decoherence 72
II.2.f Experiment picturing difference between classical and quantum superposition 73
II.2.g Transition between quantum mechanics and classical physics based on rapidity 74
II.3 There can be no discontinuities in quantum mechanics . . 75
II.3.a Can a coherent sum go into an incoherent one? . . 75
II.3.b No jumping allowed 76
II.3.b.i *The unfortunate name "quantum mechanics" leads to jumps in thought* 76
II.3.b.ii *Leaping without looking* 76
II.3.b.iii *Jumping into misleading language* 77
II.3.c Can an object go everywhere instantaneously? . . . 78
II.3.c.i *A classical analog for instantaneous momentum eigenstates* 79
II.3.c.ii *Parallels to classical physics should always be looked for* 80
II.4 Only inept use of quantum mechanics gives paradoxes . . 80
II.4.a A complete, realistic experiment is needed to show a paradox 80
II.4.b Superposition 81
II.4.c Half-integer spin 82
II.5 Quantum field theory 82
II.5.a What is a field? 83

II.5.a.i *What is the primary ontology?* 84
II.5.a.ii *Expressions for fields* 84
II.5.a.iii *Other expressions* 85
II.5.a.iv *Justifying these expressions* 85
II.5.b What do we mean by "single particle"? 86
II.5.b.i *How one photon differs from two* 87
II.5.b.ii *Product states and single states are distinguishable* 88
II.5.b.iii *How functional forms of single and multiple Poincaré states differ* 89
II.5.b.iv *Amplitudes, energy and particle number* . . 90
II.5.b.v *Wavepackets* 90
II.5.b.vi *Why discreteness?* 91
II.5.b.vii *Why is energy change equal to an integer times the frequency?* 92
II.5.b.viii *Does a photon interfere with itself?* 93
II.5.b.ix *Double slits* 93
II.5.b.x *Linear polarizers* 94
II.5.c Gauge transformations 95
II.5.c.i *Gauge transformations act on space coordinates* 95
II.5.c.ii *Gauge transformations are not space-dependent coordinate transformations* 96
II.5.c.iii *Gauge invariance and charge conservation* 96
II.5.c.iv *What field is the electromagnetic field?* . . . 97
II.5.c.v *Duality* 97
II.5.d Misconceptions about quantum field theory 97
II.5.d.i *Second quantitization* 98
II.5.d.ii *Quantum field theory is not a result of relativity* 98
II.5.e What are particles and what are forces? 98
II.5.e.i *Product states must be symmetrized, giving the real difference between bosons and fermions* 99
II.5.e.ii *Fermions are more real only intuitively* . . . 100
II.5.e.iii *What can we see and feel?* 101
II.5.f Is field theory consistent? 101
II.5.f.i *Inconsistencies of classical physics are irrelevant* 102
II.5.f.ii *How do we tell if a theory is consistent?* . . 102

III Statefunctions and Probability, What and Why **104**
III.1 Some fundamental concepts of quantum theory 104
III.1.a Is quantum mechanics complete? 105
III.1.b Is quantum mechanics mysterious? 105
III.2 Necessity for probability 106

III.2.a Experiment requires probability 106
III.2.b Required knowledge does not exist 107
III.2.c Geometrical reasons for probability 107
III.2.d Quantum mechanics requires ensembles 108
III.2.e How do we deal with individual systems? 109
III.2.f What is probability really? 110
III.2.g Is there a wavefunction of the universe? 111
III.2.h Uncertainty principles 112
III.2.h.i *Ensembles and the uncertainty principle* . . 112
III.2.h.ii *Uncertainty produced by a microscope* . . . 113
III.2.h.iii *The uncertainty principle is a consequence of the expandability of statefunctions* 114
III.2.h.iv *Uncertainty and individual measurements* . 115
III.2.i How does probability become certainty? 116
III.3 What is a statefunction? 117
III.3.a Does a statefunction report about us, or about a system? 118
III.3.b The completeness of the description given by a statefunction 120
III.3.c What we really mean by statefunctions 120
III.3.c.i *Definitions are determined by use* 121
III.3.c.ii *What is a physical object?* 122
III.3.c.iii *Physical objects and their mathematical descriptions are not the same* 123
III.3.c.iv *Physical objects must be statefunctions* . . . 124
III.3.c.v *When is a basis vector a statefunction?* . . . 124
III.3.c.vi *Are we relevant?* 125
III.3.c.vii *How an electron differs from its statefunction* 126
III.3.c.viii *How do electrons, statefunctions and formulas differ?* 126
III.3.d Multiparticle states 127
III.4 Reality is mathematical 129
III.4.a We can be replaced by robots 129
III.4.b The apparent reality of the classical world 130
III.4.b.i *Is there an external reality?* 131
III.4.b.ii *Why the world seems as it does* 132
III.4.c Artificial intelligence is completely artificial 133
III.4.c.i *The immense complexity of brains* 133
III.4.c.ii *Intelligence, lack of intelligence and language* 135
III.5 The Hamiltonian, its construction and necessity 135
III.5.a Why the energy is the Hamiltonian eigenvalue . . . 136
III.5.a.i *Schrödinger's time-dependent equation* . . . 137

III.5.a.ii *Lagrangians and action principles need not exist* 137
III.5.b Construction of a Hamiltonian 138
III.5.b.i *Relativistic wave equations* 138
III.5.b.ii *Using creation and annihilation operators for Hamiltonians* 138
III.6 Topology and geometry 139

IV Language and its dangers **142**
IV.1 Misuse of Language 142
IV.1.a The name's to blame 143
IV.1.a.i *Classical physics can be quantized, quantum mechanics not* 144
IV.1.a.ii *Statefunction is a better term than wavefunction; why?* 144
IV.1.a.iii *Defining statefunction by use* 144
IV.1.a.iv *Is a force an interaction?* 145
IV.1.b Meaningless statements often seem meaningful, so confuse 146
IV.1.c Questions meaningful and meaningless 146
IV.1.c.i *What questions actually ask something?* . . 147
IV.1.c.ii *Examples of questions lacking content* . . . 147
IV.1.c.iii *How meaningful, and meaningless, questions fundamentally differ* 149
IV.1.c.iv *Unhappiness with inability to answer meaningless questions* 149
IV.1.c.v *Questions that do not seem to have meaning but may* 150
IV.1.d Can physics be replaced by logic? 151
IV.1.d.i *Validity of self-validation* 152
IV.1.d.ii *Providing physical sense is quite difficult* . . 152
IV.1.d.iii *Is mathematics better than physics?* 153
IV.2 Examples of statements meaningless, misleading, and often nonsense 154
IV.2.a Can a system be in all states at once? 154
IV.2.b Can an object exist in many quantum states simultaneously? 155
IV.2.c Is the brain in a bizarre superposition? 156
IV.2.d Do superpositions cling? 157
IV.2.e Is there any point to point particles? 157
IV.2.f The meaning of the wavefunction 158
IV.2.f.i *How real is probability?* 159
IV.2.f.ii *What is the probability that the universe has a wavefunction?* 160
IV.2.g Can statements be in two places at the same time? 160
IV.2.h Is inaccessible information really information? . . . 161

IV.2.i How fundamental is complementarity? 161
IV.2.j Time 162
IV.2.j.i *Can time have a history?* 162
IV.2.j.ii *Is the universe bored?* 163
IV.2.j.iii *Going backwards in time, and backwards in thinking* 163
IV.2.j.iv *Pointless properties of points* 164
IV.2.j.v *Does time flow?* 164
IV.2.k Uncertainty, long may it wave 164
IV.2.l How excited are electrons? 165
IV.2.l.i *Why are all electrons the same?* 166
IV.2.l.ii *Babbling* 166
IV.2.l.iii *Photons also are overexcited* 166
IV.2.l.iv *Electron fields illustrate the harm of meaningless strings of words* 167
IV.2.l.v *Are fields permanent?* 167
IV.2.m Self-energy 168
IV.2.m.i *Are mass differences due to the electromagnetic interaction?* 168
IV.2.m.ii *What does experiment show?* 169
IV.2.n Uncertainties with analogies 169
IV.2.o Is there any value in having values? 170
IV.2.p The quantum of action 170
IV.2.q Nonlocality 171
IV.2.r Can electrons scatter other electrons? 172
IV.2.s The language of nature 172
IV.2.t Does quantum teleportation teleport denotation or connotation? 173
IV.2.u How many universes are there? 174
IV.2.u.i *The anthropic cosmological principle* 174
IV.2.u.ii *Do these questions exist?* 175
IV.2.v Are physicists conscious of the meaning of consciousness? 175
IV.2.w Physical reality; does it really exist? 177
IV.2.w.i *Must experiments be published?* 178
IV.2.w.ii *Coherence, consistency and reality* 178
IV.2.w.iii *Local realism* 179
IV.2.w.iv *Are there elements of reality?* 179
IV.2.w.v *Is the description given by a statefunction complete?* 180
IV.2.w.vi *Are fields real?* 181
IV.2.x Oxymorons 182
IV.2.y Inflammatory language 183
IV.3 Do objects have properties if these are not measured? . . 183
IV.3.a Are these the answers we want? 184

IV.3.b Do we have to look for things to be there? 185
IV.3.c Where is it? 186
IV.4 Empty words are everywhere 186
IV.4.a Qualities floating in the air 187
IV.4.a.i *Is reality really real?* 187
IV.4.a.ii *Why are atoms small and galaxies big?* . . . 188
IV.4.a.iii *Machines that think, and people who don't* 188
IV.4.b Does God have meaning? 189
IV.4.b.i *Does God have emotions?* 189
IV.4.b.ii *Does life have meaning?* 190
IV.4.b.iii *Is the universe a product of design?* 191
IV.4.b.iv *The mystery of existence* 192
IV.5 Empty words give emotional comfort not knowledge . . . 192
IV.5.a The need to explain creation 193
IV.5.b Equal emptiness in statements about quantum mechanics 194
IV.5.c Limitations of science, of knowledge 194

V Errors, confusion and quantum mechanics 196
V.1 Why quantum mechanics often seems weird 196
V.1.a Weirdness 197
V.1.b Statements stand alone 197
V.1.c A fundamental problem is the rejection of Occam's razor 198
V.1.d Be careful what you call things 198
V.1.e Quantum mechanics is not classical physics 199
V.1.e.i *Magnetic monopoles* 199
V.1.e.ii *Why gauge noninvariance does not matter* 199
V.1.e.iii *So the Aharonov-Bohm effect does not imply nonlocality* 200
V.2 There are no waves, no particles, no wave-particle duality 201
V.2.a Dots are properties of screens, not objects 202
V.2.a.i *Dots are a property of the instrument, not the incoming object* 202
V.2.a.ii *Why it is different classically* 202
V.2.a.iii *Incoming objects with large energy* 203
V.2.a.iv *What if the atoms in the detector move?* . . 203
V.2.a.v *Are there other ways of showing the object is not a point?* 204
V.2.b Quantum mechanical objects cannot be waves . . . 204
V.2.c Are leptons points? 205
V.2.d Are electrons points and protons extended? 205
V.2.e Point particles (which do not exist) cannot cause difficulty in quantum electrodynamics 207
V.2.e.i *Point-particle approximation* 207
V.2.e.ii *The inverse square law* 207

V.2.e.iii *Problems in classical electromagnetism say nothing about quantum field theory* 208
V.3 The EPR "Paradox" 208
V.3.a The EPR "Paradox" and classical mechanics 209
V.3.a.i *Classical physics is even more "nonlocal"* . 210
V.3.a.ii *There are no acausal signals, ever* 210
V.3.a.iii *Does an object have spin before being measured?* 211
V.3.a.iv *Electromagnetic radiation gives another example* 211
V.3.a.v *Where is the money in the bank?* 211
V.3.a.vi *There is no actual physical state before measurement* 212
V.3.b The basic reason for confusion about the EPR experiment 212
V.3.b.i *What does conservation of angular momentum require?* 213
V.3.b.ii *Why coherent superposition cannot refer to a single object* 214
V.3.b.iii *What might the statefunction be?* 215
V.3.b.iv *What showing coherence requires* 216
V.3.b.v *Entanglement is not possible for a single object* 216
V.3.b.vi *In classical physics there is a meaningful state before measurement* 217
V.3.b.vii *How do we know the position before measurement?* 218
V.3.b.viii *How the double-slit experiment differs* . . . 218
V.3.b.ix *The analysis is the same for photons* 219
V.3.b.x *Creation of linearly polarized beam* 219
V.3.c If quantum mechanics is time-symmetric, what is "before"? 220
V.3.d Is there a violation of causality? 220
V.3.d.i *Bell's inequalities* 221
V.3.d.ii *Quantum mechanics is local* 222
V.3.e Be careful with formalism 222
V.4 Beliefs and statements weird and wild 223
V.4.a Here, there and nowhere 224
V.4.b Nature does not depend on human beings, but confusion does 225
V.4.b.i *Our observations are not properties of systems* 226
V.4.b.ii *Confusion fluctuates, but spacetime does not* 226
V.4.b.iii *Can a system exist in all states at once?* . . 226

V.4.c Human egomania and discussions of quantum mechanics . . . 227
V.4.c.i *What the uncertainty principle does, and does not, say* . . . 227
V.4.c.ii *Discussions, but not measurements, show human hubris* . . . 228
V.4.c.iii *Was the universe designed for the pleasure of theoretical physicists?* . . . 229
V.4.d Entanglement of statefunctions and ideas . . . 229
V.4.d.i *Are all objects in the universe entangled?* . 229
V.4.d.ii *Entanglement of electromagnetic fields* . . . 230
V.4.e Quantitization . . . 231
V.4.f Classical physics and quantum mechanics . . . 232
V.4.f.i *There is no quantum-classical boundary* . . 233
V.4.f.ii *The transition from Newtonian physics to statistical mechanics is similar* . . . 233
V.4.f.iii *Planck's constant and taking a limit* . . . 234
V.4.g Constants . . . 235
V.4.g.i *The fundamental constants have no physical significance* . . . 235
V.4.g.ii *The sizes of our bodies determine the values of fundamental physical constants* . . . 236
V.4.h Tunneling . . . 236
V.4.h.i *What tunneling means* . . . 237
V.4.h.ii *Can the universe tunnel from nothing?* . . . 237
V.4.h.iii *Does quantum weirdness produce tunneling?* . . . 238
V.4.i Why do people believe in the cosmological constant? . . . 238
V.4.i.i *Why the cosmological constant must be exactly 0* . . . 238
V.4.i.ii *Obsession with subscripts* . . . 239
V.4.j General relativity cannot be quantized, it is already 240
V.4.j.i *Gravitation, coordinates and momenta* . . . 240
V.4.j.ii *Can gravity modify spacetime?* . . . 241
V.4.j.iii *What is the gravitational field?* . . . 241
V.4.j.iv *General Relativity and quantum mechanics* 242
V.5 The measurement problem . . . 243
V.5.a Does measurement actually measure anything? . . 243
V.5.b Measurement and confused superposition . . . 244
V.5.b.i *Does the double-slit experiment contradict the superposition principle?* . . . 245
V.5.b.ii *Can a superposition of states decay into a classical state?* . . . 245
V.6 Alternative theories . . . 245

V.6.a Deterministic theories that do not determine anything . . . 246
V.6.a.i *Well-hidden variables* . . . 246
V.6.a.ii *Psychological difficulties caused by quantum mechanics* . . . 248
V.6.b The many-worlds interpretation of quantum mechanics . . . 249
V.7 The vacuum . . . 251
V.7.a The rejection of nothingness in history . . . 251
V.7.b There is no vacuum, only vacuity, in the discussions of the vacuum . . . 252
V.7.c Spacetime foam . . . 252
V.7.c.i *Can violent fluctuations rip space apart?* . . 253
V.7.c.ii *Can the vacuum be ripped apart?* . . . 254
V.7.c.iii *On what scale does spacetime fluctuate?* . . 254
V.7.c.iv *What fluctuates in quantum fluctuations?* . 254
V.7.c.v *The "miracle" of particles "popping out of the vacuum"* . . . 255
V.7.c.vi *There is a difference between objects and the vacuum* . . . 255
V.7.c.vii *The Casimir effect* . . . 256
V.7.d Does the vacuum have energy? . . . 257
V.7.d.i *Is zero point energy of the vacuum pointless?* . . . 257
V.7.d.ii *Some interesting things to do with the vacuum* . . . 258
V.7.e Quantum mechanics, probability and space . . . 258
V.7.f How does the electromagnetic field affect an electron? . . . 259
V.7.f.i *Fluctuating vacuum field* . . . 260
V.7.f.ii *Quantum electrodynamics rules out vacuum fluctuations* . . . 260
V.7.f.iii *Is the uncertainty principle a superstition?* . 261
V.7.f.iv *Is the cloud surrounding particles that of photons, or of confusion?* . . . 261
V.7.g Does the uncertainty principle prevent conservation of energy and momentum? . . . 262
V.7.h Fluctuations, foam and rhetoric . . . 262
V.7.i The vacuum and vacuous statements . . . 263
V.8 Checklist for determining why quantum mechanics appears weird . . . 265
V.9 Comprehensibility of a sensible universe . . . 267
References . . . 269
Index . . . 277

Chapter I

Language, concepts and physics

I.1 Geometry, field theory, and the description of nature

The physical universe, to the extent that we now know it, is described by quantum field theory (of which quantum mechanics is a special case). Why should this be? And what is quantum field theory? What determines what it is? What properties must it have in order to describe — answer all (relevant, meaningful) questions about, give all (possible) information about — nature?

Quantum mechanics (and the laws of nature), their necessity and properties, are not arbitrary, not mysterious [Mirman (1995b,c)]. They are not assumptions for which we can have no understanding, can find no rationale, but must just accept. There are reasons for them, and these can be found and stated. Analyses supporting these views are continued, but merely continued, here; one aim is to demonstrate that we can understand the reasons for the laws of physics — with the hope that this encourages others to carry such investigations further.

These explorations are based on the view that a field — any quantum mechanical statefunction — is a representation basis function of the transformation group (not necessarily a symmetry group) of the geometry of the universe, and so physics is (largely?) based on geometry, and the groups that it requires. (Statefunction is a better term than wavefunction (sec. IV.1.a.ii, p. 144).) Geometry determines its transformation group, including its representations, giving their basis functions, among these are the fields. Quantum field theory is a property of geometry, so physics is a specific aspect of it. But geometry is a property of physics — only certain geometries, (locally) perhaps only one, can entertain physics [Mirman (1995b), chap. 7, p. 122].

From geometry and group theory we acquire much knowledge about our universe. To gain this we must understand reasons physics and geometry are so connected — through group theory, and how this gives quantum field theory as the description of nature, and determines what it is. The understanding is in language, terminology, logic, mathematics. These we consider here, in particular discussing first the mathematical and conceptual formulation of quantum field theory, and why it has these mathematical and logical structures.

I.1.a Necessity for examining foundations

But our considerations, especially at first, are more general. This is necessary to provide a foundation for more specific topics and an understanding of what our approach is and why we believe this is the necessary way of defining, and using, quantum field theory, and then later conformal quantum field theory [Mirman (2001)]. And we emphasize language (chap. IV, p. 142), not only the language of physics and mathematics, but what language is, how it should be used, and in depth how it should not be used. For it is by language that we express our concepts, our physics, our mathematics, and mathematics also is a language. But more, our concepts, our physics, our mathematics are aspects of language — they are what we express of them. Thus we must consider, not merely the laws of physics and mathematics, but what these subjects really are — for they are what we say they are. And what we say they are is bounded, but only in part, by what they are, what they can be. Many of the difficulties of these, and a large number of other topics, not only in quantum mechanics, result from misuse of language.

This, language, and how it determines our view of reality, helping us to understand, and too often causing us to misunderstand, is something that has been too little discussed, certainly by physicists and mathematicians. Thus here again we hope that this discussion, and perhaps irritation, encourage others to carry these analyses further.

Since we are considering application of groups, physics, quantum mechanics, field theory, to the real world, it is useful to start by asking whether there is such a thing?

I.1.b What is reality, and is it only classical?

It is widely believed that classical physics has trained us that science can and does describe things as they really are [d'Espagnat (1990)]. Indeed it is undoubtedly the belief of almost everyone that we do see things as they really are. But what are things really? How do we distinguish the way "things really are" and merely the "way that we perceive them"?

And if we cannot make such a distinction do these phrases actually differ? Intuitively they seem to express very different concepts. But is this distinction real or merely a fantasy of our intuition? What else can it be if there is no way to differentiate between reality and perception?

Of course, perception can, usually does, mislead. But once all possible errors are removed (in theory), and everyone agrees (to the extent possible) on what is being perceived, what more is there to reality?

These questions are fundamental to the interpretation of quantum mechanics, for instinctively there seems to be more reality in the classical reality than in the quantum mechanical one. Perhaps this instinct is a cause of much of the misleading interpretations of quantum mechanics, the errors, the confusions, the misunderstandings.

What then is the reality that, not only, science deals with? Is there a difference between classical reality and quantum mechanical (non?) reality?

Words like reality and facts appear continually in physics, but very rarely have attempts been made to clarify what they are supposed to refer to. Analyses that even make an attempt [Mohrhoff (2000)] are far too scarce. These not merely deserve, but greatly need, careful study and definitions.

I.1.b.i *What criteria should we use for reality?*

Possibly the best criteria for reality are constancy, coherence and consensus. A table is real because every time we look and touch, it appears the same (exactly, within our ability to judge). And everyone who also studies it finds the same properties (sec. I.4.c.ii, p. 18).

We can construct a simple — coherent — model for a table that predicts what we perceive with each of our senses, and what we would perceive under all conditions in which the table might find itself, under all operations that we might perform on it. And everyone else has (to the extent that it can be communicated) the same model, with the same predictions. So the table is real — our simple, coherent model agrees with the simple, coherent model of all others, and with the relevant perceptions that are independent of time, space and environment.

Not all our models are coherent: dreams are not coherent, but we would not regard these as models of reality.

That our reality is based on consensus does not mean that it can be changed by consensus. The consensus is not one that is decided by us, but one that is imposed on us. Our model is one that we are forced by nature to have, that is unique because of nature. There are laws of physics — and there is a nature, a reality — because there is a consensus, constant and coherent, that we must accept.

We can reach a consensus on the number of physics courses that a student should take, that is we develop a consensus about what we want, or what we regard as proper. However if we see someone walk off the top of a building we reach a consensus as to what then happens, whether we want it to or not. This is a consensus on what we see, and based on our knowledge, what we will see — and we have no choice but to accept this.

A social consensus is about what should happen, a physical consensus is about what does happen.

Does "reality" "exist"? We do have a model, a consensus model that is constant and coherent, and for a very large class of observations, and models drawn from them. And it is unique. What more can we want? What more is possible?

I.1.b.ii *Is there really a difference between quantum reality and classical reality?*

But isn't this definition of reality for a table also true for an electron, or a wavefunction (sec. III.4, p. 129)? Of course, for a table we can use our senses directly with no need for instruments, for electrons or wavefunctions we cannot. Do we regard a speaker we hear on radio or television less real because we need instruments to perceive the sound? Why then should we regard an electron, or wavefunction, as less real than a person on the radio? And aren't our eyes and hands also instruments that our brains use in the same way?

Thus the difference between "the way things really are" classically and quantum mechanically is actually due to no more than prejudice and unexamined habits of thought. Reality is agreed and constant perception, whether the perception is direct or requires conduits. Though reality may not always agree with our mental pictures, our experience with the "real" world is limited (certainly we have more direct experience with radio than with electrons), and cannot provide reasonable criteria upon which to construct systems, schema, models, for depictions of nature.

Perhaps then, the "reality" that we are comfortable with, and that of quantum mechanics, are not in reality different in the least. Or perhaps the "reality" that we are comfortable with is in reality not the correct "reality" at all.

I.1.b.iii *Does the universe exist if we do not observe it?*

There is a related idea that the universe does not exist "out there" independent of observation, and that reality is created by the observer. The difficulty with this view is seen by asking the question (one which is always worth asking): "How does it differ from the view that the universe really does exist, and is independent of the observer?". To be able to deal with the universe we have to assume that it does exist — try to construct a workable astronomy by assuming that stars existed, even billions of years ago, only when our telescopes are aimed at them.

To do otherwise means piling on meaningless, confusing, extraneous verbiage that makes dealing with our lives more difficult (except when it is ignored beyond providing emotional comfort). Such statements like the world is created by observation, and does not otherwise exist, say nothing. Even those who say that they accept the validity of such statements actually do not — whatever they say, they do not act as if they believe such statements, for they cannot.

I.2 Why some words have meaning

Much of our discussion is about language, how it is used, how it is misused. Too often views of the universe, of physics, of quantum mechanics, are based on words, phrases, sentences, larger collections, that seem to say something but which are completely empty. It is thus useful to consider briefly how meaning and meaninglessness can differ.

If so many words that are so often used are meaningless, why do we believe that others have meaning? In what way is a book, an electron, different from a multitude of universes (sec. IV.2.u, p. 174)? Books (in particular specific ones) and electrons have properties that are well-determined, and (especially for electrons) always the same. Thus in experiments, we can say that there are electrons, and using their properties predict (correctly) outcomes. And these properties are always the same, no matter what the experiment. Thus it makes sense — it is useful — to define an electron, give it a set of properties and use it not only directly in experiments, but to develop models of other systems which lead to experimental predictions.

And these experiments are not merely ones we do in laboratories, but which occur in stars, in other galaxies, in materials, in our own bodies, and the models of such processes using this concept of "electron" help us organize our observations in a useful way, no matter what the observation. Rather than saying that the electron is real (or that a book is real), with an undefined term "real" it is better to say "it makes sense, it is useful". It is also useful to define "real" as meaning these phrases, but we must be careful not to read into the word more than we should. And it is helpful to say "organize our observations in a useful way", but it is essential to remember that when we introduce words to simplify such phrases we should not think that they say more than they actually do.

I.2.a Constructs are needed for coherence

Electrons, protons, and such concepts are constructs that are needed to give a coherent picture of the universe, one that we can deal with — make predictions about and apply our predictions to control, slightly, the behavior of objects in the universe. There are many such concepts, for example when we talk about the radius of an atom or the size of a nucleus or a galaxy, we are using a definition different (experimentally at least) than the size of an accelerator. But these concepts also are required to be able to deal with nature. Thus such terms as electron, proton, sizes (of atoms, nuclei, Compton wavelengths, galaxies, say) do have meaning, but in this indirect sense. "Meaning" then is a fuzzy term, and can be meaningless, and misleading. It is important to be sure that words do have meaning, what it is, and in what sense.

If an electron is real, is a Δ resonance? Again, it has well-defined properties, always the same, and moreover Δ belongs to multiplets, isospin, SU(3), perhaps larger ones, these containing particles like the nucleon for which the evidence is stronger and more concrete. It thus makes sense to say that it "exists". There are other particles, and the evidence for these varies from as strong as the Δ, to much weaker, to almost nonexistent. Further there are resonances whose widths are about the size of the distance between them. Thus it becomes less and less useful to say that they "exist". This is well recognized, and when these values are listed the uncertainty is often emphasized. Unfortunately in other cases equivalent fuzziness is usually not recognized.

Another example is the construct fundamental in quantum mechanics "statefunction" which we discuss at length (sec. III.3.c.v, p. 124).

I.2.b Space and coordinates are constructs

Among the most essential constructs are space and the coordinates that give it meaning. What are coordinates?

We can define them by giving events, physical interactions, that happen at them. But this is much too narrow. It is impossible to have an event at every one of the continuous set of coordinates. Yet clearly physics does not happen in a space with coordinates that form discrete sets.

We have functions, statefunctions, and these depend on coordinates, which are continuous. We then use statefunctions for predictions, for understanding our world. They do predict events, although these observable events do not occur at every point of the continuum. Thus the continuous set of coordinates is a construct that we need to understand physics and make predictions — we use functions defined over a continuum. And this gives coordinates meaning.

It is not that we can determine every value of a coordinate — we do not have to. We do not measure at every point. However we do take the points at which we observe events to be a finite subset of a continuous set. That defines (roughly) the continuous set of coordinates.

I.2.c Language also varies from precise to nonsense

As with resonances, the same is true of language, but not so well-recognized. There are terms with very clear, very well-defined meanings. Others have definitions less clear, less explicit, but still meaningful. Eventually we reach terms with no clear definitions, and beyond, those which cannot, even in principle, be defined. It is for that reason that these are often so attractive and popular. Because there is no clearcut line from sense to vagueness to nonsense, it is easy, much too easy, to start with a statement that is completely meaningful, and quickly be saying things that not only have no content, but that cannot, that are, that must be, complete nonsense.

This is a reason that there is so much confusion, not only about quantum mechanics, not only about physics. Language is a very dangerous instrument.

I.2.d Meaning is pragmatic

Essentially, no matter what people say, or believe, or think that they believe, the ultimate meaning of words and concepts, theoretical ideas, constructs on which these are based, is pragmatic. To ask whether an electron is real, whether it actually "exists", is to ask a question that asks nothing. However to have a coherent view of nature, to be able to deal with it, we introduce a word "electron", and assign certain properties for which this word is an abbreviation, because it is useful, necessary, to do so. We do this for purely pragmatic reasons. We do not, and cannot, imply thereby that the electron, or book, is real — for the word "real" has no content in the sense it is used here, and adds nothing to the value of the concept.

Why is "electron" meaningful, useful, "real", while "virtual particle" is much less so? An electron, the name given to a list of properties (or set of statefunctions defined by these (sec. III.3.c.vii, p. 126)), appears in a vast number of different physical situations, of different physical objects, and always with exactly the same properties. These can be determined in a very large number of ways, using immensely different physical objects, stars, tracks in chambers, behaviors of cathode ray tubes, wires, and so on. It is a construct completely necessary for understanding of numerous other objects, concepts, physical situations, experiments, theories; it is needed for a coherent, even just a reasonable, view of nature.

I.2.e How electrons differ from virtual particles

A virtual "particle" is a name given to a particular term in an expansion in an approximation scheme. Its properties are determined by this approximation scheme, and by equations for systems this scheme is used for. These properties cannot be determined in any other way. The "particle" itself has no meaning outside of the approximation recipe. Not only is it not needed elsewhere, it is not even needed for this scheme. It is merely a useful device for visualization that helps in the calculations in the procedure, a mnemonic, a purely bookkeeping device. Moreover it is misleading. The word "particle" suggests that it has the status of an electron. Thus it is taken too seriously, leading to error. To say, for example, that the (quantum) vacuum is full of virtual particles is not only empty, but silly (sec. V.7, p. 251). There are terms in the approximation expansion — to describe circumstances in which objects are present, so cannot refer to the vacuum — that can be looked at as giving virtual particles popping in and out of the vacuum. That says nothing about physics or the vacuum, but is purely a picture of a bookkeeping device for a calculational method.

I.2.f Meaning comes from being a (necessary) part of a web

The meaning of a concept then, and the reality of physical objects to which one refers, comes from it being a necessary part of a web of concepts, laws and properties, necessary (at least to some extent) for others, and the entire set, to makes sense (sec. II.2.d, p. 72; sec. III.4.b.i, p. 131; sec. IV.3, p. 183; sec. V.7.g, p. 262). Such webs extend to include concepts that refer to our perceptions. Concepts making up webs have content because the sets to which they belong have, because they contribute to their meaning, so their own derives from that of the whole. And the whole has meaning because parts that refer to our perceptions obtain sense from these. We cannot understand, as fully as is possible, what we observe of a star in another galaxy, or of this book and the physical objects of which it consists, what we observe within ourselves, without such concepts as electrons and statefunctions. We cannot touch, or taste, electrons or statefunctions, but they are physically real because they are needed to provide understanding of those objects which we can touch and taste and see.

The correctness (correctly the value) of a concept is not determined by whether it is true or false, something that cannot be determined, so is meaningless, but by how necessary it is to the totality of which it is a part, and how valuable that entirety is.

I.2.f.i *Measuring instruments*

Among the necessary concepts are objects used for measurement — these are necessary to give physical value to theories. We cannot discuss whether a theory has content unless we have some way of determining how it affects our senses — we cannot so determine, but that does not imply that other physical objects cannot; the universe exists whether we notice it or not. It is widely known that there is difficulty in defining and picking these instruments, clocks, rods and so forth, especially in quantum mechanics. However quantities that we measure with them, like time, distance, mass, need not be defined by physical instruments. It is only when we have to relate the predictions of a theory to (possible) effects on our senses that we need them, and for this the difficulties may be less severe. Thus statefunctions of electrons in a star, or the temperature or pressure there, or in a crystal, depend on position. But (fortunately) we need not go into the star with a meter stick, or thermometer or ..., for these concepts to be meaningful (useful).

Seemingly, perhaps, there is a difference between temperatures and statefunctions. We can measure the temperature of an object (that we can see) so can regard the temperature within a star to be the same concept. However not only can we not measure directly the temperature in a star (or a crystal) but the quantitative difference between the temperature of a human-size object and that within a star, or for microscopic objects, even qualitative differences between these, is so vast that they are at best extensions of concepts that we have elsewhere defined. This is true even for internal temperatures (say of a crystal). To define, for example, temperature, physical laws are required, and while laws governing atoms are the same as those governing stars, in reality we use not these, but simplified models of them. And such models may (have to be) quite different for mesoscopic objects, crystals, columns of mercury, stars.

While we may be more comfortable with the concept of temperature than of statefunction, both are really given denotation by the concepts, laws, properties to which they are, to which they must be, linked, not necessarily by objects, if any, used to study them.

I.2.f.ii *Is a concept needed?*

Of course it is often difficult to tell to what extent a concept is needed to give an entire web of concepts meaning, so to tell whether it itself has meaning, whether it is needed. Difficult, quite often, but not so difficult as to excuse ones that have no content, that are not needed, that are not defined at least in terms of others. (Use of Occam's razor is vital [Mirman (1995b), sec. A.1.2, p. 178].)

However for many concepts introduced to provide alternative theories to quantum mechanics (sec. V.6, p. 245) this is quite easy. These are clearly senseless, clearly irrelevant, thus the theories do not actually provide alternatives to quantum mechanics. They can be removed and there will be no change — physically, although perhaps not psychologically. But in other cases, removing a concept will cause the theory to totter, a little, perhaps a great deal, or the concept might be replaceable by another so it is not essential, although some concept is needed in its place. There are other cases in which removing a concept will greatly weaken a theory, and for some destroy it completely, causing it to collapse (which theories, although not statefunctions, can do). These are more and more needed, so more and more "real", thus more and more physical — provided the theory for which they are essential is also essential, provided that it is necessary for us to understand (have a coherent picture of) the world, or at least that part to which the theory applies.

Although it is not always easy (even possible) to find dividing lines (regions?) between what is necessary, essential, what is important, what is only useful, what is extraneous, and what is meaningless, it is essential, important, useful and quite necessary to do so.

I.2.f.iii *Quantities defined indirectly*

Using statefunctions and the positions determined by them and the other statefunctions of particles (to the extent necessary) we can determine the macroscopic behavior of a star, thus of the light (gravitational waves, neutrinos, . . .) that it radiates. And this we can (at least in principle) sense, perhaps with instruments that we have to define. Quantities like time, position, temperature, thus need not be defined by physical objects, but only by the web of concepts of which they are parts. It is important that distinctions between needed and unneeded definitions of instruments be clear (and we need not tell how to measure time or distance or pressure inside a star — with instruments).

Properly defining the way we measure can be difficult, and extraneous (and vacuous) definitions confuse.

I.2.f.iv *What does the phrase "very small distances" say?*

An example of a concept that can achieve content as part of a web is small distance, even though it is not clear that we have to probe small distances experimentally for it to make sense. Relevant equations are differential, and these give distances meaning, even if it is not possible to measure for small values. But equations are based on the concept, and would be affected, as thus would their predictions, if concepts were wrong.

Our inability to measure distances that are very small does not imply that they are meaningless, specifically that differential equations provide the wrong formalism. Solutions of these give, say, paths of objects. Exact paths can never be found experimentally, but the solutions will be correct to whatever precision we want to (or can) measure. There is no reason to think that a path will be wrong, or that other formalisms will give one in better agreement with experiment. That it cannot be checked with absolute precision does not imply that the formalism is not correct.

Also there is no indication of a minimum uncertainty distance. The uncertainty in the distance with respect to the distance increases as the distance becomes smaller, but there is no distance at which a transition occurs. Thus it makes sense to say that the distance is $d \pm \frac{1}{10}d$, and even that the distance is $d \pm d$, but perhaps less sense. However it is useless to say that the distance is $d \pm 100d$. Then the correct distance is in the range of about $100d$. But there is no sharp dividing line.

We cannot measure small distances but the concept is needed (as far as we now know) to give content to and enable us to use quantities that we can measure. And it is these quantities, and the equations, with the correctness of their predictions, that give meaning to small distances.

I.2.f.v *Hollow rigorous formalisms*

This indicates why (many) rigorous formulations of quantum mechanics are, at best, unhelpful (sec. II.3.c, p. 78; sec. IV.1.d, p. 151). They avoid the real difficulties. It is possible to introduce terms and symbols and rules for manipulating them and rigorously derive consequences of these rules. These terms and symbols are given the same names as terms and symbols used in physics, for example in quantum mechanics. But these rigorous derivations cannot tell anything about physics unless the terms and symbols are furnished with physical meaning, say by relating them to the behavior of objects that we are already familiar with in physics (like our sensory neurons or objects that have effects on them). But it is just at this stage that problems arise (as shown by the very large number of discussions, including this one, about the interpretation of quantum mechanics). Ignoring this stage, and then deriving rigorous results does not give any that are relevant to physics, rigorous or otherwise.

Likewise there are theories with clear, definite statements of postulates, but no statements of how these are related to experiment, what these mean, or ones with postulates that are given definite physical meaning, but with it having nothing to do with any of the further discussions or developments of the theory. Their irrelevancy is not noticed (by the author or reader) so the irrelevancy of the discussion becomes well-hidden. This is often the case with such terms as "clocks" or "measuring rods" which are carefully defined and introduced, but then not used.

A physical theory is about physics (not mathematics), so to have content it must be related to physics. But this, giving physical meaning to the terms and concepts, can never be rigorous, and will always be dangerous, hazardous, precarious, but critical.

The very concept of rigor in physics is questionable, not fully defined, perhaps not fully definable [Mirman (1995a), sec. P.5, p. xv].

I.2.g Physics does have meaning

Physical concepts can be given meaning, reasonable, often quite precise, meaning, whether they directly affect our sensations or not. And they must be given it, words and sentences expressing them must have content, else they are mere spots of ink on a page, mere vibrations in air carrying sound that is nothing more than noise.

Many, if not all, of the views given here may seem strange, unreasonable, if not obviously wrong. However physics is an experimental subject, and the relevant experimental data are the ways these terms are actually used, not in how people think they are being used. How else can they be understood then with the present interpretations?

Physics makes sense, and its language makes sense, provided that we wish it to.

I.3 Quantum field theory and what we need to study it

Narrowing our focus from the way we gather knowledge of the universe, and how we do, and can, interpret it, we look at concepts underlying our study of quantum mechanics and quantum field theory, and how these are determined by geometry and the groups of its transformations.

I.3.a The need for interpretation

Examination of quantum field theories requires in depth consideration of their interpretation and so of quantum mechanics, which is quantum field theory. Is this really necessary? Clearly many think so, for this is a topic of great interest. We believe that what we do here helps justify this interest, and that it leads to, not only understanding of the foundation of quantum mechanics, but specific results and predictions, like the reason for, the necessity for, conservation of baryons, that the proton is stable [Mirman (2001), chap. IV], and the reason that all terms in an equation have the same units [Mirman (2001), chap. I]. There has been much confusion about quantum mechanics, not its formalism, but its meaning. To understand foundations we have to confront this confusion, and errors so common in the way quantum mechanics is thought about, try to show why they are wrong, and how mistakes can be avoided (chap. V, p. 196). Some are obvious, some quite subtle. This is a subject of great controversy, and many, perhaps all, readers will disagree with some, perhaps all, views given here. But we hope that this will at least stimulate, and infuriate, all readers to think about these subjects — carefully — and extend what is done here, perhaps in different directions.

I.3.b The relevance of the conformal group as a transformation group

What is the transformation group of our geometry [Mirman (1995b), sec. A.2, p. 178]? The Poincaré group [Mirman (1995a), sec. II.3.h, p. 45] certainly, and this thus has fundamental significance. But there is a larger group, of which it is a subgroup, the conformal group. We try to analyze reasons for its relevance in some depth [Mirman (2001), chap. I]. Having done so we examine the relevant mathematical aspects of the group and some of its representations. Ones that we consider are of an unfamiliar form. To find representations of a group we decompose it into subgroups, and build its representations from those, say the relevant little group [Mirman (1995c), sec. 2.2, p. 12; (1999), sec. VI.2.c, p. 284]. In particular the states are labeled by representations and states of the subgroups. There are different decompositions, and the form of representations and states that they give thus differ. Here states of the conformal group are labeled by its Poincaré subgroup. Thus we construct representations of a noncompact simple group with states that are basis states of an inhomogeneous subgroup [Mirman (1995a), sec. XIII.4.b, p. 382].

Comprehension hopefully gained can be applied to quantum field theory, obtaining restrictions on that and knowledge of how these lead to the description of nature. Quantum field theory so restricted by the conformal group becomes conformal field theory. Clearly this is of much interest, and we start a study of it, and its applications.

While there has been much work on conformal field theory [Ketov (1997); Schottenloher (1997)], usually it has been quite limited, generally considering only two dimensions, and massless objects. But there is far more to the conformal group, and it is thus quite interesting to consider why that is relevant and whether it might lead to results not apparent from more restrictive cases. It is the study of this that we wish to start, but only start since this is a vast subject, with many novel aspects, as we see.

I.3.c Nonintuitive, thus confusing

These subjects, quantum mechanics, quantum field theory, and then the conformal group, conformal field theory, are nonintuitive. There is thus often much confusion in discussions about them. But to understand these we must try as best as possible to clear away misunderstandings. Many come from language — it is very dangerous, and the dangers of language are usually not evident, a reason it is so dangerous. This is frequently strikingly illustrated by discourses on quantum mechanics, quantum field theory, and their meanings. What we are, not only as students of nature, but as creatures capable of thinking, is based on language, on our possession of it, our ability to use it, and unfortunately our ability to misuse it.

Many statements about (not only) quantum mechanics actually have no import, they are combinations of words, themselves meaningful, but which cannot be put together giving thus expressions without content. Because each word has meaning — in different contexts — the vacuity of sentences and phrases in which they are misused is well hidden. Studies of language therefore need careful analysis and must form a major part of our considerations (chap. IV, p. 142). Otherwise it will be impossible to understand what we are doing.

I.3.d Why do we disagree?

While readers might disagree with what we say here, with our view of nature and physics, many (most?, all?) may find that what we are asserting is just what they are saying. The only difference is that they also have mental pictures present while they are saying these, but which really are irrelevant. But such pictures confuse, and make it unclear to them, and others, what they are really stating. And if upon thinking what they are, and have been, saying they still disagree, then they can clarify explicitly why they disagree and how our views really differ, not how they would like them to differ, or how they seem — superficially — to disagree, perhaps because of language that is not thoroughly analyzed.

Here then we try to assess foundations, logical, mathematical, linguistic, perhaps even metaphysical, of quantum mechanics, quantum field theory, conformal field theory. But this we wish to do to understand physics. Thus to illustrate that such study is useful, and to illustrate meanings of what we have harvested from it, we then consider some applications.

I.4 Frameworks and theories

The idea of a scientific theory is quite broad, actually too much so. We therefore consider narrower ones, frameworks [Mirman (1995b), sec. 1.3.1, p. 3] and theories. A framework (for theories) is a broad view of its domain. It has certain underlying concepts upon which its theories are based. There is no clear distinction between a framework and a theory, some pictures are close to one of the extremes, others more equivocal. Frameworks shade into theories, but the extremes, of a general framework and of a very specific theory are qualitatively different in such ways as what they are, how they are used, how they are judged, what their value is, how they are related. Thus it is useful, and clarifying, to distinguish them, a reason that we use different names.

I.4.a In science there are many frameworks, and far more theories

It is helpful to start this analysis of frameworks and theories by illustrating what we are saying with the terms. While our interest here is of course physics, these ideas are not limited to that. Examples from other fields can clarify concepts, and their differences.

I.4.a.i *Classical physics and quantum mechanics as frameworks*

Classical physics is not really a theory, it is a framework. It is based on the view of objects as waves and particles, these localized, idealized as point particles; with all relevant values determinable in principle with arbitrary precision; variables like position, velocity, momentum, force, and so on. Newton's laws are more a theory, but still also a framework, Maxwell's equations form even more of a theory, but still somewhat of a framework. Descriptions of behaviors of charges within materials are theories, being quite, yet not necessarily fully, specific.

Quantum mechanics tends also to be a framework rather than a theory. It is based on the underlying concepts [Mirman (1995b), sec. 1.4.4, p. 13] that all physical objects are described by (actually are) functions of space and time and perhaps other variables, and are representation basis functions of the relevant group (they lie in the domain of its operators — these can act on the functions).

I.4.a.ii *Ptolemic and Copernician views and evolution are frameworks*

The Ptolemic view of astronomy is a framework, while the sets of cycles and subcycles assumed provide theories based on that. Eventually it did not prove useful. The Copernician view was more useful, better theories could be built on it, and it also lead to a broader framework, classical physics. Of course now we recognize that the ancient view is totally useless, the Copernician better, but really only the next step. Yet it is actually not correct to say that one is true, the other false. One is almost totally useless, the other is of some value.

The theory of evolution is not a theory, it cannot be falsified, but is rather a framework in which it is possible to organize, not merely large amounts of data, but many more specific frameworks, and within these build theories. It allows us to think about much of biology in an organized, heuristic, fruitful manner, to understand, so to be guided. And there are no other frameworks that allow any of this, that are useful (except perhaps emotionally to their proponents). Evolution is not so much true or false, but rather extremely useful.

The big bang view of the universe is perhaps more toward a theory than quantum mechanics is, but more of a framework than, say, descriptions of supernovas or superconductivity.

I.4.a.iii *Quantum field theory is a framework*

Most of what we do here is really a study of a framework, quantum field theory. Of course there is no clear dividing line, and some parts are more of a theory. The assumptions, to the extent they can be determined, or even exist, thus may vary. The usefulness of quantum field theory, and our view of it, can be considered objectively, as we try to do. But it is in the end, subjective. It is helpful to be aware of this, to try to see assumptions, differences between parts that are a framework and those that are theories, parts, concepts, that are necessary and those that may be superfluous, perhaps misleading.

The more these are understood, the more likely subjective judgments will be, not so much correct, but sound, the more likely they will provide fruitful, not misleading, guidance.

I.4.b Theories are formed within a framework

It is within a framework that specific theories are constructed, these given by specifying the constituents of the system, electrons, protons, holes, rotons, unit cells [Mirman (1999), sec. I.3.b, p. 7], for example, with their properties, and the realizations of the group operators, such as the particular form of the Hamiltonian, and perhaps which operators are diagonal, plus various other details as may be appropriate. These constituents, properties, realizations, and so on form the assumptions of the theory. From it experimental predictions can be developed and tested. But te,sting the underlying framework is more difficult. That is less specific, more general, with the view much broader. And if there are difficulties, say experimental, these are more likely due to the specific theories that lie within the framework. But if there are many problems developing theories, preferably testable ones, than the framework is not so much wrong as useless. Of course to determine whether it is useful enough may not be possible unless there are competing theories, competing frameworks. A framework may be poor, of limited value, as was classical physics in important areas of physics just before the development of quantum mechanics, but might be better than none at all.

Theories can have subtheories, which can have their own subtheories, and so on. Thus for example we have theories about crystals, and also ones about specific crystals, and also subtheories about the behavior of objects within them [Mirman (1999), sec. XI.2, p. 569]. Subtheories about behavior can belong to general theories about crystals, or subtheories of these about specific ones. Thus a theory about the behavior of objects in specific crystals is a subtheory of both a theory of the behavior of objects in crystals, and of a theory of the structure of a particular type of crystal. Theories and subtheories form not so much a tree, but more of a net. Often they are part of a web, giving content to each other.

I.4.c Frameworks are not true or false, but useful or not

One of the difficulties in determining the underlying assumptions of a particular description is that the concept "assumption" is less applicable to frameworks than to theories; a better word than "assumption" would be "concept". We can test whether a theory is true (that is useful) or not, so find whether assumptions are correct. It is generally felt that for a scientific theory to be reasonable, it must be falsifiable. But this can be too narrow; it certainly is for frameworks. A framework is neither true nor false, but rather useful or not. Thus it is not so much whether assumptions are correct as whether they form a useful foundation for theories that can be tested.

It is not that classical physics is incorrect but rather no useful general theories can be built using its concepts. Of course, extremely useful theories, but limited and phenomenological, are based on it. Quantum mechanics is useful because it forms the foundation of theories that can be tested and found to agree with experiment. If there were disagreements, say nonlinearities, it is likely that theories could be changed without changing the most basic, underlying concepts, the framework. Quantum field theory is neither true nor false, but (so far) quite useful.

A theory of superconductivity is a theory, it based on certain assumptions about charges within materials. It can be checked, and an experiment can falsify it (provided the experiment is done correctly, which can be quite difficult, provided all confounding factors are known, it is fully understood, and the results properly interpreted). It cannot be proven true, no theory can, but rather shown more and more likely to give correct predictions and guidance — or not.

Science then consists of frameworks, theories and concepts chosen, perhaps accepted is a better term, because of their simplicity, coherence and predictive power. Although this is well-known it is useful to emphasize it explicitly [Maxwell (2000)].

We can never be sure that we understand nature completely or properly. Thus the correct terms for frameworks and theories are not true or false, terms of questionable meaning anyway, but extremely useful, very useful, useful, less useful, not too useful, of very little use, totally useless.

The word "useful" is very broad here. Science is useful because it has lead to the transistor, to the many instruments that so many find useful.

But it is also useful in explaining (predicting our observations resulting from) the appearance of a supernova, and behaviors of electrons in crystals and in stars. It is useful in replacing a jumble of sensations with coherent pictures, and these with a smaller number of coherent pictures, continuing with a smaller and smaller number of broader, more inclusive pictures. It is useful in enabling us to understand the world — develop coherent pictures that we can use to predict and so control our environment. Theories and frameworks are useful, thus "true", if they satisfy criteria such as these.

Frameworks (and theories), even if almost useless, are used until and unless better — more useful — frameworks and theories are developed.

I.4.c.i *Criteria differ for theories and frameworks*

Why distinguish between theories and frameworks? A very specific theory, say of the conductance of a certain material under given conditions, can be checked experimentally. The experiment, if done and interpreted properly and all confounding variables are accounted for, will tell whether the theory is correct or not, that is whether its predictions agree with experiment. If not there will be little flexibility in the theory and it will have to be discarded. But it is unlikely that quantum mechanics can be shown to be incorrect. If there is disagreement with experiment, specific theories within that framework have to be modified. And there likely will be much flexibility. But as more and more modifications are needed, as more and more different theories are required, as these become more and more forced and artificial, the framework becomes more and more useless, and finally must be abandoned.

Thus for the extremes, very general frameworks and very specific theories, there is such a great contrast between how they are judged and how they are used that these are essentially different concepts. So it is helpful to distinguish them.

I.4.c.ii *Usefulness and correspondence to reality*

Of course we might regard frameworks or theories as useful because they correspond to reality. But what does "correspond" mean here? What is the difference between "a theory corresponds to reality" and "a theory is useful"?

It is reasonable to say that a concept corresponds to reality if others also believe that it corresponds to reality (sec. I.1.b.i, p. 3), that other people agree with our concepts and views of the "external world". For a concept, a theory, a framework to be useful it must ultimately be found useful by others (sec. IV.2.w.ii, p. 178). Again even with criteria like these is there any difference between "correspond" and "useful"?

I.4.c.iii *Tools, not truths*

Frameworks and theories are in reality tools, in a certain sense like microscopes or accelerators or enzymes. They need not be perfect, perhaps cannot be, but they are not only useful but essential. We use the best ones that we can, even if they fall far short of what is really needed. The best ones are used until better ones are developed. We cannot abandon a theory because it does not always work, just as we cannot give up a microscope if it does not always provide the needed resolution. We abandon these tools, no matter what type, only, and we can abandon them only, if there are better tools available. We must have theories, and imperfect ones are better than none at all, if that is the choice.

Fundamentally science is nothing more than a tool. There is a difference between science and religion, one we believe, one we use.

I.4.c.iv *Decisions about value are subjective*

Tests for whether a framework is useful, or a theory is true (actually useful), are basically pragmatic. We can never be sure, we can never study enough cases, enough different topics, do enough experiments. But life is short, time and funds are limited. We accept a theory or framework, stop doing an experiment, when we feel that it is more valuable to do so than to continue, when we feel that it is better to do other things, study other subjects. Although this is based on objective knowledge, the decisions are essentially, necessarily, subjective.

Science is an aspect of life. Scientific decisions are like any others in life: which theories to believe in (that is to use); which parts of a theory, which concepts, are really necessary, to what extent; which experiments to do, when to stop; what career to chose; which job is best; who to marry; what to eat and wear each day — these are among the many decisions we must make. In some cases the answers are clear and obvious, strongly based. In others making the best choice is almost a matter of guesswork and hope. But always, no matter how strong the objective reasons, there are subjective decisions involved. A theory is never proven true or false, just as a choice of what to eat is never proven correct or not. Decisions about them are based simply on our best judgment — hopefully.

Classical physics provides a valuable, necessary, framework for the theories on which buildings are constructed. But buildings sometimes collapse. Of course, we can be quite confident about classical physics in this realm, but never completely, absolutely certain. However while this is the framework, specific theories, about the behavior of materials, of the forces exerted by and on these, and so on, are required. While we can be confident of these (in general), we can never be completely sure, and definitely less so than about underlying concepts of classical physics. And these theories must be tested, carefully checked and studied. But eventually decisions have to be made about their correctness, otherwise no buildings would ever exist. And sometimes these decisions are wrong. Thus it is in all of science, all of life. Eventually decisions, judgments, have to be made, and judgments, even those based on objective considerations, are necessarily subjective.

Theories are never proven true or false, but rather subjective evaluations are made whether to accept or reject them, whether to look for better ones or go on to other areas, and these decisions are based not merely on the experimental evidence, but, necessarily, on practical, pragmatic considerations. The usefulness of frameworks is perhaps even more subjective.

I.4.c.v *Theories, physical and phenomenological*

It is important to distinguish between complete physical theories (of which none are known), and phenomenological ones — incomplete, partial pictures, not fully consistent, not able to explain all (relevant) phenomena, needing extra assumptions for specific subsets of the phenomena. Phenomenological theories are needed, absolutely essential, as the history of science demonstrates. Classical physics is an excellent example of a phenomenological framework, friction of a phenomenological theory.

But their weaknesses must always be remembered, otherwise they become seriously misleading. Their heuristic values are undisputable, their intuitive appeal often seductive (typically because they are constructed from guesses that are affected, determined, by our intuition), so there can be great danger of them deceiving. Heuristic value should never be confused with consistency. Often it is easy to see the difference, but too often it is not.

If we try to explain the scattering of nucleons by nucleons by picturing little green men standing on nucleons with baseball bats, and this leads to useful equations describing the scattering, then the picture is useful, helpful. But no one would take it seriously as a real theory of scattering. Unfortunately there are too many theories similar to this, but with their intrinsic absurdity much better hidden.

I.5 Underlying assumptions

From geometry we find the necessity of quantum mechanics, the dimension of space, and much else [Mirman (1995b,c)], and hope to find more. What assumptions lead to these? How much freedom do we, and nature, have? Could we change assumptions and obtain different physical laws? The answers that we get are perhaps disconcerting, and perhaps incorrect. Obviously considering underlying assumptions is fundamental, and greatly adds to understanding. But in addition, we present them here — as well as we can — to challenge the reader to find whether our list, or lack of it, is correct, and whether there are others not stated. We leave it to the reader to decide whether the extent of the — apparent — freedom in these assumptions is valid, whether others are possible and reasonable, and lead to possible physics.

It might seem that symmetries, invariance transformations, particularly of space and time, are important. While discussions are often phrased using these, very little actually needs them. The groups are the transformation groups of geometry, not necessarily symmetry groups (although it is interesting that they turn out to be that also). Most of what we obtain holds even if symmetry were badly broken, and often require no more than that there be invariance in some limit, as when a field becomes zero, or the curvature of space vanishes, so space becomes flat (and often not even this). Whether symmetries are actually broken is generally not relevant.

I.5.a Are there assumptions if there can be no other universe?

One assumption is that there are objects in the universe and that these interact. But this is necessary, it would be a void universe unless there were objects and interactions. This illustrates the problem. We can assume, for example, that a system is frictionless, because there are systems with ranges of values of friction, and this assumption picks out the subset of objects that we wish to consider from a larger set. But we cannot assume that the universe consists of only objects that do not interact, for then there would be no universe, nor can we consider that there are some objects that do not interact, for then they would not exist. We thus have no freedom in our choice of this assumption.

Likewise to seek assumptions on which derivations of the dimension of space, of quantum mechanics, of much else, are based [Mirman (1995b,c)] is to ask for universes with reasonable, but different, physics and then ask for particular postulates that select from the set of possible universes the one that we are considering (necessarily, because we live in it). But there are no other universes known to be conceivable, and apparently no way for there to be other realistic possible laws of physics, complete and describing a sensible universe. We also thus seem to have no freedom in our choice of these assumptions.

We must assume that the universe is "reasonable", an undefined term, but here it does not matter how it may have to be defined. While we cannot (now?) give all requirements for a reasonable universe, we can say that others are not.

What is a proper universe, what are the criteria? Our universe is clearly one, but a universe that consists only of an ideal gas, a universe without interactions, is not. Thus we cannot say that the requirement of interactions is an assumption. Rather it follows from the necessity that the physics in the universe (so the universe) not be vacuous.

To find the assumptions we use we must compare them to ones giving other reasonable universes. But we do not have a complete theory of even one reasonable universe — not even of ours. Thus we cannot compare ours to other reasonable ones, or even say that such are realizable, conceivable — and it seems unlikely that others are. Conditions for a reasonable universe seem to make different fundamental laws impossible.

Therefore it does not appear that assumptions underlying our basic (perhaps all our) results can be found. To do this we would need another example of a sensible universe, even one that we can just conceive of. None are known, nor even approached. So it may be that there are no assumptions, there is no freedom — which much here seems to imply. It may be that only one nontrivial universe is possible. Everything (fundamental?) then follows from the fact that our universe actually has, can have, physics going on within it.

But this is a question worth much further exploration, if possible.

I.5.b The necessity for regularity, and the need for inertia

A fundamental belief (not only here) is that there are laws, regularities. This may seem obvious, trivial, trite, vacuous. And it would be unfortunate if it were not true. It is a fundamental requirement, but apparently a necessary one — how could we exist if it were not satisfied, how could a universe exist? Yet there are those who seem to deny this, of course inconsistently. For example, some believe that inertia is caused by the presence of distant matter in the universe (as many historical reviews emphasize [Norton (1993)]). This would mean that if there were only one object in the universe its motion would be completely haphazard, jumbled, its velocity changing randomly. Of course, if this were the only object in the universe such a statement would be without content — there would be no way of checking it. However if there is an additional object, unless we make the silly assumption that its presence determines the behavior of the first, this second object, the observer, would see this disorganized, random motion — this unlawful motion. We can take the motion of the observer to be the same as that of the first object, random and disorganized in the same way, and regard the second object as seeing the first behaving lawfully. However the concept that inertia is determined by the objects in the universe then becomes empty. This can be extended to any number of objects.

Inertia thus is a reflection of, or requirement for, existence of laws of physics. Also as we see [Mirman (2001), chap. I], there is a basic geometrical difference between straight lines (like paths of particles with constant velocity) and of circles or hyperbolas obtained from circles by making one coordinate imaginary (paths of objects with constant acceleration), and other curves. Thus to say that inertia depends on the presence of other matter in the universe, implies that existence of physical laws, and properties of geometry, also depend on its presence, because otherwise the motion of an object would be totally random, obeying no laws.

The necessity for inertia comes from the existence of laws, and from straight lines being distinct (due to the geometry of our world) in that they are generated by the members of the group over space. Also equiangular hyperbolas, as we see, have an intrinsic connection to geometry, leading to the relevance of the conformal group [Mirman (2001), chap. II]. This property of geometry and its transformation group does not depend on observers or physical objects, near or far. And to look for the reason for the law of inertia, perhaps the simplest law and one that is based on a fundamental property of geometry, as a consequence of other laws, is to try to understand the simple, the obvious, the basic, in terms of laws that are, at least, no more fundamental, usually (apparently) far less so. If motion can be random, why should we expect that there are other laws from which we can obtain the nonrandomness of motion?

To assert that inertia is caused in some mysterious way by other objects, is really saying nothing — it is a sentence without content, and another of the many, many examples of such (chap. IV, p. 142).

Yet if the law of inertia is one of the least mysterious of laws, it does point to a more preeminent mystery. Why are there laws at all, why are there regularities? Why is the universe lawful, simple, knowable? However there is no way of answering these questions. The reasons for the knowability of our world are forever unknowable.

I.5.c Assumptions, formalism and language

Of course, this all follows from the formalism that (we must?) adopt. But why must we use it, can we express the (required) transformations in other ways? Yet this formalism does work, it does give correct answers. Why? And so have we obtained the answers about assumptions underlying our work, and the freedom that we have in picking them — or have we just raised the questions that need to be studied to understand why the universe is governed by the laws that we find, and why these are (apparently) so definite?

I.5.c.i *Assumptions come from language*

Many assumptions here come from language that we use, thus tend to be hidden. Words that we use to describe nature make many presumptions about it. The question, which we cannot consider here, is whether it is possible to use other language, other terms, so use other assumptions, thus giving a different view of physics, of the laws of nature? Or perhaps we have no choice, the language we use is the one that we must use, for that is the only one that describes, can describe, nature. This is another question worth exploring, if it is possible to explore it.

I.5.c.ii *Where are the assumptions and what can we do about them?*

One reason that it is difficult to find assumptions is that they are hidden in subtle aspects relating symbols to physical operations, and without such relationships symbols have no meaning. But physical operations are in fact quite complicated, symbols are simple. Thus understanding all parts of an operation, and all its subtleties can be impossible (sec. I.2.f.v, p. 11). This is often mentioned below for it is a fundamental difficulty in trying to comprehend the meaning, not only, of quantum mechanics — but of physics, of life.

Often explicit rigorous systems are set up, with clearly stated postulates, and these are used to show the correctness of the relevant physical theory, such as quantum mechanics. The problem is that the postulates avoid the main problems, which are thus not considered, the physical meaning of symbols and operations that they refer to (in the postulates, but often not in the following discussions). Thus the rigorous mathematics which fully proves the results wished for is irrelevant to physics.

And postulates do, in fact must, leave out important requirements, they must because not all physical requirements are known (and perhaps can never be known). Yet even when these are known, and fundamental (like the required transformation properties [Mirman (1995b), chap. 5, p. 85]), they are often simply ignored giving rigorous mathematical systems that have, can have, nothing to do with physics.

I.5.c.iii *What can be postulated about quantum mechanics?*

In addition postulates can be inconsistent, not mathematically, but physically. There can be physical requirements that some postulates, which may be unnecessary, are in conflict with and thus cannot be included.

For example, there is the belief that measurement of a variable forces a quantum mechanical system into an eigenstate of the corresponding operator. While this is a fundamental abstract postulate of quantum mechanics, it is not clear that it can be so postulated in a consistent system.

Suppose that we measure the position of an electron by scattering an object from it, and assume that we can know the position and momentum of the scattered projectile exactly (simply to avoid imprecision due to it, otherwise we would know even less about the electron). The original state of the electron is a wavepacket. From a study of the outgoing projectile we can find the final position of the electron with infinite precision, and thus have forced it into a position eigenstate (and similarly for a momentum eigenstate). Is this really true? Is the final statefunction of the electron a δ function, rather than an extended wavepacket? From Schrödinger's equations, for the electron and the projectile, with interaction terms, and the initial statefunctions, we can calculate the final statefunctions. It is certainly not obvious that the calculation will always, or ever, give a final state of the electron that is a δ function, rather than an extended wavepacket, when that of the outgoing projectile is such as to allow a precise determination of the position of the electron. Given a set of differential equations, Schrödinger's equations, initial conditions, and some information about final conditions (the solution for the projectile) we cannot assume, without calculation, the rest of the solution. So can this postulate that the measurement forces the system into an eigenstate ever hold?

Statefunctions must be solutions of Dirac's equation (for half-integer spin), or its nonrelativistic limit [Mirman (1995b), sec. 6.3.3, p. 118], Schrödinger's equation. We cannot in general impose extra conditions on these solutions (such as that they be δ functions) without introducing inconsistencies.

As the basic rules governing quantum mechanics are known, like Schrödinger's (or better Dirac's) equation, it is possible to determine (in principle) behaviors of systems under all circumstances. Thus requirements like a measurement forces a body into a momentum eigenstate become questionable. Of course the word "measurement" is rather fuzzy, and it is necessary to specify exactly how the measurement is carried out, and show how it leads to the hoped for results. This then becomes an extra assumption, but perhaps an inconsistent one.

Often assumptions are stated as providing the foundation of quantum mechanics. However it is not (always) clear that the set of assumptions is consistent, and often, since the basic equations are available, whether all are necessary. If quantum mechanics is based on an inconsistent set of postulates then it would not be surprising for it to predict strange results. But this has nothing to do with whether nature is strange (it certainly is not), but is purely a consequence of the wrong set of premises.

Actually the postulates of quantum mechanics are not arbitrary [Mirman (1995b)], and if chosen arbitrarily can lead to systems that are inconsistent and wrong.

This is often not carefully studied, and not only in quantum mechanics.

I.5.d Is physics based on our biology?

One problem in trying to understand assumptions is that it is not clear to what extent they (if they exist), and the theories based on them, are themselves based on our physical and biological properties.

How much do our theories, our views of the universe, depend on our own biology? Certainly the languages we use are fundamental (in what way?), including (the accidental nature of?) mathematics — an essential language — which is related to our ability to count, the way we count, even the base of our numbers, this determined by the number of our fingers (so by the existence of fingers and of numbers). If dolphins constructed a (complete?) world view, and they have no fingers, would they have a reasonable base for their numbers, would they have numbers, language? Their language would be very different — to what extent is this (fundamentally) relevant? Would they be able to describe the physical universe, the world around them (as fully as — we think — we can)? How different would their frameworks, their theories, be? How different would their world view be? How different could these be? How deep could such differences go? Of course, they have a very different view of nature, at least from living in the sea rather than on land. Do dolphins see the world that we do? (Of course the word "see" implies eyes, like that of mammals(?) which are not universal, and vision; "learn of" might be better, emphasizing the difficulties.) Physical theories constructed by intelligent dolphins would be expressed very differently. Would they be the same as ours at some fundamental level? What is that level? And what does "same" mean here?

Of course if something jumps off a building, or drops an object (rigorously in a vacuum), it will fall no matter what organism does that, or how observations are expressed (although dolphins, who are familiar with objects supported by water, may not include that in their theories). And this is true of all laws of physics — that we know (?) are actually (?) laws. But determining differences between what is and what we view and communicate about nature, finding boundaries between laws of physics and their expressions, are not likely to be trivial.

If then other creatures, with completely dissimilar properties, views of the world, ways of viewing it, construct theories, what would they be like? These would be different in language, techniques, mathematics. But could they be fundamentally different? And what is the meaning of fundamental here? Could frameworks be different? Could theories? Might it be that frameworks must be the same, but theories could differ?

These are some of the major questions in trying to understand the extent to which our understandings really reflect the physical world. Do these questions actually ask anything (sec. IV.1.c, p. 146)? Of course if a chimpanzee were to see something become loose from a tree it would know that it falls, as would much lower forms of life. A worm would not, but we do not regard worms as having different physical laws, but rather lacking in the ability to have any. All animals capable of learning from such experience undoubtedly agree that the object falls. Yet can agreement with our fundamental views of nature go much deeper? Must it?

But perhaps the universe cannot be fundamentally viewed — completely and correctly — in ways that are greatly different from ours. It is one of the implications of much of the considerations here, perhaps rather disturbingly, that a fundamental, complete, correct — properly abstracted — picture of the universe cannot be really different from the one that we are apparently being lead to. Nature must be what we find it to be, although perhaps not what we describe it to be.

I.5.e Why must the world be what it is?

If the world must be the way it is, energy and momentum conserved say, the dimension 3+1 [Mirman (1995b), chap. 7, p. 122], quantum mechanical, and so on, must it be so because there is no other way of constructing a consistent physical universe, or because our biology requires that we see it that way? The indications are that there is only one possible universe, only one way of achieving a consistent (comprehensive) physical system. But it would be interesting to find if dolphins, or inhabitants of other worlds, other planetary systems, other galaxies, would disagree, if they could. Certainly language, mathematical descriptions, ways the universe is discussed, would be vastly different, but how deep would, could, such differences really go?

Scientists believe that the laws of physics are the same everywhere, in every part of the universe, that all alien creatures would find the same laws. But what does this say? What does "same" imply here?

These questions are impossible to answer, but perhaps they are worth pondering — maybe hints can be found, hints that would clarify assumptions, views, understandings, descriptions, languages, mathematics, biology, on which we construct our beliefs about nature.

I.5.f The wonder of simplicity

What is the greatest wonder of all? Eugene Wigner pointed out the remarkable effectiveness of mathematics in physics. Perhaps more remarkable is its effectiveness in mathematics. That it is consistent is far from trivial, but quite necessary [Mirman (1995a), sec. IX.2.c, p. 253]. For example the dimension of space is determined by several equations. None can be expected to have integral solutions, but all do, and all the same. Thus the dimension must be 4, and can be 4. But this is not unique. However for dimension 4, and this only, the orthogonal-group algebra is not simple, but only semisimple. Thus there is only one dimension that is possible, and fortunately it is possible: 3+1 [Mirman

(1995b), chap. 7, p. 122]. This result, that it is possible (quite accidentally?) and unique (quite accidentally?) is stunning. This is a miracle (?) that should be far better appreciated (sec. IV.4.b, p. 189).

We see more and more that laws of nature, and reasons for them, are simple and understandable. Yet why this should be so, even why the universe is possible, is, as our study impresses on us, a deep, even startling, mystery. The extraordinary simplicity of our universe, that laws of nature are knowable, understandable, can be known and understood in such a simple way, leaves us amazed. Yet no matter how extraordinary these seem, there is a puzzle larger still: Existence is the greatest mystery.

I.6 Fundamental definitions and their rationales

What concepts do we base these analyses on, and why? To discuss this, and of course anything else, we must define the terms we use. But here, as is usual, terms also contain underlying assumptions and formalism. Thus it is necessary to examine what we are actually trying to say, why that, and what the mathematical and physical meanings of our language(s) actually are. This is normally not done, certainly very explicitly. That lack leads to error and confusion. Here then we try to be very explicit and precise — and hope that errors, imprecision, ambiguity will be noticeable so that readers can correct them. Thus properly we have to start with terminology and language, this forming the foundation, and the core, of our discussion.

Since much, if not all, of what we do is based on geometry, that is the subject to begin with.

I.6.a What is a geometry?

Many arguments here come from geometry, especially from its transformations (thus group theory). How do we determine a geometry? What do we mean by geometry, by a geometry? Start with sets of sets of numbers, real or complex, perhaps others. Each set of numbers specifies (that is defines) a point, and the first property of a geometry is the size of each of these sets, that is the dimension of space (always assumed identical for all points) and the type of numbers in these sets — real, complex, or others. These numbers may not have continuous ranges — we might, say, consider points of a lattice or crystal and define the geometry using them [Mirman (1999), sec. I.3, p. 5].

Over these sets of numbers we specify functions, distances, angles, a metric, curvature, and so on. It is these functions that determine the geometry. They allow transformations, some — symmetry transformations — perhaps not necessarily all, leaving the geometry invariant (where the meaning of invariant may need careful specification). Taking these transformations, symmetry ones or not, to form groups, we obtain representations and their basis states (whose form depends on the realization [Mirman (1995a), sec. V.3.c, p. 157]), and thus further objects defined over this geometry, vectors, different types of products of these vectors, perhaps, especially for less familiar groups, additional objects [Mirman (1995c)]. As we will see the nature of the realization is crucial (and because there are only few types that are common) usually not noticed — so at times emphasized. Thus, starting with a set of sets of numbers, we have postulated a structure over it. Of course these definitions are not arbitrary, they must at least be consistent, and hopefully form a minimal set (the axioms of the geometry).

For physics these geometrical objects are identified with physical ones, that is we give them names that we have decided to use in physics. This identification gives a physical theory. But assignment of names is not just a matter of arbitrarily picking labels, but an essential, and difficult, physical procedure. We have to consider, in depth, what this is comprised of — or rather start to consider it, so at least to raise some of the questions and point out some of the difficulties involved.

Geometry has intrinsic properties, like dimension or the sum of angles of a (small) triangle. These do not depend on physics, but they are restricted by physics — it is only in a (the?) geometry with (the?) correct properties that physics is possible [Mirman (1995b)].

It is important to understand this process of developing a geometry, and giving it physical meaning. Possibly we could get equally correct physical theories using different geometrical definitions and names, and different physical names for the geometrical objects. For example, the sets of numbers that we started with might be regarded as a Euclidean space, and objects like curvature and the metric giving it taken as simply functions defined over that space (sec. III.6, p. 139). Hence we could choose our language to describe the physics over the space, either say that it is flat, with objects like curvature only functions over these sets, and with physics then depending on these functions, or we can say that space is curved, with physics taking place in the curved space. It is possible that only one view could give correct physics, or it might be that both languages are useful, maybe one more than the other, maybe not. This is something to consider for each theory in which properties of geometry are fundamental. Physical concepts may not come from the nature of the geometry, but rather from the language we have chosen. If we are aware of this, we might be able to choose a better language, giving a theory, perhaps just as correct, but simpler. But it would at least help us to know what is required, and what is chosen.

Understanding assumptions underlying a theory is always useful — although rarely easy.

I.6.b Why space is flat

Given a space, we can arbitrarily label its points (assuming its dimension is fixed). Geometrical properties of the space are determined when a distance function $d(x, x')$ is specified. That the space we live in has a certain geometry, it is Euclidean, is not required by the laws of geometry (whatever these are), but is in fact a physical assumption, something observed from experiment. The distance function is an experimental result. Why is it Euclidean? The reason for the dimension and (so) signature are known [Mirman (1995b), chap. 7, p. 122]. It would probably be difficult to have a (reasonable) physics if the space had singularities. Thus we assume a manifold, and the question becomes why is it flat?

Being a manifold it must be flat in a small region. If it had curvature then there would have to be a distance built into the geometry in order to tell the size of the region in which the curvature becomes apparent (no matter how precise we choose to measure, once the precision is fixed, a distance is necessary). That is, there has to be a radius of curvature. There are clues about why there is no distance. What distances are there in physics? The largest intrinsic one is the Compton radius of the electron (or a few magnitudes larger using neutrinos). The implication is that (except perhaps for gravitation) these do not affect geometry. The properties of our geometry are not affected by the objects within it (except that they limit the types of geometry possible), and the distances attached to these objects.

It might seem that there is an absolute distinction between spaces that are flat and ones that are curved. A sphere however is given by a function over a flat space — giving the points of the spherical surface. But the flat space is of higher dimension than the sphere. However we can also change the radius of a sphere, making it eventually into one with infinite radius — a plane. And we can reverse the transformation to change a plane into a sphere, which can be deformed (by nonlinear transformations) into other shapes. Thus we can regard the surface of the sphere as determined by a function over the plane, the function being the transformation that takes the surface of the plane into that of the sphere. And the sphere can be pinched, and parts cut out, to give surfaces with holes. The effect of transversions, a subset of the transformations of the conformal group [Mirman (2001), chap. I], on lines and circles suggests some of the possibilities, as we see [Mirman (2001), chap. I,II]. These transformations are nonlinear, perhaps with singularities. But they do provide functions over the plane (or hyperplane for larger-dimensional spaces) that take it into geometrical figures of very different form.

These suggest that we can take space as flat and objects like curvature as functions over it — or perhaps that we can use a language of flat space with such functions. Or perhaps these say the same thing.

I.6.c Space, time, now, then and causality

Space and time are taken identical in the theory (except for aspects related to the signature being non-definite). Yet intuitively they are very different, and while it is clear that this intuition is purely subjective, and not part of a correct fundamental theory, it can cause problems, so we should be aware, and careful, of it. To an external observer (who, of course, cannot exist) events in our brains are given by a statefunction (quite complicated) defined over space and time, with space and time having the same status (just as they do if we observe, say, an atom) except for the signature of the metric.

While we might feel that there is a difference between the past and now, past and future, now and the future, there is not — in terms of the physical laws and the complete physical description of the system. Statefunctions of systems, including our brains, can be defined as depending on space and time, so never changing (there is nothing to change with respect to). Distinctions between space and time are purely subjective, but can cause confusion in discussions of quantum mechanics.

Thus quantum mechanics is completely reversible — there is no fundamental distinction (in this sense) between past and future, just as there is none between the positive and negative z axis. There can be no irreversibility in a quantum measurement — it can always be run backwards. However the system (statefunction), perhaps including that of our brains, at one time (subjectively after the measurement) can be vastly more complicated than at a different time ("before" the measurement). Such aspects as complexity can be used to determine a direction of time. But these are not inherent in the laws of quantum mechanics — however they are very relevant in discussions, for one reason, because they cause mistakes if not handled carefully.

Is physics causal? How can it be if there is no distinction between past and future? What causality means is that laws of physics (but not our computations using the laws) make exact predictions (of a statefunction) on any space-like surface, given the statefunction (actually of the universe — this meaning that everything relevant must be known) on any other space-like surface, in the past or future (retrodicting might be more difficult because the complexity of statefunctions might change with time, but that is not here relevant). Knowing boundary and initial conditions (completely) at one time (on one space-like surface) for a statefunction of a system determines it at all other times ("before" and "after").

It is in this sense that causality has meaning — it is a statement about the nature of laws of physics, and boundary and initial conditions they require and allow. Our subjective impressions, or feelings of time, are irrelevant.

Related to this is what we mean by a law of nature. It may be defined as having the ability to describe all initial states of physical systems in such a way that there is a rule allowing the final states to be predicted uniquely from the initial state using the nature of the system and its interaction with its environment. The absence of laws then implies that identical physical systems may start from the same initial state and end up in different final states. This statement may seem consistent with our observation of quantum phenomena, in that we can only give probabilities for final states. Quantum mechanical laws are, in this sense, different from classical ones. Quantum mechanical systems definitely obey laws, specifically the laws of quantum mechanics. But these give, not definite final values, but definite final statefunctions, so definite final probabilities.

I.6.d What is a field theory?

A field is a function, of both space and time [Mirman (1995b), sec. 1.3.2, p. 4]. Since it can be viewed by all observers, no matter how transformed (by Poincaré-group transformations), it is a representation basis state (sec. I.8.b, p. 48) of the group (and perhaps others, like that of isospin). This is true for any statefunction. Thus there is no difference between quantum field theory and quantum mechanics. Generally quantum mechanics has been taken to mean the theory with a fixed number of particles, quantum field theory with a variable number. But there is no reason to make this distinction; quantum mechanics (with a fixed number) is simply quantum field theory, but for the special case of the number of particles fixed. Quantum mechanics is thus quantum field theory (sec. II.5, p. 82) with multi-particle states (sec. II.5.d.i, p. 98) restricted to only a fixed number. The formalism, done properly, does not, cannot, make a distinction between a field theory and ordinary quantum mechanics (although the Poincaré group does distinguish between single basis states and products [Mirman (2001), chap. IV]).

Thus there is no independent concept of field operators. The operators are actually ones used to construct basis states of representations of the relevant group, which is what fields are.

What is a field theory [Mirman (2001), chap. IV]? We wish to determine the behavior of a field. Since it is a representation basis state, we first give the representation and state labels [Mirman (1995a), sec. I.6, p. 25]; for the Poincaré group, these are mass and helicity (for a massive object the spin in its rest frame), with states labeled by the z component of spin and momentum values (for representations with states that are momentum eigenstates [Mirman (1995c), sec. 1.1.1, p. 2] — this may not be possible for all representations [Mirman (1995c), sec. 2.7, p. 28], so that the set of representations may be far richer than generally realized). Then we have to give equations governing the field. Actually we have just done that, at least for the familiar groups and representations for which this is possible, for these are the group invariants, giving representation labels, and the diagonal operators, giving state labels. If there are other groups, corresponding operators and eigenvalues are also needed. These give differential equations that statefunctions, fields, basis vectors, must satisfy. The real problem is that since the field has interactions, these operators contain nonlinear terms — the interactions, which must be stipulated or determined. However they link different fields, so each must be specified in the same way, that is its labels and interaction terms given. But these interaction terms may introduce further fields (as with hadrons), so the process continues, perhaps indefinitely.

Once all realizations, representations [Mirman (1995a), chap. V, p. 146] and basis states, so the differential equations, are given, the theory is complete, and all that is necessary is to solve the equations for cases of interest. Obviously neither this step, nor the preceding ones, are trivial.

I.6.e What do we mean by coupling constants?

Since interactions are fundamental, their strengths, given by coupling constants, must also be. What does "coupling constant" mean?

Unfortunately there is a great deal of confusion, caused by using the term in different ways, perhaps even in the same sentence. For example, it is sometimes said that it can change with energy. Or we have statements like ones [Schweber (1962), p. 555] claiming that if the charge of a body is observed far from it, the value is Q, but close to the object the charge is Q_o, a different value. That is charge Q_o surrounds itself with a cloud of electrons and positrons, and this changes it to Q. One problem is that Q_o is unmeasurable, so meaningless — thus the entire statement has no content (sec. V.7.f.i, p. 260).

The behavior of an object is given by its Hamiltonian — or equivalently for spin-$\frac{1}{2}$, Dirac's equation [Mirman (1995b), sec. 6.3, p. 114]. In these equations there are interaction terms, with numerical constants, the coupling constants. These are universal constants, defined by their appearance in these equations, so cannot vary with energy, nor with how they are observed. The fine-structure constant, for example, is $\frac{1}{137...}$, no matter what the interaction, the particular process, or parameters, or how charge is observed. This is what we mean by a coupling constant — the value that appears as the coefficient in a nonlinear term in an equation governing an object. Solutions of the equation are functions of momentum (including energy), but the equation itself cannot be.

In phenomenological theories, which are used when it is impossible to calculate using correct ones, phenomenological coupling constants may replace actual coupling constants. These, being purely phenomenological, and not physical, can have various properties that universal constants cannot, like energy variation. But it is important to distinguish between universal numbers, of fundamental physical significance, defined by correct (although perhaps unknown) Hamiltonians, the actual physical interactions that they contain, and phenomenological functions which are purely calculational (coming from models, simplified, but, usually fundamentally inconsistent).

What does "cloud of electrons and positrons" mean? This, of course, is also purely — only — phenomenological, a picture perhaps helpful but with no real meaning. We want eigenstates of a Hamiltonian, say that governing the electron and electromagnetic potential field and their interaction. Finding these exactly is impossible. Thus we use models and pictures. One is obtained by considering, say, the interaction to first-order in a perturbation expansion. Higher orders can then be depicted, and studied phenomenologically, with this picture of a cloud. But the charge on a particle does not, and cannot, vary with distance from the observer. What might vary is a function used in place of charge, with no real physical meaning, but one that allows calculation, something not possible with the physical charge. A better term, instead of charge, is effective charge [Cao (1997), p. 294], but, unfortunately the word "effective" is often omitted.

A variable coupling constant marks a theory as phenomenological. It cannot be fundamental.

These phenomenological concepts can be useful, necessary, essential, perhaps, but it is important to remember that they are not physical concepts, only aspects of models. Otherwise great confusion can, and has, occurred. Phenomenology is an indispensable part of physics, but often quite misleading because it is forgotten that it is purely phenomenological.

Originally the definition of coupling constant meant that it is constant, but, as emphasized with this, if definitions are changed without that being explicitly stated, there is a strong suggestion that befuddlement is desired.

I.7 Transformations and groups, their meaning and use

Perhaps the most important concept of much of the present discussions is that of transformation. The underlying philosophy, which clearly has strong experimental support, is that transformations, transformation groups of geometry, and likely other transformation groups, strongly restrict, perhaps completely determine, physics. One set of transformations that will be emphasized are those of the conformal group [Mirman (2001), chap. III], but the basic considerations and results are more general, relating physics and geometry. Thus we must consider what transformations are, what these transformations are, and what their relevance is, and why.

I.7.a Reference systems, geometry and physics

Transformations, external ones at least, are between coordinate systems. What are these systems [Mirman (1995b), p. 203]? This question is more subtle than it may seem, so we here just mention several points, and emphasize the need for further analysis.

Many definitions are based on observers, and physical objects they use. To some extent this is reasonable, for if objects and processes cannot be observed, at least indirectly, in principle, they do not exist. However underlying the concept of observer is the suggestion of human being. But laws of nature cannot depend on whether they are found by human beings, or whether human beings exist or not. We are new to the universe, and the laws of nature did not change as prokaryotes appeared and then evolved into eukaryotes, into chordates, into humans.

Thus an observer is a physical object [Mirman (1995b), sec. 5.1.3, p. 88].

I.7.a.i *Coordinate systems, and how they are given*

Coordinate systems that underlie our discussions are those found by assigning numbers to each point of the geometry. But how are numbers assigned if there are no geometers to assign them, and even if there are, how do they do so? And how do we determine the geometry to assign these numbers to if the points do not already have numbers attached?

The geometry is given by the sets of numbers, the coordinates, that define it, these having certain properties that are determined by, or perhaps rather determine, the geometry, which though limits them. The dimension is one such property, the sum of the angles of a triangle another, and so on. There are many ways of assigning these numbers, all giving the same geometry. Different assignments of numbers cannot change the dimension, or the sum of angles of a triangle. Many of these different ways are related by the transformation groups (as for example by rotations — but not those between Cartesian and polar coordinates?).

These thus are postulated: properties of the geometry, and coordinates systems assigned to it. Then it is the agreement with experiment of the deductions from them that justifies the postulates. Geometry is a fact of physics.

I.7.a.ii *Physical coordinate system are needed*

Of course to do experiments physical coordinate systems are needed, and these must have the properties (somewhat like, or perhaps required by, those?) of postulated ones. Why do we not just start from them, rather then introducing two steps, postulation and physical definitions to relate these postulates to the nature that we observe? In fact, we cannot. For to provide physical definitions, we must make assumptions about geometry and coordinate systems defined over it, and how they are related, and also related to physical properties and systems (physical objects or procedures giving the systems). To do otherwise is not to do anything differently, but rather only to hide what we are doing.

Of course, these assumptions are not arbitrary; except for they way they are expressed, they are greatly restricted, perhaps unique.

Many discussions of these topics start by defining coordinate systems, say using sets of rigid rods and clocks. But then they do not use the definitions any further, except perhaps at the end when specific experiments are considered. Thus they do what we are doing, postulating and — perhaps — only at the end giving definitions physical meaning, relating them to physical objects and processes. We are trying to make this distinction between the postulates and the physical assignments explicit. And it is essential to do so — in order that physical assumptions and postulates are not vacuous, and do not confuse and distract.

I.7.a.iii *Defining geometry, mathematically and physically*

What is done, here and usually in discussions of subjects like relativity, is best found by observations of analyses, rather than by taking literally definitions that are then ignored. A geometry is defined, and geometries can be defined without having physical meaning, and usually do not have, often cannot have, physical meaning. This is done by specifying properties and those functions that give — that are required to give — them, such as distances and transformation groups, with perhaps certain aspects, like symmetry emphasized. Also there are other functions over the geometry, like statefunctions, needed for later physical interpretation. From these postulates and axioms, conclusions are drawn (which we then wish to interpret physically). None of this, although physically motivated, has anything to do with physics — it is purely mathematical.

After this the geometrical objects, distances, curvature, statefunctions, and so on, are given physical meaning, and rules are stated to relate them to observable objects and measurable quantities. Determining and specifying this process is a difficult and subtle problem. The danger in trying to start with it — say by describing coordinate systems (such as stating that they are sets of clocks and rigid rods), which are then not further used — is that concepts introduced, such as distance, have intuitive meaning. But this makes it difficult to properly relate them both to the underlying geometry being studied, and to actual techniques that are used to determine them. Defining coordinate systems first helps hide the lack of proper explicit definition of physical objects — by the means for actually determining them, for giving them values that we can find experimentally. Yet it is just such definitions that change a purely mathematical theory into a physical one. It can be the most difficult part of developing a physical theory, so it is where the greatest chance of confusion, error, and of being mislead, can occur.

I.7.a.iv *Physical theories must make physical sense, not be rigorous mathematically*

An unfortunate tendency in much of recent physics is development of theories that are highly rigorous mathematically, but nonsense physically. Theories about physics can have no sense if their objects, axioms, postulates, have no physical sense, and are not correct physically. Mathematical rigor, and definitions of physical properties that are then not used or have no meaning, perhaps because they are not given explicitly enough, help hide the physical senselessness of the theory, so make such senselessness more likely.

I.7.a.v *Starting with mathematics, then going to physics*

Here then what we start with is purely mathematical, although physically motivated, and with objects often given physical names (so requiring care for these have intuitive meaning which may not correspond to the way they are actually used, or the way they will eventually be defined physically).

Then we consider how these correspond to our intuitive ones, how they are determined and measured by actual physical processes. But this is a quite difficult topic, undoubtedly the discussions are highly skimpy compared to what is necessary for a sufficiently thorough analysis and description — and perhaps must always be, for complete ones may be infinitely long. It is important to be aware of this, to see gaps, weaknesses, to consider how to improve associations, justifications, demonstrations, analyses, how better to relate concepts to physics.

I.7.b What is a transformation group?

Our understanding of physics, at least as considered here, is based on transformation groups of the (required?) geometry of the universe, for it is these that connect physics — objects and observers (which are identical) — to the geometry. So this (most) fundamental concept must be defined and analyzed.

Symmetries are not the most fundamental aspect of geometry and physics, transformations are, even if they do not leave space invariant.

The groups of geometry, the rotation, Lorentz, Poincaré groups, perhaps more general ones, are symmetry groups, but more fundamentally transformation groups, as is, perhaps in a slightly different sense, the conformal group. They are properties of the geometry, whether it is symmetrical under them or not — that geometry, so physics, is invariant under these groups is an additional property, and that such invariances are also present is interesting, but not understood, perhaps not (fully) understandable, but certainly to be noted.

What is a transformation group?

I.7.b.i *Transformations define the points of the geometry*

Points of space are related by a set of transformations, given by, and (largely?) defining, the geometry, this set forming a group. It is the transformation group of the space, whether space is invariant under it or not — for that is how points are related. Thus for (locally) Euclidean space, the transformation group is the inhomogeneous rotation group. To go from one point, or direction, in such a space to another, transformations of this group are used — and no other ones are or can be. For the pseudo-Euclidean space (with dimension 3+1) the transformation group is the Poincaré group. Curvature, say, may be different at different points, and axes may be singled out by the curvature tensor — it might be simplest with one choice of axes, for example — or the sum of angles of triangles may vary with position. Space is then not invariant under the group, but it is still transformed by it (its points are related by it) — that is a property of the geometry of the space.

I.7.b.ii *Defining the Poincaré group*

How is the Poincaré group defined? Given the geometry (locally) that is determined the dimension of space, that the coordinates are real (space is not complex or symplectic) and continuous (it is not, say, a lattice [Mirman (1999), sec. I.3, p. 5]), there is a set of transformations that leave these properties invariant. It is thus a purely mathematical concept — given the mathematical definition of the geometry, the group leaving that definition unchanged is fixed. It is the affine group (thus leaving the linear structure fixed) of a vector space that leaves the Lorentz form fixed.

While this seems to imply that the transformation group is defined so as to leave invariant certain properties of the geometry, it is not clear that this requirement is necessary, or can be imposed. Thus it is questionable that there is a group that changes the dimension of the geometry, or one that changes a space with real coordinates into one with complex ones. Also the set of classical groups is limited [Mirman (1995a), sec. XIII.3.b.ii, p. 379]. In particular transformations of a real space with signature (k,l) form inhomogeneous pseudo-orthogonal group IO(k, l). We leave open the question whether just giving these most basic properties of a geometry is sufficient to specify its transformation group (up to isomorphism), yet there does seem very little (if any) freedom once the type of coordinates, the dimension and (so) the signature are specified.

I.7.b.iii *Inhomogeneous groups and solvable groups*

The Poincaré group [Mirman (1995a), sec. II.3.h, p. 45] is inhomogeneous. Here we define an inhomogeneous group by just giving the definition of its Lie algebra [Mirman (1995a), sec. XIII.4.b, p. 382]. An inhomogeneous algebra is a semi-direct sum [Mirman (1995a), sec. III.5.c, p. 102] of a semisimple algebra and an Abelian one that transforms according to a representation of the semisimple subalgebra. The Poincaré algebra P is a semi-direct sum of the semisimple Lorentz algebra, L, and an Abelian part A, transforming under its defining representation [Mirman (1995a), sec. XIV.1.c, p. 404], often interpreted as momentum. Thus

$$P = L \oplus A, \qquad \text{(I.7.b.iii-1)}$$

and its commutation relations are

$$[L,L] = L, \quad [L,A] = A, \quad [A,A] = 0; \qquad \text{(I.7.b.iii-2)}$$

these are the general commutation relations for an inhomogeneous algebra.

The simplest such algebra is the two-dimensional Euclidean algebra, SE(2), which has elements M, A_1, A_2, with commutation relations

$$[M, A_{1,2}] = A_{2,1}, \quad [A_1, A_2] = 0. \qquad \text{(I.7.b.iii-3)}$$

However this is not only inhomogeneous, but solvable [Mirman (1995a), sec. XIII.3.a.i, p. 376]. The Poincaré algebra is not solvable. All its generators, unlike the SE(2) algebra, appear on the right.

That SE(2) is solvable is a (the?) fundamental reason for the properties of massless representations, electromagnetism and gravitation [Mirman (1995c)].

I.7.b.iv *The relevance and implications of transformation groups*

Why are transformation groups relevant, and why do they impose restrictions on geometry, thus physics? They impose limitations on a geometry because they define it. Why on physics? There are different observers — for interactions, if nothing else, produce them, as for example when an object decays. Their observations [Mirman (1995b), sec. 5.1.3, p. 88] must be related, and relationships between observers, thus observations, are transformations, and since observers have different coordinates, momenta and angular momenta, these are transformation of geometry. But if there is a transformation from one observer to another, there must also be one from the second to the first, the inverse. The product of these, which undoes them, is the unit transformation, which thus exists. And transformations can be from one observer to another, or from the first to a third, and then to the second, or can be of various other products — all must give the same results; each transformation is a product of others, so products must be in the set of transformations. (Also for the coupled, nonlinear, equations governing the behavior of physical matter to have solutions, there are strict conditions [Mirman (1995b), chap. 5, p. 85] which again leads to such requirements.) All these imply that the transformations form a group [Mirman (1995a), sec. I.4, p. 14], and since observers can have any real coordinates, a continuous group. But the sets of such (classical) groups are extremely limited. And as the group defines the geometry, this means that (local) geometry is extremely limited.

Geometry is thus limited by algebra, and physics is limited by geometry so by algebra. The universe, it seems, may not have much of a choice about what it is, what it is like [Mirman (1995b,c)]. Mathematics will simply not let it; algebra and geometry conspire to tell nature what it must be.

That the definition of the geometry is invariant does not mean that physics need be invariant under these transformations. If we model noninvariance with a magnetic field, then all directions are not the same, but observers are still transformed by the group. And we might conceive (perhaps) of a geometry in which the sum of angles of a triangle depends on the directions of its sides. The geometry itself would not be invariant, but the group would still give the transformations of coordinate systems (these transformations define the different coordinate systems).

This is true in general, giving the definition of the transformation group: the set of transformations, allowed by the geometry, between points and between directions, and so between possible observers.

I.7.b.v *Invariance and transformation groups*

The foundation of our discussions is that there is more than one observer (physical object) in the universe, their observations (the effects of other physical objects on them) must be related and that these relationships are transformations between the coordinate systems that objects define. These transformations are determined by the geometry, they form groups that are the transformation groups of geometry.

This is different from symmetry or invariance — that the laws of physics are unchanged by the transformations. That is an additional assumption. It is highly interesting that it does hold, but most of our discussions do not need it. Laws of physics to a large extent are determined by geometry, and would not change, or would change little, if they were not invariant under its transformations. The laws are deeply rooted, being based on the most fundamental properties of geometry.

I.7.b.vi *Invariance, covariance and gravitation*

Thus, for example, there has been much disagreement about foundations of general relativity [Norton (1993)], whether it is based on general covariance or not, and what general covariance means. Actually this is irrelevant. Einstein's theory really says that gravitation is a mass-zero spin-2 representation of the Poincaré group. This is necessary and sufficient for general relativity [Mirman (1995c), chap. 8, p. 135]. The assumption that it is a mass-zero spin-2 representation completely determines the form of the gravitational field, and gives Einstein's equation [Mirman (2001), chap. IV].

What is significant is not general covariance, or the principle of equivalence, but the representation under which the field, the connection, transforms (with no implication that it is a symmetry transformation). And general covariance is (usually) nothing more than gauge covariance, a property of the Poincaré group for massless objects, and possible for these only [Mirman (1995c), sec. 3.4, p. 43].

All disagreements about underlying postulates and their meaning are unnecessary and irrelevant, including such questions as whether the name "general relativity" is correct, whether Einstein's theory of gravitation is really a theory of relativity. The use here of the name "general relativity" for Einstein's theory of gravitation does not carry any implication that gravitation is, or is not, a theory of relativity. This is the name by which it is usually called, and while, like the name "quantum mechanics" it can be misleading (sec. IV.1.a, p. 143), it is used because that is the name by which people know it. History often dominates logic in nomenclature, and often in (too) much else.

The term "relativity" is relevant to the understanding of neither the special nor general theories. Special relativity is merely a statement that the dimension of space is 3+1, as it must be [Mirman (1995b), chap. 7, p. 122], while general relativity is a theory of gravitation that is (for free gravitation at least) completely determined by gravity transforming under a massless helicity-2 representation.

I.7.b.vii *What is the effect of translations with interactions?*

Momentum operators, exponentiated, translate an object. While the meaning of this is familiar for one momentum component, the Hamiltonian (or perhaps the statement is so familiar that it is never thought about, thus people who use it do not know what its meaning is), what about the other three components? These are related by Lorentz transformations so their properties are closely linked. If interactions appear in the Hamiltonian, as they must, they have to also appear in the other three momentum components.

With H the Hamiltonian, and $|t_1\rangle$ the state at time t_1, $exp(itH)|t_1\rangle$ is the state at time $t + t_1$. It, written in this form, is defined over all space. If at t_1 it is a one-particle state, and that particle can decay, at a different time it is a sum of states of one-particle and of states of the decay products [Mirman (2001), chap. IV].

If the statefunction is a wavepacket, then the probability of decay is different at different points of space. Thus the three-momentum, operator $exp(ip_\iota x_\iota)$ takes the state at one point (which we might want to consider being defined for all time), to that of another. The states, including the fraction of the state for an undecayed object, and the fractions for the different decay products, are different, at any time, for different points of space. Thus all momentum operators contain interaction terms, as required by their being Lorentz vectors. There is no difference here between the Hamiltonian and other momentum components.

I.7.b.viii *Why transformations determine physics*

Physical transformations imply that there are different observers, for example the reader and author, who look at the same event. Observations must be related, and by the transformation group. This places very strong restrictions on physics. Many, perhaps all, (fundamental) laws of physics are determined by the requirement that observations be consistent.

What physically is symmetry, and how can space be, and not be, symmetric under transformations? First consider transformations of an object being observed. If, say, the spin of an electron is rotated with respect to an external magnetic field its behavior changes. Space does not seem symmetric. Of course here we are rotating one part of the system with respect to another. But we can consider an intrinsic direction in space, with the behavior of objects depending on their orientation with respect to it. If a complete system is transformed and its behavior does not change, and this is true for every complete system, then space is symmetric under the transformation.

What if we transform the observer? A hydrogen atom, with the proton the observer [Mirman (1995b), sec. 5.1.3, p. 88], the electron the object, provides an example. The electron might be in an excited state, and fall to a lower one, producing radiation which is intercepted by the proton affecting its behavior. This is the observation. If space is not invariant under the transformation, simulated let us say by a magnetic field, and the spin of the proton were rotated with respect to the field, its observation of the electron would differ. Thus it would see space noninvariant both if the electron were rotated and if it were. So for space to be invariant under transformations, behaviors of all objects and all observers (which are the same) must remain invariant under transformations (of complete systems).

Of course symmetry of space is determined experimentally.

We can thus say that transformations are physics. The choice of a coordinate system may be made according to the convenience for the problem under study, thus giving, say, transformations between Cartesian and spherical axes. Gauge transformations are similar [Mirman (1995c), sec. 3.4, p. 43]. Sometimes it is unclear whether transformations are translations of one descriptive language to another or they are really physical processes. But there are transformations or else there can be no physics.

I.7.b.ix *Consistency and group theory*

One essential requirement (for consistency under transformations) is that all terms in an equation transform as the same basis state of the same representation (of perhaps different realizations) of the transformation group. This is not merely a matter of symmetry. It would be required even if symmetry were badly broken. Thus if there is a magnetic field, states of a representation go into sums of states of the same representation of the rotation group under rotations of the coordinates. If states in an equation were to transform as different basis states (say a scalar equaled a vector), then there would be (at least) questions of mathematical consistency. More basic than symmetry, is whether a theory is mathematically consistent. If terms were to transform differently, consistency would be extremely implausible. This gives, for example, that classical physics is inconsistent and quantum mechanics necessary [Mirman (1995b), sec. 1.4, p. 7], that the cosmological constant (sec. V.4.i, p. 238) must be zero [Mirman (1995c), sec. 8.1.4, p. 139], and much else.

Consistency requires more than that indices match, which is a condition that is neither necessary nor sufficient [Mirman (1999), sec. IX.8, p. 505]. Physicists often seem so obsessed by indices that they are unable to see absurdity even when it is glaring (sec. V.4.i, p. 238).

Most results come then, not from symmetry, but from the necessity for mathematical consistency [Mirman (1995b), chap. 5, p. 85]. It is very difficult, if not impossible, to imagine how classical physics could be correct, or the dimension of space other than 3+1 [Mirman (1995b), chap. 7, p. 122], even if space were not invariant under Poincaré transformations.

Theories violating such requirements, would be, not merely noninvariant, but either vacuous, or inconsistent.

I.8 Why states belong to representations, even without invariance

A fundamental postulate — perhaps better necessity — is that statefunctions of objects are basis states [Mirman (1995b), sec. 5.3.1, p. 101]. Why is this required (and not by us)?

Consider, at any time, two observers, related by a group transformation, for example two relatively rotated objects [Mirman (1995b), sec. 5.1.1, p. 86; (1995c), sec. 6.3.7, p. 109]; group operations transform the system, or equivalently the observer, at any time, so whether states vary with time is irrelevant. The basis functions that the observers see are different of course, but they are related by transformations. And by definition of a representation, group transformations take a state of any representation to a sum of states of the same representation; from any state the group transformations generate a set, and this set forms the representation.

While this is clear for linear realizations of semisimple groups, including the breaking up of these states into such sets, nonlinear realizations still need much further study.

Because of this necessity the rotation group forces statefunctions (that depend on coordinates) to be sums of spherical harmonics. Why? One reason is that, because of our geometry, these form a complete set for (well-behaved) functions of angles, which statefunctions must be (they cannot have singularities). And they must be functions of angles for we can use different axes, that is observers — and there are possible observers for axes with any orientation (with respect to an arbitrary fixed set). Thus statefunctions do, and must, satisfy the requirement that they be expandable in terms of spherical harmonics, so they must be such sums.

I.8.a Statefunctions are basis states

Statefunctions are transformed by operators forming representations of the relevant group — they are basis states of it. To see the implications of this consider the rotation group, for which basis states seen by observers are related by angles of rotation [Mirman (1995a), sec. X.4, p. 277]. A basis state of one observer is a linear combination of those of another, but the terms in this sum are all of the same representation, and all observers see states of the same representation. Rotation-group transformed states of total angular momentum l are functions of the untransformed states — of the same l. The states of each l value, for one observer, depend on, and only on, states of the same l of the other observer. While a statefunction may be a sum of states of different representations, each term in the sum goes into states of the same representation; terms in the sum from other representations are irrelevant to the transformations of it.

So in a rotated system statefunctions are still sums of spherical harmonics — necessarily — but the coefficients are different since the coordinate system is. The group transforming the coefficients is the rotation group, because 3-dimensional real space is transformed by the rotation group — a necessary (defining?) property of this geometry. This group transforms spherical harmonics, its basis functions, into themselves, into spherical harmonics of the same representation.

The crucial aspect is that the coefficients in one system are linear functions of those of any other, with the functions relating coefficients depending on the angles between the two systems. But, whether space is symmetrical or not, whether the Hamiltonian is invariant under rotations or not, the functions relating the coefficients for these axes are determined — by the group — and are such that the coefficients of any spherical harmonic l in a system depends on, and only on, the coefficients in another system for that l. Basis states change but the representation remains the same. Rotations cannot change the representation of the rotation group, for they are transformations of it.

And this is true for any group whose transformations take a possible observer (object) into another possible observer. Statefunctions must be sums of functions of the transformation parameters, these defining the space, thus must be representation basis states of the group. And statefunctions seen in one system belong to the same group representation as those seen in any other (at least for linear realizations of semisimple groups, for which representations are most clearly defined).

So in general functions that are in the domain of group operators, are sums of group basis states. And statefunctions, since they can be transformed, must be in these domains.

I.8.a.i *Statefunctions are basis states even without invariance*

Although it may be clear that if a system is invariant under a group its statefunctions form sets that are the basis states of the representations, if there is no invariance, why must they still belong to representations? One reason is that representation basis states form complete sets for statefunctions, so they can be expanded in terms of these. However the representation to which each term in such sums belongs is not changed by transformations of the group — a state of a representation goes into only states of the same representation, which follows from the definition of a representation.

The rotation group relates statefunctions. However ones defined with respect to different axes may have different properties. In a magnetic field a statefunction giving spin along the z axis parallel to the field belongs to a different energy state (state of different expectation value of the Hamiltonian) than does one giving spin along some other axis z'. And one state can be constant in time, the other not. But two observers, with axes z and z', looking at the first statefunction agree on its rotation representation, as they do looking at the other statefunction. (There is an assumption here that states break up into sets — representations. This is true for linear representations of semisimple groups. However for inhomogeneous groups, and especially nonlinear representations, we have to be careful about what we assume [Mirman (1995c), sec. 1.2.1, p. 7]. Our experience with familiar cases might mislead if we do not remember what it is restricted to.)

As another example putting a crystal in an external field breaks symmetry. Basis states then have different energies, so are no longer degenerate and may not be time-independent — the system can go from a state to others (of different energy, exchanging energy with the field). However they are still basis functions of a single representation [Mirman (1995b), sec. 5.3.1, p. 101; (1995c), sec. 6.2, p. 98].

Space group transformations [Mirman (1999), chap. III, p. 132] also illustrate this. For an electron in a crystal states of a representation can have different energies — the degeneracy is broken [Mirman (1999), sec. XI.2.h, p. 579]. But at special positions [Mirman (1999), sec. IV.6, p. 239], some degeneracy remains because there is symmetry so statefunctions are basis states of the symmetry group there. They must be representation basis states because, by symmetry, they can be transformed into each other. And they are states of the full symmetry group of the crystal even though the Hamiltonian is not invariant under it at all (or even most) positions — the Hamiltonian depends on the lattice, but the lattice does not depend on the Hamiltonian (physics depends on the geometry, but the geometry does not depend on the particular physical system).

So it is with all geometrical transformation groups (which need not be symmetry groups). That is why, even though states have different energies say, they still belong to a group representation, and the same one as that for all energy differences zero. Transformations relate statefunctions as seen in different coordinate systems, this is purely geometrical, so whether the Hamiltonian is invariant is irrelevant. This is one way of giving the meaning of a group representation.

I.8.a.ii *Statefunctions are physically required to be basis vectors*

Thus statefunctions are sums of basis states of the group of transformations over space. Why physically must this be so? That a function is a basis state of a group representation means that it lies in the domain of its operators — these are able to act on it. And for the transformation group of space, they have to be able to act on it; there are in the universe different observers (sec. I.7.b.viii, p. 42) — the universe does not consist of a single object (fortunately) — and these can observe (interact with, or with the objects like radiation produced by) any object [Mirman (1995b), sec. 5.1.3, p. 88]. But observers are different, and are transformed into each other by (perhaps the group containing) the group of space, the Poincaré group. Thus their views of the statefunction of an object are also related by group transformations, so these must be able to act on statefunctions, requiring them to be basis states of a representation of the group. Views of statefunctions must be related by the same transformations as those of observers because they also observe each other and each other's behavior, but these are determined by statefunctions of objects that they measure, and unless the views of statefunctions are properly related, the entire set of observations of behaviors would be inconsistent (the equations governing them would be inconsistent [Mirman (1995b), sec. 5.2, p. 90]) — physics would be inconsistent so impossible.

I.8.a.iii *Why angular momentum values are limited*

Why do particles have spin 0, $\frac{1}{2}$, 1, $\frac{3}{2}$, ..., only? Why not spin $\frac{1}{3}$? The answer is that only functions labeled by these (that are eigenstates of the rotation group invariants) transform under representations of the rotation group. And it does not matter whether it is a symmetry group. If a particle with spin up had a different energy than one with spin down, or one of these states were stable, the other not, there still could not be a particle with spin-$\frac{1}{3}$. Relatively rotated observers see different statefunctions of a spin-$\frac{1}{2}$ particle, but these are related by the rotation group. If one sees a linear combination of states, the other sees a different linear combination, but both are of only two states (one may be zero).

There is no way of getting three states with angular momentum other than one. An object with three states has twice the angular momentum as one with two. And this is true whether angular momentum is conserved or not, whether space is invariant under the rotation group, whether energies of states are the same or not. A transformation between two systems can be done in one step, in two, in any number, in an infinite number of ways. These must all give the same result for physical observations cannot depend on how we carry out the mathematical steps to correlate them [Mirman (1995b), sec. 5.2, p. 90]. But only representation basis states (here of the rotation group) have this property. Only (sums of) these then can be statefunctions of physical objects, and only integral and half-(odd)-integral angular momentum values are allowed for these states — by the rotation group, because it is a transformation group (of this space), not a symmetry group, although — interestingly — it is that also, an additional property.

Thus the rotation group is the transformation group of Euclidean 3-space, whether that, or physical objects, are invariant under it or not. Properties of objects (like angular momentum values) defined over this space, and their transformations, are given by this group, that is by the geometry of the space, expressed by the group.

And it has representations of all integral and half-(odd)-integral dimensions, but no others. Total angular momentum (as is well-known) is given by (or gives) the dimension of the representation (for this group). This determines the allowed angular momentum values.

That geometry forces statefunctions to be basis states of its transformation group places strong restrictions on them; that these, and only these, are the possible angular momentum values is an example.

I.8.b Why fundamental objects are given by irreducible representations

Statefunctions are representation basis states. But of what type of representations? Those of fundamental objects, electrons, protons, and so on — fields (sec. I.6.d, p. 32) — are basis states of irreducible representations of the Poincaré group (or perhaps a larger one of which it is a subgroup). Why a representation, and why must it be irreducible?

I.8.b.i *Representations of what groups are required to be irreducible?*

Elementary objects are given by irreducible representations. But of what groups?

Representations of the Lorentz and Poincaré groups for massive objects are determined by their states in the object's rest frames so only SU(2) is relevant. Massless representations are so constrained [Mirman (1995c)] that these questions are irrelevant for them. For example it is impossible to have a superposition of (so an object consisting of) a photon and a "graviton". But an electron is defined in its rest frame, determining its statefunction in all frames, and when translated.

States of different Poincaré-group irreducible representations cannot be a single object. The Hamiltonian must be of a momentum-zero representation [Mirman (1995b), sec. 1.4.1, p. 9], and a rotation-subgroup scalar. That would be impossible with an interaction keeping a sum of states of different irreducible rotation representations unchanged.

A particle in an irreducible representation remains in it, unless it changes to a different type of particle. We must regard a fundamental object as one described by an irreducible representation of the Poincaré group (and perhaps larger groups), and ones described by reducible representations as not fundamental, but composite.

With internal symmetry there are interaction terms that are not invariant under SU(3) or SU(2). So the requirement that an elementary object belongs to an irreducible representation becomes less clear. We often define an elementary object as transforming under an irreducible representation of the SU(2) subgroup of SU(3). Thus we consider the nucleon as a single object, the Λ a different one. But we can also take all particles of an SU(3) representation as states of a single one. It is reasonable to consider the proton and neutron as different, while it is not for a proton with spin up and spin down. And it makes sense to consider the nucleon and the Λ as a pair (or triplet) of objects rather than a single one. But clearly it would not make sense to consider an object that can have either spin-0 or spin-1 as a single object. It would be taken as (at least) a pair.

Thus for internal symmetry whether an object transforms under an irreducible representation is to some extent a matter of definition; however in fundamental ways neutrons and protons are really different, and considerations below apply to them. While it is at times convenient to take them as states of a single object, that of isospin-$\frac{1}{2}$, this cannot hide their dissimilarity, nor that between two isospin states and two spin states.

For the rotation group (or SU(2)), the question is if there were a particle with states of say spin-$\frac{1}{2}$ and spin-$\frac{3}{2}$, but identical in all other ways (including of course mass) could it be regarded as a single object, or must it be taken as two? Clearly a single particle transforms according to an irreducible representation.

I.8.b.ii *How reducible and irreducible representations describe objects differently*

One difference between sums from a single irreducible representation of a compact semisimple group, and from more than one, is that for a single representation, but not for a sum from different irreducible representations, there is a coordinate system in which the sum is a single state, so describes only one type of object in that system, therefore in all (the labeling operators can be diagonalized). There is always a z axis along which the angular momentum points, no matter what its direction in other coordinate systems. The statefunction thus is a single term in that system, along that axis. But there is no coordinate system in which the sum of angular momentum 1 and angular momentum 2 states is just a single term in general.

It might seem that for isospin it is not possible to find a coordinate system for which a statefunction has only one component, so the analogy does not hold. However, as seen for charge [Mirman (1995b), p. 213] similar analyses apply for the two different interpretations of the SU(2) group (and likewise for larger such groups).

I.8.b.iii *The representation is irreducible*

Most systems generally have statefunctions that are sums. Why for an elementary one do we require that the representation be irreducible? If the statefunction of an atom is a sum, a single experiment finds a particular angular momentum value but the value is different for each repetition — the statefunction gives the probability distribution for the total angular momentum. But an atom is composite. If we found an electron with different spin values, we would not consider it as a single type of object.

This can be clarified using the rotation group and a statefunction that is a state of a reducible representation, a sum from, say, total angular-momentum representations $l = 0$ and $l = 1$ (the sum of a scalar and a vector). The vector has, in some coordinate system, a single nonzero component, and this component in another system is 0. One term in the sum varies with rotations, the other not.

For total angular momenta $l = 1$ plus $l = 2$, a statefunction is (schematically)

$$|s\rangle = a|1\rangle + b|2\rangle. \tag{I.8.b.iii-1}$$

Measuring the spin component, say by finding the magnetic field produced by the particle, gives a sum of terms from different angular-momentum z components. Changing the axis along which measurements are made changes the sum. The two angular-momentum states transform differently, and a series of measurements would show that the sum state belongs to a reducible representation. What is wrong with this? If the object were always in this sum of states, with the same coefficients, if this were possible (which is certainly not true for the rotation group, and similarly for ordinary-type representations [Mirman (1995c), chap. 2, p. 12] of any semisimple group), would there be any real difference from it being in a state of a single, irreducible representation?

How do elementary objects and composite ones differ? Why if the statefunction is a sum from different representations do we regard it as composite?

I.8.b.iv *Differences between elementary and composite objects*

Terms from different representations in the sum giving a statefunction of an atom with different energy expectation values would lead to decay, so over time the statefunction becomes a term from only one representation. Clearly the original state was composite.

For the nonrelativistic hydrogen atom, for each principle quantum number there is a set of degenerate angular momentum states. Why do we regard such a system as composite? Some reasons are discussed below for the deuteron (sec. I.8.b.vii, p. 53), which apply for this system also, but there are others. While a statefunction of an isolated atom can be a sum of degenerate states, this is not true for one in a field. Then energies differ, and not all states are stable. Measurement of energies of a collection of atoms gives not a single value, but a distribution. We would not then think of the system as elementary, but compound.

The distinguishing property then is the response of the system to interactions. A set of free electrons, or pions, all have the same energy. But this is not true for a set of nuclei with statefunctions that are sums over different representations (or particles with states of different internal quantum numbers).

I.8.b.v *Interactions lead to irreducible representations*

The Hamiltonian is a scalar — it is the fourth component of the momentum generator of the Poincaré algebra [Mirman (1995c), sec. 1.2, p. 7], so invariant under rotations in 3-space. Consider an interaction of one object with total angular momentum l_o and another of angular momentum l_e to give a system with total angular momentum l_f, perhaps an atom interacting with a photon. Statefunctions of these objects (which themselves may be products) are sums over angular momentum states. The Hamiltonian must act on each term of the statefunctions, so must be a sum, thus of a reducible representation, that has to be reduced. The reduction is to a scalar — it has a scalar in its decomposition [Mirman (1995a), sec. XII.2, p. 340], and only a scalar.

To act on a sum the Hamiltonian has to be a product of the statefunction of a (possibly multi-component) object in state $|l)$ with the states in the sum. A first order term in $|l)$ could not do so for $|l = 2)$ states, but could for $|l = 1)$ states. Thus such an interaction would project out the $|1)$ state, and the statefunction of the object, originally a sum of states of different representations, would then belong to an irreducible representation. While higher powers could allow interactions with both terms, Clebsch-Gordan coefficients [Mirman (1995a), sec. IX.5.b, p. 266] would be different. The coefficients in the sum would have to be unchanged for every interaction, every possible scattering, else the state is composite. At best this is extremely unlikely. Otherwise there would be no definite sum of states that could be regarded as giving the object. Also by proper choice of values, after many scatterings the coefficient of one term could be made almost zero, so that the statefunction would essentially belong to an irreducible representation. Such an irreducible-representation state is thus the proper one to pick as the statefunction of a fundamental object — the representation is invariant under all transformations, and under all interactions (else we would regard the object produced by an interaction as different).

Probability distributions of values of coefficients of different irreducible representations after a scattering depend on variables like momentum, energy, angle, say. Interactions therefore change such sums, so they could not be taken as giving fixed, elementary objects, such as electrons or protons.

In general, except for (almost certainly impossible) representations of very unusual types, states of different representations would behave so differently, in several ways, that they could not be taken to describe a single, elementary, object.

Perhaps most important a sum would not be time invariant. An atom whose statefunction is a sum over different states decays, so coefficients of states in the sum go to zero (an example of this argument about the effect of interactions). A state that is a sum over different angular momentum states at one time thus becomes a different one later — it is not elementary. For a proton, which cannot decay [Mirman (2001), chap. IV], the spin is fixed for all time. as is true for a neutron, although it can decay. But for it a single object becomes several. Certainly one criterion for an elementary particle is that its parameters, mass, spin, isospin, interactions and any others, be well-defined (which its momentum or position or angular momentum z-component are not) and constant in space and time. Thus its statefunction must be a sum over basis states of a single irreducible representation only.

I.8.b.vi *Translation groups and how they differ*

There is one exception to this requirement that the representation be irreducible. A representation of both the Poincaré group, and its translation subgroup, is given by a single momentum value (for those representations in which momenta are diagonal [Mirman (1995c), sec. 2.7.2, p. 31]). Yet we regard an electron given by a wavepacket as elementary, even though the wavepacket is a sum of states from an infinite number of irreducible translation (so Poincaré) representations. Why? It is not possible to go to a frame in which the statefunction is a basis state of an irreducible representation. However for each momentum value in the wavepacket, there is one frame in which that momentum is zero, the object's rest frame for that momentum. And all frames are equivalent. Thus the wavepacket consists of a sum of states of elementary particles, all equivalent, except for the frame in which they are at rest, but there is in principle no way to distinguish these frames. They are equivalent, but different, so the particles are. There are essential differences between the states in the sum of different total angular momenta, but unlike these, there is no fundamental distinction between states making up a wavepacket.

An interaction Hamiltonian (which really decides what an elementary particle is), for example, treats all these wavepacket states the same, but it does not so treat different angular momentum states.

Perhaps most fundamentally, there are (in principle) states of an electron, say, with only a single momentum value, so a rest frame. And there is no difference in physical laws seen from the different rest frames. It is upon this that the definition of an elementary particle such as an electron is based, its representation in its rest frame (that is using the little group [Mirman (1995c), sec. 2.2, p. 12; (1999), sec. VI.2.c, p. 284]) and how that transforms under the full group. Representations of the full group are obtained from little-group representations, these providing the labels that are the parameters defining the object (and the values of the labels are the same for all these representations). A wavepacket is then a sum of states all of the same Poincaré representation, so describing the same elementary object, but referred to different coordinate systems.

I.8.b.vii *Why the deuteron is composite*

Why do we regard the deuteron as composite, even though it has but one state? Neutrons and protons are observed as free objects, and can be scattered from each other, and from the interaction observed the properties of the deuteron can be calculated (to an adequate approximation) to obtain (reasonable) agreement with observed properties. We can also calculate the interaction of the deuteron with electrons, protons, neutrons, and other objects, using the known interactions of the neutron and proton with these, and obtain agreement with observation.

Perhaps most significant, the deuteron has integer spin, thus is a boson. But this is not completely correct. Deuterons very close to each other, say (in principle) in high-energy scattering, actually see the particles of which they consist, and these are fermions. There must be corrections to Bose-Einstein statistics for deuterons taking into account their internal structure, that of systems of pairs of fermions. This shows that a deuteron has internal structure, and is not elementary. And scattering of a proton or neutron must take account of the exclusion principle for the particle and the corresponding one of the deuteron. In an interaction these elementary states participate, not that of the deuteron, except phenomenologically.

Thus a deuteron behaves as a system of two fermions, not as an elementary system. It is described by the product of a pair of irreducible representation basis states, and this effects its statistics, interactions, and other properties. The creation operator of the deuteron produces a group representation basis state with no implication of elementary particle properties, because to properly describe it and its behavior its state must be a sum of products.

Chapter II

Foundations of quantum theory

II.1 Comprehensibility of quantum theory

Physics is too often presented as a mysterious set of laws beyond understanding whose only justification is that they work. But why are these the laws, why are quantum mechanics and quantum field theory necessary, what determines their nature, their character, the way they describe the universe? Once we go too deeply into these questions they become unanswerable [Mirman (1995b), preface, p. vii], as seen at times here. Yet we can understand much about the reasons, why they hold, and why they are what they are, as strongly emphasized here and previously [Mirman (1995b,c)]. Nature must be quantum mechanical, and the nature of quantum mechanics, quantum field theory, must be what it is.

So now we continue the analyses to understand in greater depth properties of quantum mechanics and quantum field theory. We find that mysterious features that we just have to accept because they agree with experiment are not mysterious at all. They, and their necessity, can be understood, often quite simply. However too often they are not understood, not because they are strange or incomprehensible, but because they are not thought about carefully. So in the course of our study we have to often mention errors that lead to many of these misunderstandings.

Quantum mechanics and quantum field theory have to be the framework for physical laws, and they have to be what they are. Thus we turn to some of their properties to explore these arguments.

Although it is quite clear how and why quantum mechanics and its properties are required, there are attempts to explain it in other ways [Brukner and Zeilinger (1999); Zeilinger (1999)], such as using concepts like information, but these are often based on assumptions that do not necessarily have to be true, and in many cases have no real (physical) rationale.

II.1.a The reasons for physical laws are based on logic and experiment, not history or the occult

There is a belief among many physicists that quantum field theory is a relativistic extension of quantum mechanics, although it is merely quantum mechanics with the many-body aspects explicit (sec. II.5.d.i, p. 98). This is purely a result of its historical development. Too many physicists base their views, not on reason or logic or experiment, but on the way that physics has developed historically, as with the belief that spin-$\frac{1}{2}$ is a relativistic phenomena although it is actually a well-known property of the rotation group [Mirman (1995a), sec. X.6, p. 287]. They believe this merely because spin-$\frac{1}{2}$ is predicted by Dirac's equation, which is Lorentz invariant. But that does not mean that Lorentz invariance is needed, only that spin-$\frac{1}{2}$ happened to be predicted by a Lorentz invariant equation — although it could have been found from the rotation group. Because these physicists understand physics in terms of its history, even though the laws of physics have been around far longer than have the people who created the history, they are often mislead and confused (sec. II.4.b, p. 81; sec. II.4.c, p. 82).

There is also, for some reason, a belief that the TCP (CPT, PCT) theorem is connected to quantum field theory [Streater and Wightman (1964), p. 142]. In reality it is little more than a trivial consequence of elementary geometry [Mirman (1995c), sec. 4.2.6, p. 60].

And too many physicists look for the most complicated, most esoteric explanations rather than the simplest (sec. V.4.i, p. 238). But the simplest is most likely to be correct, and the most helpful [Mirman (1995b), sec. A.1.2, p. 178]. This can be seen again and again, and not only from examples appearing here.

II.1.b Complex numbers

Quantum mechanics is based on the complex number field, not on the real number field [Mirman (1995b), chap. 2, p. 25]. This raises two questions, why is the number field complex rather than real, and why is it complex rather than another such as those of quaternions and octonions? Complex numbers are fundamental and needed, but it is questionable whether others are (although it is difficult to show this rigorously since it cannot, apparently, be ruled out that a complete, consistent, physical theory can be found based on them that is not simply a rewriting of quantum mechanics, although this seems quite unlikely).

Undoubtedly other number systems can be used for some aspects of quantum mechanics [Khrennikov (1995)], but it is not clear that they can give a full theory, especially the required group properties. Certainly those who believe they can should investigate this, and perhaps will develop new insight.

Complex numbers cannot be replaced with real ones. There is no real number that is the square root of -1. A pair of real numbers is needed, requiring a new number system. Related to this is the one-dimensionality of the representations of Abelian groups over the complex numbers but not over the reals [Mirman (1995a), pb. V.5.a-4, p. 163; (1999), pb. VI.4.a-1, p. 297]. With real numbers representations must be two-dimensional, but a two-dimensional real number is a complex number, thus merely a rewriting of complex numbers in a different notation.

II.1.b.i *Complex numbers from group generators*

The eigenvalues of momentum generators are real numbers — these are measurable, and if they were not real that would present serious problems [Mirman (1995b), sec. 2.1.2, p. 26]. Momentum generators — generators of translations — are (thus) realized as

$$P_{\mu} = i\frac{d}{dx_{\mu}}; \qquad \text{(II.1.b.i-1)}$$

with eigenfunctions (schematically) $exp(ipx)$, complex numbers. The argument is the same for angular momentum generators, and so for the Poincaré invariants. This is a reason why quantum mechanics is complex; the complex field gives the proper representation basis states.

Why does P have this realization? Coordinates of space are real (so transformations over them have to be complex [Mirman (1995b), sec. 2.2, p. 32]), and P (exponentiated) is the translation operator. Thus P is defined as (schematically)

$$exp(iaP)x = x + i\Delta x + \ldots = (1 + iaP + \ldots)x; \qquad \text{(II.1.b.i-2)}$$

the limit gives P as this derivative. Hence that quantum mechanics is complex follows from the properties of transformation groups on real space.

This emphasizes again the role that geometry plays in determining laws of physics. That space is real makes quantum mechanics complex.

II.1.b.ii *Implausibility of other number systems*

Other numbers, like quaternions and octonions, and their algebras [Mirman (1995a), pb. III.2.e-5, p. 75] are unlikely to appear because they can be expressed as complex numbers — represented by matrices with complex entries — so do not add anything. This is likely true for other number systems (if there are any that can possibly be considered).

Statefunctions of quantum mechanics (group basis vectors) are complex functions. They cannot be real; might there be further problems if they were quaternions [Chevalley (1962), p. 16; Gilmore (1974), p. 4; Peres (1979)]? The transformations include translations, rotations, and boosts, giving the Poincaré group. This and its Lorentz subgroup are complicated, thus can have unusual representations [Mirman (1995c), sec. 2.3.2, p. 17], making a rigorous answer difficult, but perhaps not impossible since requirements for a consistent physical theory are quite strict. We take basis states of physical representations to be eigenfunctions of momentum, or angular momentum — labeling operators — or expandable in terms of these, as complete sets. This presents additional requirements on other number fields making a consistent physical theory different from quantum mechanics more difficult, if not impossible.

(Any function expandable in terms of basis states is itself a basis state. The phrase implies that there is a chosen complete set of mutually orthogonal basis states, spherical harmonics for example, and other functions are written in terms of them.)

For example, suppose that basis states were quaternions. We would have for momentum eigenstates, with i, j, k the quaternion units,

$$P_{\mu}\psi(x) = \frac{d(\psi_1(x) + i\psi_i(x) + j\psi_j(x) + k\psi_k(x))}{dx_{\mu}} = p_{\mu}\psi(x), \tag{II.1.b.ii-1}$$

so each ψ would be of the form $exp(ipx)$. This we could factor, giving

$$\psi(x) = exp(ipx)(1 + i + j + k), \tag{II.1.b.ii-2}$$

with the same coefficient for all basis states. It would be meaningless.

There is an important distinction between real and complex numbers, and other number systems — certainly geometrically. There are two infinite series of classical groups (besides sympletic groups whose connection to geometry and number systems is less clear), those of transformations on real spaces, orthogonal groups, and ones that are transformations on complex spaces, unitary groups; transformations on others are more restricted. But real numbers cannot form a proper basis for physics.

Also these groups lead to the dimension of space being 3+1 [Mirman (1995b), chap. 7, p. 122], and it is questionable whether other spaces can satisfy the necessary conditions giving the dimension. These indicate that it would be very difficult, quite likely rigorously impossible, to have a fundamental physical theory based on any numbers but complex ones.

The difference between going from real numbers to complex ones, and going from these to other number systems, is that a consistent formalism cannot be constructed over real numbers, transformations of space require that they be realized over complex numbers, but this field does allow a consistent formalism. Hence there is no need to go further. This strongly implies that correct formalisms over other fields can be converted to ones over complex numbers.

II.1.b.iii *Hermiticity and unitarity*

Why must the operators of quantum mechanics be hermitian, so the finite transformations are unitary? For position and momentum the reason is clear. Otherwise their eigenfunctions become infinite [Mirman (1995b), sec. 2.1.5, p. 29], and it is very difficult to believe that there could be a possible world of physics in which that happens (although anyone who disagrees may contribute to our understanding by constructing one). Also there are regularity conditions on eigenfunctions needed for a possible physics, such as those of the Lorentz group, and this forces hermiticity. We shall not attempt a general answer here (but it may be that one can be found), for that would probably require examination of every operator, and of course there can be many different ones in specific theories. It is likely that regularity requirements are similar for each.

Space is defined by the translation operators, these acting on a basis vector at one point take it to all other points, so defining the points (sec. I.7.b.i, p. 38). If momentum operators were not hermitian, so translation operators not unitary, that would imply that coordinates of space (so space) are complex, not real. And this, as we have seen, would give results that are quite unphysical [Mirman (1995b), sec. 2.2.1, p. 32].

II.1.b.iv *Why probability is relevant to hermiticity*

An essential requirement of a fundamental theory is unitarity of the translation operators, giving the momentum operators hermitian. This follows from conservation of probability (if the total Hamiltonian were not hermitian, so the time-translation operator not unitary, the total probability would vary with time). This raises the question why must probability be conserved? Of course, it would be very difficult, likely impossible, to have a reasonable, consistent physics if it were not.

For a set of free objects the total energy of a system is the energy of each multiplied by the total probability of finding it. But the total energy, and energy of each, are time invariant because the Hamiltonian is the time-translation operator, and also space invariant since it commutes with all momentum components, being a Poincaré inhomogeneous operator. Otherwise space would not be Poincaré invariant. For free objects probability must be conserved. This can also be used in a definition of probability, giving conservation by definition.

Could the Hamiltonian H depend on time t so its eigenvalue would be time-dependent and energy not conserved? Time is defined by the time-translation operator, $exp(iHt)$, but H is a sum of terms (for free particles) each acting on one of the particles. Thus it actually defines different times, one for each object (each "clock" is independent). If laws of physics were time dependent we could take the time given by one clock (free particle) as the correct one and then Hamiltonians for others could depend on this. Thus their eigenvalues would also vary with time, and the total energy would not be conserved. For example in a system of several particles, if one disappears the total energy and probability change. The point of time when the object vanished would be distinct, so laws of physics would not be time-independent. This does not seem to happen, perhaps fortunately, energy and probability are conserved, the Hamiltonian is hermitian, and the time-translation operator unitary.

It is often believed that time-translation symmetry ensures energy conservation, and space-translation symmetry give momentum conservation. But this cannot be true. Energy is the eigenvalue of the Hamiltonian, which is the time-translation operator, and the Hamiltonian must, of course, commute with itself so its eigenvalue is constant — energy is conserved. Thus energy is conserved by definition. But there really is a physical assumption here, although it is not correct to state it as time-translation invariance. The assumption is that there are statefunctions that are eigenfunctions of the momentum operators (including the Hamiltonian), and that physical systems can be described by such statefunctions. In fact, the statefunction of a total system is such an eigenfunction. For these momentum is conserved. But this would not be true if particles could appear and disappear.

Does the relationship between conservation of probability and of energy change if there are interactions? What do we mean by conservation of probability if an object decays becoming more than one [Mirman (2001), chap. IV]? At the initial time the probability of finding the object is 1, that of finding sets of decay products is 0. The probability of finding the single object decreases, those of the sets increase. At all times the total probability of finding something is 1. Further if an experiment at some time finds the original particle, there is no possibility of finding decay products, and if decay products are found, the original object will not be. Of course, if both the original object and its decay products were found together, energy would not be conserved. Conservation of energy gives conservation of probability here.

With a change of particle number, constraints due to conservation of probability are less straightforward than for a fixed number of objects. Hermiticity could be a stronger condition for it restricts the form of momentum operators more than simply requiring them to annihilate one object when they create another, and conversely.

II.1.c Probability and the absolute square of statefunctions

Probability is given by the absolute square of statefunctions. Why? There are two questions here (beyond the already discussed need for probability [Mirman (1995b), sec. 3.3, p. 47]), why the absolute value, and why the square? Are these just incomprehensible postulates that just accidently happen to be true, or can we derive them, understand them, from more basic requirements?

Probability usually refers to the "probability of a particular classical outcome", that is if we perform experiments and find a mark on a film, a bubble in a chamber, a pointer, say, all in a certain position, the probability (distribution) is the function of position that gives the fraction of the experiments in which the macroscopic object is found at that position (or within a small interval about it). We use experiments such as these to illustrate the arguments. The term probability might appear in other ways (so for example in interference experiments, if we wish to discuss probability for them, phases are quite relevant). But the use here is the most common, and consideration of why any aspects of these arguments are irrelevant for other cases aids understanding of them. Words are often used in many ways, so it is impossible to discuss all types of experiments in which a word like probability might be used. However these illustrations should be sufficient to show why such rules of quantum mechanics are necessary, and how they arise.

II.1.c.i *The phase cannot appear*

There are several reasons why the phase of the state of the incoming object (of the cloud chamber experiment, for example [Mirman (1995b), sec. 3.2.1, p. 43], but the considerations are general) is irrelevant, so only absolute values give observed results. The initial value of the phase is arbitrary, for example when it depends on the origin of the coordinates, as with $exp(ikx)$ or $exp(i\omega t)$, and this origin is arbitrary. Also to determine statefunctions we must repeat experiments many times, and phases average to zero.

Also what is relevant is not the phase (the phase of a single object cannot be measured, so has no meaning [Mirman (1970), p. 3356]), but the phase difference between the incoming object and, say, the struck molecule. But we do not measure the latter, so neither phase can affect observations. And if we did measure the phase of the molecule, we would change it. The statefunction of the atom resulting from the breakup, so of the system which the bubble in the chamber or the mark on a film is for example, does depend on this phase difference. But even if we knew it, the dark spot is altered by random motions of objects in its environment, and these average out phases of statefunctions of the spot. Also the phase of a statefunction of this dot determines details of its makeup. But we just notice the dot, and do not examine details, thus do not learn anything about phases.

Since a dot, say, is macroscopic it consists of a large number of particles, another reason its phase averages to zero (sec. II.2.c.ii, p. 69).

II.1.c.ii *How probability comes from the square*

Hence only the amplitude is relevant in determining properties of the dot, and how many form. But why the square, why does the number of dots depend on the square of the amplitude of the incoming object?

What does it mean to say that probability is proportional to the absolute square of a statefunction? For electrons hitting a screen and producing dots, and a beam containing two electrons, the statefunction of the beam is twice that for a single electron, but it produces two dots, not four. Thus for this, the number of dots — the probability of finding an electron or dot — is proportional to the sum of probabilities, not to the square of the statefunction of the system.

If an experiment is repeated many times (causing the phase to average to zero) the distribution of dots is proportional to the square of the statefunction — it peaks more strongly than the statefunction. This indicates why it is proportional to the square: interference. However for two (incoherent) electrons interference terms average to zero, thus the square of the amplitude is the sum of the squares of the amplitudes of each, rather than the square of the sum.

Another way of understanding this is by conservation of probability. If the probability of finding two electrons is one (and zero for any other number) and each electron produces a single dot, the probability of finding two dots is one, and zero for any other number (ignoring overlap).

II.1.c.iii *Probability is proportional to energy*

This interference is the same as in classical electromagnetism, for example with light that passes through two slits giving an interference pattern. The number of dots produced by the light, or the strength (or brightness) of the interference bands, is proportional to the intensity of the light, the square of the amplitude. Why? The creation of a dot, or bright line, requires energy, so the number of dots, or brightness of a line, is proportional to the energy density, and energy, in classical electromagnetism, is proportional to the amplitude squared, thus the number of dots, or the brightness, is.

The same is true for quantum mechanics. Producing a dot on a screen requires energy — if we repeat this many times the number of dots in a (small) region is proportional to the energy in that region (the greater the energy the more often — the greater the probability — that an atom is excited and a reaction triggered) and this is proportional to the amplitude of the statefunction squared. But whether slits are there or not, whether there is interference or not, the total energy, like the total probability in quantum mechanics, is unchanged. Energy, and probability, are merely redistributed.

Intuitively we can consider n electrons, and by the same argument giving the energy in classical electromagnetism, the amplitude of the statefunction is n times that of a single electron, so the energy is proportional to the square of this amplitude — the more electrons present the more work that has to be done to add an additional one, thus increasing the amplitude (or classically, the field). Since this is true for n electrons, by consistency it must be true for the statefunction of a single electron, so the probability of finding it is proportional to the energy density, which is proportional to the amplitude of the statefunction squared. And this is true then for any system, again by consistency.

Why is the number of atoms, say, proportional to the statefunction? If all are in the ground state this is true by normalization. It really does not tell much about physics. But if we ask why the number of excited atoms is proportional to the square of the statefunction of the excited states, the answer is again because of the energy. The number of excited atoms (ignoring irrelevant possible multiplicity of excited states) is proportional to the energy needed to excite them, which equals the total energy of the excited atoms (minus that of the ground state). But this equals the expectation value of the Hamiltonian, so is proportional to the square of the statefunction of the excited state.

II.1.c.iv *How the formalism gives energy, so probability*

The formalistic reason that energy density is proportional to the square of the amplitude is that it is given by the expectation value of the Hamiltonian H — because energy is defined as the expectation value (often the eigenvalue) of the Hamiltonian, which is one of the momentum operators of the Poincaré group [Mirman (1995b), sec. 6.3.3, p. 118]. This value is then invariant under the group, and being a constant it is important, so is given a special name, energy. Also it is a representation label of the group (for representations with momentum operators diagonal [Mirman (1995c), sec. 2.2.1, p. 13]), therefore is left unchanged by group operations. To find representation labels we consider diagonal operators acting on a basis state of the representation which gives the same state times the (invariant) eigenvalue, and taking the product of the (normalized) state with its conjugate — the expectation value of an operator — gives the eigenvalue, and for the Hamiltonian (the timelike component of the momentum) that is named energy.

This, conservation, is the reason that energy is important so is given a name. Of course this is true for other quantities, momentum, angular momentum, mass, charge and so on.

State $|s\rangle$ has expectation value (schematically) $\langle s|H|s\rangle$, which is proportional to the square of the amplitude of the state, so energy is proportional to the amplitude of the state squared. In such an expectation value the statefunction and its conjugate appear, another reason for the absolute value.

As H is an operator, and energy, probability, transition probability, are numbers, expectation values are needed. In general to get numbers from an operator, matrix elements of the operator between two states are necessary. This is the formal reason for the appearance of the square of the amplitude.

Thus essentially, probability is the absolute value because of the method that we (are forced to?) use to realize group transformations, by means of operators requiring expectation values to obtain numbers like probability.

II.1.d Gleason's theorem

There is a completely rigorous proof relating probabilities to absolute squares of statefunctions, Gleason's theorem [Cooke, Keane and Moran (1985); Gleason (1957); Pitowsky (1998); Varadarajan (1993), p. 1821]. However it is not entirely convincing. It uses the properties of Hilbert space, but experiments are not done in Hilbert space. The present vague discussion is meant to supplement this.

Rigorous mathematical proofs have no meaning if they do not properly correspond to and describe physical situations that they profess to explain. If it is not fully clear how symbols and concepts of a theorem relate to actual experiments then no matter how rigorous the discussion is mathematically, its physical content is doubtful. Thus it is useful to study how properties of quantum mechanics are obtained from what we know about experiment.

For example, the requirement that absolute squares be used means that phases are irrelevant. This is certainly not true if we consider a single event. For an object that is scattered, which is the measurement, and then is involved in an interference experiment, the phase is crucial. But for large numbers of objects, or experiments, phases average to zero, so are irrelevant. The above discussion is meant to illustrate why this occurs and so why the absolute value (of the square) is the relevant quantity. It is useful to understand physical reasons, as well as mathematical ones. They do not supplant mathematical reasons they supplement them — and they help in the difficult checking of the relevance of the mathematical arguments. Mathematics may be completely correct but if it is irrelevant, it has nothing to do with physics.

However the physical reasons are less clear in explaining why probability, and energy, are related to the square of statefunctions. The explanations are more formalistic. One reason is that the physical meaning of the statefunction in this context is itself less clear if it is not linked to the concepts of probability, energy, and transitions. Thus physical explanations seem to necessarily be somewhat circular. But this is worth thinking about further.

II.2 Wavefunctions don't collapse, oversimplifications do

One idea that has caused much confusion is "collapse of the wavefunction". Some people have even gone so far as to say that gravitation is the cause of this "collapse", that is (presumably) when human beings make measurements, or merely look at an object, this causes a gravitational field to form that results in a change of a system's statefunction. This implies that human beings (or at least physicists) are able to cause (a change of) a gravitational field merely by observing the behavior of objects, say particles going through slits — just by looking at the dot produced by such an electron, for example — and even by observing macroscopic objects (for these too have statefunctions that must "collapse" when we look at them).

Perhaps what collapses is not the statefunction, but common sense.

While some have felt that this terminology is awkward [Kayser and Stodolsky (1995)], very rarely, if ever, have the processes and the meaning of the terms been considered with care.

The difficulty, as in so many other cases, is that there is not a full description of what is happening. People (unfortunately including physicists) just assume that in some mysterious way a system goes into some final state, and do not consider the steps in which it does so, thus are mislead into believing that the process is discontinuous. Actually what is discontinuous is the way we consider the transformation. If we look at just the initial and then the final states of course the process is discontinuous. We can only see the continuity if we follow the process fully and carefully.

Discontinuity cannot be true, and it is not. But carelessness unfortunately can be true and too often is, and certainly can make discontinuity appear true.

Another way of expressing this belief [Leggett (1999)] is that while a body can be in two or more eigenstates at one time that when a variable is measured, a definite value is always found! Shocking isn't it? It would be interesting to learn how an experiment can possibly give more than a single value at once. And in what circumstances is a system not in several (perhaps an infinite number of) eigenstates of operators at any time (sec. IV.2.b, p. 155)?

II.2.a Processes are always continuous

There is no "collapse of the wavefunction". Schrödinger's (properly Dirac's) equation describes the system at all times. Statefunctions of systems after measurement or collision or fragmentation are found from the initial ones using the equation (in principle), and they evolve continuously; there is no sudden change.

If for example we consider an object striking a screen forming a spot, the statefunction of the system after the formation, the product of that of the struck atom plus all objects attracted to it and the scattered object, is found from the initial one using Schrödinger's equation, and if so found would be seen to vary continuously. In principle it is possible to calculate final (perhaps extremely complicated) statefunctions from initial ones, and the entire transformation from one statefunction to another is completely continuous. Never is there a sudden change or collapse. Any such appearances result from ignoring the (continuous) intermediate stages by regarding these as happening instantaneously [Mirman (1995b), sec. 3.2, p. 42].

And the spot seems classical only because we do not examine its details to find its statefunction. Regarding the final object as "classical" also makes it appear that collapse occurs. So there seems to (but in principle cannot) be a difference between the microscopic, quantum, objects we start with, atoms, molecules and incoming particles, and the macroscopic, classical, objects (like spots) that we take as the final object.

An analogy is given by a ball hitting a wall. The velocity changes instantaneously, there is a "collapse of velocity" from one direction to the opposite, Newton's laws do not hold for the collapse. If misunderstanding classical physics were as big a fad as misunderstanding quantum mechanics is, there would be a large number of learned papers discussing the breakdown of Newton's laws and the mysterious nature of the discontinuous process. Of course there are forces on the ball, no matter for how short a time, these produce an acceleration which results in the reversal of velocity. The process is completely continuous, Newton's laws always hold, there is no mystery. Measurement is never instantaneous [van Kampen (1990)]. If intermediate steps are left out certainly transformations look discontinuous. It is no different for quantum mechanics.

II.2.a.i *How we make it seem as if wavefunctions collapse*

The process of a beam forming a dot on a screen (sec. III.2.i, p. 116) can be considered to be a "collapse of the wavefunction", even though the statefunction of the (entire) system always varies smoothly, and can be found from Schrödinger's equation (provided that we are very good at calculations). An incoming beam triggers the process, and then we combine all possible final states that give a similar macroscopic appearance. Taking different statefunctions as equivalent when they give the same macroscopic object (the dot) is really what is often meant by "collapse of the wavefunction". The statefunction of the incoming particle, the triggering particle, changes very quickly, thus appears to change discontinuously ("collapse") but of course does not — and there is no reason to think that it does. We could calculate the scattering of a particle by an atom, for example, thus tracing the change of statefunctions through the scattering process. Then we would understand that it is continuous — and would not think otherwise. But here the particle triggers a set of reactions among a very large number of objects — which occur after it has finished the triggering process and left, so giving a large change, a dot. The change therefore appears to happen discontinuously since it is fast and occurs after the initiating object has left — but is in fact continuous, as can be seen with a slight bit of thought.

This is true in general. If we look at an electron through a microscope, the probability distribution (that is the probability of finding it in some region) is mapped to the probability distribution of the scattered photon — knowing the latter (which we do, since that is what we measure) gives the former. The photon distribution is measured (using a large number of experiments) by having it interact with a screen (perhaps a retina, which is a screen) resulting in a dot.

In measuring spin direction (say by sending a particle through a magnetic field) we look again at the probability of its forming a dot (or the brightness of a point on the screen).

In all cases the argument is the same.

II.2.a.ii *Measurement and collapse*

Actually when we measure (where "we" might be a set of electrons) there is an initial state of the system, a product of the statefunction of the object that we are studying, with that of the rest of the system (the measurement apparatus), and a Hamiltonian (exponentiated) which acts on it (this action is the measurement), to give a final state — a sum of products of statefunctions of the object with those of the apparatus. We study the latter, and the particular one that we find is the result of the measurement — which might be a wavepacket of an electron centered at some point taken as the position of the object. Quantum mechanics does not tell which statefunction we find, but gives the probability of each possible one in the sum. This whole process is continuous, there is no collapse.

II.2.a.iii *What is a measurement, and why they occur continuously*

As described here a measurement seems to be a rather complicated process involving macroscopic objects interacting with microscopic ones. But these are examples used to illustrate the arguments, and are used because they are what most people think are measurements.

What is a measurement? Actually it is the interaction of objects resulting in the change of the statefunction of any one (called the observer). This change is the measurement. Thus electrons, protons, ..., are always making measurements [Mirman (1995b), sec. 5.2.4, p. 94]. These are the observers, and observers do not have to be humans. The number of our laboratory experiments is of measure zero compared with the number of all measurements in any period of time, no matter how small, being made in the universe.

Thus if the wavefunction were to "collapse" upon measurement that of every object in the universe (sec. V.4.c.ii, p. 228) would be "collapsing" continuously.

II.2.b Wavefunctions must be of complete systems, including brains

There is another way of putting this. There is no "collapse" that does not fit into the Schrödinger dynamics, provided the entire system, including our brains, is included.

Consider an outside observer who sees a system being studied, say an atom, a device to measure it, with a pointer, the environment, plus our brains. Assume it can make a huge number of copies and so measures the statefunction of this entire system. The part of the statefunction describing our brains shows that we view the pointer in a rest position (of course what this brain-statefunction is like is totally unclear, but in principle we can consider one). At a later time the atom decays, the pointer moves, and the statefunctions of our brains describe them viewing the pointer in the final position. The development of this entire statefunction, from the initial to final states, is given by Schrödinger's (really Dirac's) equations (plus those for the electromagnetic potentials, and of the other particles involved), exactly, in principle, just as is a statefunction of an atom. It is completely quantum mechanical (in principle), and continuous.

The difficulty comes from trying to understand in the approximation (and that is what it is) in which we ignore, or in some ways average over, say, the statefunctions of brains. But this is due to the approximation (which is of course necessary) and not from any peculiarities of quantum mechanics — as is true in general. There is a measurement problem only in this sense; it arises solely because we do not include part of the system.

Since these transformations are unitary it would appear that a transformation of a statefunction of a system that is a superposition to one that is an eigenstate cannot be described by quantum mechanics. The problem is the word system. Transformations of statefunctions of entire systems are always unitary. However if we consider only part of a system, say a particular object, than the transformation of that need not be unitary. Probability need not be conserved for only a subsystem. And the entire system includes the observer. Forgetting this makes nonunitary transformations seem, fallaciously, to disagree with quantum mechanics.

Such errors are similar to those leading to the belief in superselection rules [Mirman (1995b), p. 203].

This is something that needs much more work, considering various systems, but carefully and completely. Too often important aspects of systems are ignored, and then various paradoxes arise. Decoherence is part of the answer, but not all (sec. II.2.e, p. 72).

II.2.c Macroscopic states do not distinguish different microscopic wavefunctions

We generally regard a measurement as giving a macroscopic state. How does this occur? The initial statefunction describes a microscopic object, say an excited atom (plus of course its environment). It decays and the energy emitted affects many different atoms and molecules, causing chemical reactions say, but many. The effect of this large number of reactions is macroscopic, perhaps viewable as a black dot. Thus the process leads to a statefunction describing a collection of many objects.

A statefunction of the apparatus after the Hamiltonian is applied, after the apparatus and object interact, gives the probability that affected objects, which are all close to each other, are near a particular point. But for any point there are (very) many different statefunctions that give the same macroscopic state — the blackened dot at that point, or the position of a pointer. Each of these gives a probability that the dot is caused by a specific set of reactions, and that a specific set of molecules is involved. But since all we see is a dot, not individual molecules, we add these to give the probability of the dot at that point, which equals 1 (since we have observed it and know that it is, about, there). We obtain the probability of a dot being at any point from the statefunction of the apparatus right after an interaction, so statefunctions of individual molecules are irrelevant — all are regarded as the same. After we know where the dot is, probability is meaningless. But there is no collapse, the statefunction varies smoothly at all times. Collapse seems to appear when we regard all these statefunctions as identical — and for this reason.

II.2.c.i *Statefunctions for microscopic and classical systems are fundamentally equivalent*

In principle there is no difference between statefunctions for microscopic systems, like atoms plus incoming electrons, these combining into excited atoms, and a macroscopic system, say dark regions. It is only the way we treat them, considering different states of the former as actually different (for example, taking different basis states in the sum of which a statefunction is constructed to be for different atoms excited), while for the latter we consider many statefunctions to be the same, ignoring such distinctions as those between different sets of dissassociated molecules or different reaction routes to the same dot. Also for macroscopic systems statefunctions are products of a large number of basis states — there are a large number of molecules that are changed in forming the dot — while for microscopic systems there are only a few. Because of the large number distinctions between states described by different basis vectors in the sum are impossible to discern, so all are lumped together.

Likewise for a pointer the argument is essentially the same. It is described by a wavepacket, just as for an electron, centered at some particular point, whose probability is given by the statefunction of the microscopic object being measured. The calculation of this wavepacket can, in principle, be done using Schrödinger's equation, starting with the statefunction of the original state. The pointer consists of many smaller objects, and it is their statefunctions that are calculated. The statefunction of the pointer is then a sum of products of statefunctions of individual objects, with interference terms canceling because of the vast multitude of these subsystems, and indistinguishable terms in the sum lumped together. The sum is then a set of sums, each taken as a single term, and each is for a different position of the pointer. The probability due to a term in this sum can be calculated, in principle, from the original statefunction.

Statefunctions of pointers and electrons differ only because phases of the former vary so rapidly that they cannot be taken to have definite phases, and because different ones are not distinguished.

II.2.c.ii *Macroscopic objects prevent interference*

It is true that quantum mechanics does not say what the final position of a pointer is. Rather it is possible to find the final statefunction of the pointer, this consisting of many terms each giving the pointer in different positions, and since it is macroscopic there will be (essentially) no interference terms. If an experiment is repeated many times, in some fraction of cases (given by the original statefunction) the pointer will be near one position, in other cases near different ones. Probabilities of these various positions can be found, but not (in general) the results of a particular experiment. Considered this way, there should not be (in principle) a measurement problem.

Schrödinger's cat is classical [Mirman (1995b), sec. 9.3.2, p. 167] not because of decoherence, in which the relative phase between states is destroyed giving an incoherent superposition. For it words "alive" and "dead" each refer to a huge number of states, each a sum of an immense number of states of elementary particles (sec. II.2.f, p. 73). So

$$|a) = \sum \prod |i, j), \tag{II.2.c.ii-1}$$

and likewise for $|d)$, where the product over j is that over the product of states of all particles making up the cat, while the first over i is over all possible sets of these states of particles giving "alive". Each state of a particle is time-dependent and (since the particles have different energies) the time dependence of a product is extremely complicated, so the time variation of each term in the sum is different, thus the time dependence of the sum is extraordinarily complicated. If there is a superposition of $|a)$ and $|d)$, for it to be coherent, there must be a fixed relative phase. But clearly, with the (huge number of) phases of the terms in the sums $|a)$ and $|d)$, and these varying differently, this cannot be. A classical object therefore cannot be in a coherent superposition of states. This is true even if it is completely isolated, even if it could, in some way, be prepared in such a superposition, for the relative phase (a concept of doubtful meaning since there are so many) would be immediately washed out. A coherent superposition of states [Mirman (1995b), chap. 4, p. 58] — which requires a fixed phase — of a macroscopic object is prevented, not (only) by the environment, but by the vast number of particles of which it consists.

These points are sometimes recognized [Englert (1999); van Kampen (1991,2000)], but not often enough.

II.2.c.iii *Does observation force a quantum mechanical system into a state?*

Also we cannot see this superposition of dead and alive because our observation forces the system into one particular state — it causes the statefunction (or the cat?) to jump into one of the states. But this is true of classical physics also — the entire experiment can be done in exactly the same way, with exactly the same possibilities of dead and alive until we look, when the state (or the cat?) is forced into (or found to be in) one of these two situations. The only difference between classical physics and quantum mechanics is that in the latter there are cross terms between various states, giving an interference pattern. And these are important for systems in which, say, phases do not vary rapidly. But for complicated systems phases, so cross terms, (extremely) quickly become zero. This is the reason for the difference between classical physics and quantum mechanics.

II.2.c.iv *When does a recording occur?*

One reason that wavefunctions seem to collapse is that events (seem to) take place suddenly. Thus if an atom decays and the emitted photon is captured, that capture, the measurement, occurs at one specific time. Does the statefunction change suddenly at that time? The statefunction refers to an ensemble (sec. III.2.d, p. 108), so it gives the probability of the capture as a function of time. If it is used for a single event, misinterpretations result.

Also, to consider the process completely rigorously, the statefunction of the entire system, including the film recording the arrival of the photon, and our brains, must be used. What often causes confusion is considering the statefunction of only part of a system, say a decaying atom, and applying it to a single event, rather than an ensemble (sec. V.3.b, p. 212).

A statefunction cannot refer to a single event; it just gives the probability of each specific event in the set of possible ones.

II.2.c.v *How a measurement takes place*

Consider the cloud-chamber [Mirman (1995b), sec. 3.1.2, p. 40] to outline how a measurement occurs. This has an advantage that there need be no brains involved. We can use a set of molecules which when struck by an incoming object fall apart, thus giving irreversibility (in the sense that the final state is far more complicated than the first one). Resultant atoms then interact with other objects in the chamber, drawing some toward them, causing interactions between these, forming a bubble, or a spot on a film. It is this bubble or spot that is the observed macroscopic object.

The statefunction for the initial system, incoming particle plus supersaturated vapor, can be given (in principle), and the subsequent development of the system fully computed from Schrödinger's equation (perhaps not in reality, but certainly in theory, although how to compute statefunctions giving formation of droplets seems unclear). This gives a macroscopic, visible, track. There are no discontinuities, no collapses of the wavefunction. And for this there will be no measurement problem.

It is true that quantum mechanics does not say what the final position of a pointer or dot is. Rather it is possible to find final statefunctions of these. They consist of several terms each giving the pointer or dot in different positions, and since these are macroscopic there are (essentially) no interference terms. If an experiment is repeated many times, in some fraction of cases (given by the statefunction) the pointer or dot ends up in (close to) one position, in other cases in different positions. The probabilities of these can be found, but not (in general) the results of a single experiment.

Statefunctions of pointers or dots are actually sums (integrals) of statefunctions (going with different positions), each a product of statefunctions of (microscopic) particles making up the (macroscopic) object, with the various statefunctions giving the same position different because the states of microscopic particles are different. It is important to remember that the pointer is macroscopic, and what the statefunction of a macroscopic object is — that should forestall problems.

II.2.d Does consideration of brains lead to an infinite regression?

If we consider a brain as part of a system, with the statefunction being that of the whole system, do we not need another external observer to study the statefunction, thus leading to an infinite regression? This question, which is common, is an example of basing quantum mechanics, and its meaning, on human beings, human observers. Statefunctions of entire systems, including the brains, have meaning even if there are no other observers to measure them, just as statefunctions of parts of a star, and of a star itself, have meaning, even if there are no human beings inside the star. What gives these meaning is that they are (essential) aspects of a theory, a coherent whole, a complicated intertwined net, and this theory, and that it is coherent, provides and is needed for an understanding of the reality that we experience (sec. I.2, p. 5). Each component of it, whether we experience it specifically or not, is necessary. This connection to the other parts, necessity for coherence, necessity that the part be present for coherence, is what supplies content to a constituent. We can know that the theory has meaning from our experience, and from its coherence, its interrelatedness (sec. I.2.f, p. 8), but the laws of nature do not depend on whether we are aware of them or not, whether we experience nature or not, whether we exist or not.

When we add brains of observers so to make a system closed we are illustrating the meaning of concepts, for we have to understand definitions and concepts, but laws of physics hold whether we understand them or not, whether we have observed states of brains or not, whether we have provided concepts that we need or not, but that the universe does not.

Nature exists independent of our interpretation of it.

II.2.e Decoherence

Decoherence provides part of the reason that a pure state becomes a mixed one. But it is not the complete reason for classical states. It is similar to a box filled with (classical) gas. It may start in a "pure state", that is with positions and momenta of all molecules exactly specified.

External influences will quickly give a mixed state. But the important aspect is that a macroscopic state is equivalent to an immense number of microscopic states (for the gas, ones describing molecules with different positions and momenta, but with densities — about — the same for all).

This is likewise true for a classical state that is really equivalent to an immense number of quantum mechanical states. It is not only that external influences produce a mixed quantum mechanical state from a pure one, but it is this equivalence of the many quantum mechanical states that makes a classical state different from a quantum mechanical one.

Decoherence is not the reason (although at times it may play a part) that classical and quantum mechanical states are different — a better way of saying this is that classical objects differ experimentally from quantum objects, especially in that they do not allow coherent superpositions. Statefunctions of classical objects are very large sums of a very large number of products of statefunctions of single objects. But phases of the various terms in the sum vary extremely rapidly and differently since products in the sum are different. Thus interference terms average to zero far faster than they can be found. A sum then is an incoherent superposition, which is what a classical statefunction is. The concept of coherent superposition cannot be applied to classical objects.

II.2.f Experiment picturing difference between classical and quantum superposition

This can be illustrated by an experiment [Mirman (1995b), sec. 9.3.2, p. 167]. Consider a closed box containing a pile of papers with nonsense about quantum mechanics written on them (sec. II.2.c.ii, p. 69). Above the pile is a set of blades attached to a counter near a radioactive atom. If the atom decays it activates the counter, causing the blades to drop tearing all the nonsense to pieces. Thus the state of the system is a coherent (?) superposition of nonsense, $|n)$, and obliterated nonsense $|no\ nonsense)$, that is (schematically)

$$|system) = |n) + |no\ nonsense), \qquad \text{(II.2.f-1)}$$

so the probability of finding these states after the box is opened is

$$\begin{aligned} P &= (n|n) + (no\ nonsense|no\ nonsense) \\ &\quad + (n|no\ nonsense) + (no\ nonsense|n). \end{aligned} \qquad \text{(II.2.f-2)}$$

The first two terms are the same as in classical physics. For a small object the last two terms change probabilities. But for a macroscopic object they vary extremely rapidly in time, so immediately (whatever this means) average to zero. Thus for a large object, like piles of nonsense, the results in quantum mechanics and classical physics are exactly the same.

But there is a more basic problem (sec. V.3.b, p. 212). What does the equation for $|system\rangle$ mean if we cannot measure it, or study it in any way? How can we tell whether the system is in a coherent superposition, or any superposition, if we have no information about it? Does the equation have more sense than simply a set of letters and symbols? Are words describing this experiment anything more than sequences of letters, sentences anything more than sequences of words?

Thus not only is the pile of paper filled with nonsense, but all too often discussions of such experiments are also. Unfortunately there seems no sets of blades hanging over them.

II.2.g Transition between quantum mechanics and classical physics based on rapidity

One explanation for the transition from quantum mechanics to classical mechanics is based on the rapidity of these phase changes. At one extreme, say for an isolated electron, the phase difference between states of a superposition is constant, thus it is coherent. At the other, for an object made up of a large number of particles, phases vary so rapidly that the concept of coherent superposition cannot have meaning. As the number of particles or their interaction with the environment increases, such phase changes occur more and more quickly. Then quantum mechanics becomes a less and less useful description, classical mechanics becomes more and more so. But fundamentally there is no difference, nor is there a sharp boundary.

There are experiments that show coherent superposition of statefunctions of complicated objects. Although these objects are referred to as being equivalent to Schrödinger's cat, they are really mesoscopic, so are not classical objects. What then should we consider as macroscopic — classical — objects? One criterion is that the cross terms in superpositions vary so extremely rapidly that they average to zero far faster than they can be measured (eq. II.2.f-2, p. 73). Thus mesoscopic objects for which this is not true are very different from classical ones, and experimental results for the former cannot be applied to the latter.

It is also possible to produce "paradoxes" in the application of quantum mechanics to classical objects by using objects that are not classical, for example in the sense of interference terms averaging to zero very quickly, but calling them classical, as with the magnetic flux in a SQUID with a Josephson junction [Leggett (1984), p. 592] say, which is certainly not classical. Quantum mechanical behavior of such objects is not more spectacular than quantum mechanical behavior of an electron. Why should it be, why should anyone think that it is? Regarding such objects as classical seems no more than a desperate attempt to show that quantum mechanics is "weird".

II.3 There can be no discontinuities in quantum mechanics

Solutions of the fundamental equations governing quantum mechanical statefunctions are analytic; they cannot be discontinuous. However strange misinterpretations have arisen, some due to the unfortunate name quantum mechanics, that imply discontinuities, even though these contradict the basic principles of the theory. Here we just mention a few.

II.3.a Can a coherent sum go into an incoherent one?

Transformations of quantum mechanics are unitary, thus it is argued that they cannot take a coherent superposition into an incoherent sum of states, which is what seems to happen when we measure. The argument is of course correct. Schrödinger's, and Dirac's, equations do not allow this.

A system of a measuring instrument and, for example, objects 1 and 2, with two states in a coherent superposition, labeled up and down, has initial statefunction, schematically,

$$|system, i) = \{|1, u)|2, d) + |1, d)|2, u)\}|a). \qquad \text{(II.3.a-1)}$$

Here $|a)$ is the statefunction of the measuring apparatus, which is macroscopic so is a sum of a vast number of microscopic states (which is a coherent superposition although the phases of the states in the sum vary so rapidly with time, and so differently, this cannot be experimentally shown). The final state is

$$\begin{aligned}|system, f) = (&|1, u)|2, d)\{|a_1) + |a_2) + |a_3) + \ldots\} \\ &+|1, d)|2, u)\{|a_1)' + |a_2)' + |a_3)' + \ldots\}, \qquad \text{(II.3.a-2)}\end{aligned}$$

and is a coherent superposition, as it should be. But if we only consider the two microscopic objects then their superposition is incoherent. That however is because we are looking just at the subsystem.

Such is also true for scattering say by a hydrogen atom with total angular momentum 0. The initial state of the atom is a coherent superposition of proton up, electron down and proton down, electron up (times that of the incoming object). After scattering the entire system is in a coherent superposition, but not the atom alone.

Time translations, like other such transformations, are unitary, but only for complete systems, some parts of which may be "classical".

II.3.b No jumping allowed

Another concept, related to these, is that of quantum jump. But there is no such thing [Zeh (1993)]. If an atom is in an excited state and goes to the ground state it does not do so suddenly but rather its statefunction is a sum of statefunctions of these two states, with the coefficient of the first decreasing in time, that of the second increasing [Mirman (2001), chap. IV], and these always do so continuously; the derivatives of coefficients with respect to time are always continuous. There is no sudden change.

There are no jumps in quantum mechanics, there is nothing discontinuous. There is no state vector reduction, no collapse. Wavefunctions do not collapse, their behavior is always completely continuous, not governed by fiat, but by Schrödinger's (actually Dirac's) equation, always, everywhere.

Errors leading to such misconceptions require discussion.

II.3.b.i *The unfortunate name "quantum mechanics" leads to jumps in thought*

Beliefs that there are discontinuities in quantum mechanics, that there are jumps, comes in part from the very unfortunate name "quantum mechanics" (sec. IV.1.a, p. 143). But discreteness is not fundamental, and arises, but only in some situations, from regularity or boundary conditions, not from the properties of quantum mechanics.

There are no sudden jumps in quantum mechanics, no changes of state in zero time. The statefunction for one state decreases in time, for the others increase, and all continuously. While a statefunction is a sum of discrete states, coefficients are continuous, so every statefunction is continuous.

There is no experiment that in any way indicates a sudden jump. The name quantum mechanics is unfortunate, since it indicates discreteness and discontinuity; the latter violates physical laws (like the Dirac and Schrödinger equations), the former is not general, but only incidental, merely a particular property of special cases.

II.3.b.ii *Leaping without looking*

This doctrine of quantum jumps manifests itself in such statements as an electron can take a quantum leap from one orbit to another without traversing the space between. It would indeed be an interesting experiment that shows the electron is never in the space between, that it can never be found there. Does quantum mechanics require that this be true, or does it forbid it? Those who believe such things can undoubtedly easily explain how they are consistent with the uncertainty principle (although they are undoubtedly consistent with a principle requiring uncertainty in understanding of those who say things like this). Of course, there is always a nonzero value of all statefunctions at all points of space, even for ground states, thus a nonzero probability of finding an electron anywhere, at every time. Leaping without thinking can be dangerous.

To even speak of jumps it is necessary to explain how an experiment can show that a statefunction is strongly peaked at, say, that describing an electron in an excited state, and then an extremely short time later it is peaked at a product of the electron in a lower energy state times that for a photon moving away from the atom. But such an experiment would result in an interaction that itself causes a rapid change of state, thus it cannot be correct to say that the electron, in an unexamined atom, jumps from one state to another.

We can also measure the time a photon arrives at a detector, and then say that the electron jumped at that time. However this is not what the formalism gives, there is no sudden change of statefunctions, rather they are sums of terms, one (itself generally a sum) describing an excited atom, the other (sum) describing an atom in a ground state plus a photon, and the terms in the sum have coefficients that are time dependent, the first decreasing from giving probability 1, the second increasing to 1. Absolute squares of coefficients give probabilities (so requiring ensembles) at any time of particular experimental results, such as the probability that a counter registers a photon in some specified time interval. But this probability is a continuous function of time. Of course, if anyone can find a probability that is (exactly) discontinuous, that would be fascinating.

In an ensemble of atoms (large in number, so the observation of a few will have little effect on the total), an observation at any time shows that a certain number are in excited states, at a later time that number is less, and the number of atoms in the ground state more. There are no sudden changes, jumps, in these numbers, they vary smoothly with time. Probabilities for observing photons at any time also vary smoothly.

II.3.b.iii *Jumping into misleading language*

This is another example of how language can mislead. Because terms like "jump" have been used, in some cases thoughtlessly, perhaps in others as merely an abbreviation, perhaps sometimes maliciously, some people, without sensing how these are actually meant, or without understanding what those who use the terms are saying, take them seriously, as if they actually stated a physical fact. Of course those who say things without knowing what they are saying are at fault, but misleading terminology makes it too easy for both writers and readers to be careless and confused. It is necessary to be very careful in introducing vocabulary for even if the person who does so fully understands its meaning and its limitations, others are likely not to (chap. IV, p. 142).

The idea of jumps probably originated in Bohr's theory of the atom, an inconsistent (thus of course, phenomenological) theory. Jumps provided a means of representing something, but since the theory is phenomenological it is not clear what. The concept of a jump is a phenomenological idea that forms part of a phenomenological theory. It is quite surprising that now when we have (apparently) a consistent formalism, quantum mechanics, this no longer necessary phenomenological idea still persists.

II.3.c Can an object go everywhere instantaneously?

One of the "paradoxes" that the idea of instantaneous state vector reduction (jumping) seems to lead to is that if an object is localized, and then a measurement sends it into a momentum eigenstate, it instantaneously has an equal probability of being anywhere, since its position must be completely unknown. Thus it was in the laboratory before the measurement, but has as much chance of being in a galaxy a billion light-years away as being in our galaxy (and essentially no chance of being in the laboratory since that is so small compared to the size of the universe). Of course the measurement is not instantaneous. But even so the object is very quickly sent into another galaxy. So if we measure the momentum of an object and find that it is in an eigenstate, there is really zero probability of ever seeing it again. This does appear somewhat implausible.

If an object is studied instruments for doing so must also be in the laboratory, thus (to some degree) localized (it would be difficult to perform the experiment if the object were in a laboratory on earth and the instrument in another galaxy), and not in a momentum eigenstate. Since the measuring instrument is localized its momentum before measurement is not completely known. It should be clear that such an instrument cannot force an object into an exact momentum eigenstate (anyone who disagrees can show just how this is done).

The problem is that the formalism that we use is an idealization. Even though it may in principle be rigorously correct it cannot represent actual physical situations completely (sec. I.2.f.v, p. 11). This is especially true for real physical objects, which must be idealized. Thus we can write equations giving the actions of instruments on objects using symbols as abbreviations for physical interactions. With this formalism the symbols act on representations of statefunctions to give other statefunctions, say taking a wavepacket to a momentum eigenstate. But of course what they describe is not fully the same as what actually takes place. And what they describe may well be unphysical leading to paradoxes that are really nonexistent.

For a statefunction that is reduced to a momentum eigenstate there are several problems. There would be no difficulties or paradoxes with an object that goes into a momentum eigenstate if we could not determine that it did. But to show that it went into such an eigenstate we have to measure it again so that it remains in the laboratory (or not too far away), which means that it is somewhat localized and thus not in a momentum eigenstate. Hence the experiment that leads to the paradox is self-contradictory.

And statefunctions cannot go from a wavepacket to a momentum eigenstate instantaneously. Essentially the object is scattered, with initial state a wavepacket, the final state (assumed to be) a momentum eigenstate. But scattering takes place over a finite period of time. This is usually ignored, as we ignore the continuity of the change of momentum of a ball hitting a wall (sec. II.2.a, p. 65). In classical physics we recognize that the change of momentum does not happen in a length of time that is zero. We only treat it that way as an approximation. For some reason this is often overlooked in quantum mechanics, where a more careful treatment is required; that strange ideas arise from this oversight is hardly surprising.

In reality a measurement is an interaction, and if an object is studied with an instrument it is a complicated interaction. To analyze it correctly Schrödinger's equation, with the complicated interactions, is solved (in principle) to find the final state, and this solution can never be an eigenstate. This shows the dangers of adding arbitrary postulates to what is a complete theory, with statefunctions determined by Schrodinger's equation.

There is a further difficulty. If we apply the formalism of quantum mechanics to study the interaction of an object and instrument it gives the probabilities of the different outcomes. What is the probability of the object going into exactly a momentum eigenstate — a state of measure zero? Of course, zero. And the probability of it going into a very narrow wavepacket in momentum space is very small (taking account of localization of measuring instruments, and the complete details of measurements, likely, almost, exactly zero). Objects remain somewhat localized. Happily they remain in our local group of galaxies.

Superfluous assumptions are of course unnecessary and likely inconsistent.

II.3.c.i *A classical analog for instantaneous momentum eigenstates*

The classical analog of an object going instantaneously into a momentum eigenstate is light in the form of a wavepacket (as it must always be) that instantaneously goes into a state with an exact wavelength, and thus is instantaneously spread over the entire universe. By conservation of energy the amount in a laboratory is so small that this light will never be observed. Undoubtedly those who believe that a quantum mechanical object can go instantaneously into a momentum eigenstate also believe that light can go instantaneously into a state with an exact wavelength, and therefore be spread over the entire universe.

II.3.c.ii *Parallels to classical physics should always be looked for*

Too often experiments are invented to show that quantum mechanics is weird, paradoxical, outlandish. But it often does not really differ from classical physics. Classical physics is intuitive, so we do not think about what is being done. We think about the unfamiliar quantum mechanical cases, but sloppily. It is usually helpful to compare classical cases with quantum mechanical ones. Then we may find that there really is no essential difference.

Classical physics may be even more weird, paradoxical, outlandish, than quantum mechanics. But being familiar we do not notice it, we do not think about it. Understanding quantum mechanics is almost always advanced by comparison with classical physics, where possible.

II.4 Only inept use of quantum mechanics gives paradoxes

Quantum mechanics does not predict paradoxical results. It is only improper idealizations, incorrect assumptions about what it does predict, or the use of abstract mathematics with (often at best) only tenuous connections to actual physical processes — and carelessness, thoughtlessness — that give paradoxes.

II.4.a A complete, realistic experiment is needed to show a paradox

In order for anyone to claim that quantum mechanics gives paradoxes, or other problems, it is necessary to describe completely a realistic physical experiment, showing all measurements and that their results are paradoxical. Abstracting the experiment to mathematical formulas, and (thus) leaving out essential parts, is not sufficient. It is expected that this type of abstraction does lead to difficulties, paradoxes, and often nonsense. That says nothing about quantum mechanics, but about the incorrectness of the abstraction procedure, its interpretation, and often about the care, or lack of it, taken in developing theoretical schemes.

II.4.b Superposition

There is a belief that the superposition principle is one of the fundamental postulates of quantum mechanics. The meaning of superposition, and when it is relevant, may be more subtle than is generally believed [Mirman (1995b), chap. 4, p. 58]. There are historical reasons for this belief, which are not the same as physical reasons (sec. II.1, p. 54; sec. II.4.c, p. 82), as with the belief that quantum mechanics is linear [Mirman (2001), chap. IV]. Thus it can become, not a physical law or deduction, but a superstition (sec. V.7.f.iii, p. 261), as often happens [Mirman (2001), chap. I]. It should be considered carefully in specific cases, not just assumed thoughtlessly.

The principle, when relevant, holds for linear group representations, in particular those of the rotation, Lorentz and translation groups. It applies to electromagnetic interactions because the photon does not interact with itself — equations governing that part of quantum electrodynamics dealing with free fields must be linear [Mirman (1995c), sec. 10.2, p. 173]. These are the most familiar parts of quantum mechanics, and the physics it deals with: rotation, Lorentz and Poincaré groups, electromagnetic theory. And these are the first (and even in present times almost only) subjects to which quantum mechanics is applied. Thus their linearity has prejudiced people about superposition. But physics is based, not on prejudice, but logic, mathematics, and experiment — and hopefully on thought.

However, while the most familiar aspects are linear, gravitation is necessarily nonlinear [Mirman (1995c), sec. 4.4, p. 67]. Sums of solutions of Einstein's equation are not solutions — superposition does not hold. A gravitational field produces a gravitational field, so the resultant field due to the overlap of two gravitational fields is not the sum of the fields. Yet general relativity is a (the) quantum theory of gravity [Mirman (1995c), chap. 11, p. 183]. There is in fact no way of getting a theory of gravity that is linear, that allows superposition (in this sense). No matter what changes are made (which unless purely formal, would give an incorrect theory), gravitation must be nonlinear.

Perhaps more important all theories are nonlinear, all objects interact (else we would not know of them so they would not exist). If the statefunction of an electron and its associated electromagnetic potential field, say, is $|e)|\gamma)$, the statefunction of two electrons and their fields is not the product of these, for the presence of one electron causes the other to accelerate, producing an additional electromagnetic potential field. So it is incorrect to write the statefunction of a system, with electrons 1 and 2, as

$$|system) = |e)_1|\gamma)_1|e)_2|\gamma)_2 - |e)_2|\gamma)_2|e)_1|\gamma)_1, \qquad \text{(II.4.b-1)}$$

for this does not include the correct field.

Superposition, and its content, especially when applied to a new domain, have to be considered with care [Mirman (2001), chap. IV].

II.4.c Half-integer spin

For half (odd) integer spin, a rotation of 2π changes the sign of statefunctions [Mirman (1995b), sec. 8.4, p. 155]. But such a rotation brings systems back to their original states, so should have no effect. The standard argument explaining this lack of an effect is that physical quantities are given by absolute values (of squares), so a sign change is irrelevant. However this is not quite true. If we take a beam of electrons, split it, send one beam through a potential, thus changing its phase, and then recombine the beams, there is an interference pattern, depending on the potential (and time in it). If the potential rotates the spin by 2π, the change of sign alters the interference pattern, thus is measurable [Mirman (1995b), p. 210].

We can only use, and need, the argument about absolute values if there is interference between statefunctions of objects with half-integer and integer angular momentum (that is these are coherently superposed). However the relative phase of two beams, of half-integer, and of integer, objects cannot be determined. Thus rotating one by 2π does not result in an experimentally determinable change of sign of its statefunction. Hence even without appealing to absolute values here, the change of sign is unobservable.

It is often stated, and even believed, that half-integer spin is a result of relativity, or field theory. Of course this is not true, it is rather a consequence of the rotation group being homomorphic to SU(2) [Mirman (1995a), sec. X.6, p. 287; (1995b), chap. 8, p. 146]. Historically spin-$\frac{1}{2}$ first appeared theoretically in Dirac's equation, the relativistic generalization of Schrodinger's equation. Thus many assumed that it is a consequence of relativity. But the reason for a physical fact is not determined by when, or how, we learn of it (sec. II.1, p. 54; sec. II.4.b, p. 81). This is essentially the well-known logical error (which physicists should know enough not to make) of believing that because something precedes something else, it causes it (*post hoc ergo propter hoc*). The physicists' version is that because a feature is discovered in the study of ..., that feature is a consequence of ..., where the ellipsis is quantum field theory, relativistic quantum mechanics, and undoubtedly many other topics (probably not only in physics).

II.5 Quantum field theory

The main topics of discussion here are quantum field theory, and later the conformal group and restrictions obtained when it is applied to

quantum field theory, giving conformal field theory [Mirman (20...
But to consider quantum field theory we first studied quantum mechanics, for quantum field theory is based on it, and actually the same as it (sec. I.6.d, p. 32; sec. II.5.d.i, p. 98). Understanding, and misunderstanding, of one gives understanding, and misunderstanding, of the other. Thus having considered fundamental properties of what is referred to as quantum mechanics, and unfortunate fundamental errors in thinking about it, we now turn to quantum field theory, to define it and to see why quantum mechanics is, or is at least a special case of, it.

II.5.a What is a field?

The basis of our approach is that physical objects are (given by) statefunctions, and that these are representation basis vectors of the group of geometry. A field then is, and is no more than, such a basis vector — so a function of the group parameters. Usually quantum mechanics is considered as describing single objects, quantum field theory as describing multiple ones. But the set consisting of the number 1 is merely a special subset of that of all positive integers. Thus we do not (and really cannot) distinguish between quantum mechanics and quantum field theory.

However in quantum field theory [Mirman (2001), chap. IV] two aspects are emphasized. The number of objects is variable, and these interact. Of course quantum mechanics would not be interesting if there were no interactions. However they are considered to be due to potentials, rather than nonlinear terms in equations governing fields. But there are no such things as potentials, these are purely phenomenological expressions used to simplify procedures for working with interacting statefunctions. Even the Coulomb potential has no fundamental meaning [Mirman (1995c), sec. 7.2.2, p. 126], but is purely phenomenological. And there is no such object as an external electromagnetic field; this is merely a phenomenological representation — it is an aspect of an approximation scheme. Correctly Dirac's equation (or those for the Poincaré invariants, plus the labeling operators, for other spins) should be written for all charged particles in a system, plus equations for the electromagnetic potential. Then the concept of an external field vanishes. This is in reality too complicated, so effects of all but one (or a few) particles are replaced by this function.

There are no such things as virtual quanta (sec. I.2.e, p. 7). These are no more than a way of picturing, so keeping track of, certain terms in an approximation scheme, which being no more than such a scheme, has no fundamental significance.

Thus we have to consider statefunctions that are governed by equations that contain nonlinear terms, ones in which statefunctions appear in products with others. And we have to write statefunctions in a way to show that they describe physical situations in which the numbers of objects change [Mirman (2001), chap. IV].

II.5.a.i *What is the primary ontology?*

A question that is regard as fundamental in discussing a theory is its primary ontology; what are the basic objects? Often this fundamental question is deeply analyzed, but it does not matter what answers are decided on — nothing about the theory, how it is used, or how it is interpreted change. It is worth a few remarks, but no more than that. Too many might mislead.

Fields (that is functions of both space and time) are the primary objects, and that is the reason that quantum mechanics is necessary and classical physics impossible [Mirman (1995b), sec. 1.4, p. 7]. However we should not jump from this to a meaningless concept. Particles are not emergent properties — particles do not exist (sec. V.2, p. 201). Thus the basic objects (of quantum mechanics and quantum field theory) are not particles, point or otherwise, for theories cannot be built upon meaningless notions. If it is necessary to discuss basic objects the best ones, the ones on which quantum mechanics and quantum field theory field rest (with, however, help from other concepts), are really group representation basis states. It is the physical objects that they describe (actually are) that are closest to a primary ontology. However we cannot distinguish in meaningful ways physical objects from mathematical objects that we use to describe them (sec. III.3, p. 117).

Fundamentally it is misleading to discuss a primary ontology. The word ontology implies that objects are in some sense real, something that we can touch and feel (sec. III.3.c.ii, p. 122). The fields of quantum mechanics are not classical fields, which we might picture as say waves in water, but only functions of space and time. They have no existence beyond this. They are not objects out of which other objects are built. Thus people think of classical particles as in some sense being "real", while electromagnetic fields are something used only to transmit forces between particles, thus being less real (sec. II.5.e, p. 98). Quantum mechanical wavefunctions (fields) are not real in this sense. There is no such type of ontology in quantum mechanics.

II.5.a.ii *Expressions for fields*

We thus write a field as a infinite sum of terms each a sum of products of single particle (say plane wave) states, summed over all numbers and types of particles, and integrated over momentum [Mirman (2001), chap. IV]. This is required to be a Poincaré basis vector (which is somewhat unclear for a nonlinear representation). However it must be an eigenfunction of the invariants (these must be invariant because of the algebraic structure of the group, independent of representations and realizations), and of the momenta, and (here) these contain nonlinear terms — there are interactions. Momentum operators have to transform properly, and commute [Mirman (1995c), sec. 6.3.8, p. 110]. Also fields must commute for spacelike separation.

So we define a set of Poincaré basis vectors, for linear representations (free particles), say plane waves (for representations with momentum operators diagonal). An eigenstate of the Hamiltonian with interactions would then be a sum of products of these, with integrals over momenta, limiting the sum to particles that actually exist. This assumes that these form a complete set for such states.

And all requirements must be satisfied simultaneously. It is possible that their compatibility puts limitations on possible interactions. This is not considered here, but it might be useful to do so.

II.5.a.iii *Other expressions*

Another approach is to start not with plane waves, but other states, say coherent states, and form such a sum of products. Most of what we do is independent of the form of states, and what operators they are eigenstates of, but study of the relevance of this could also lead to useful conditions.

A Fock representation can be characterized by the existence of a number operator. In other representations this is not the case; they do not have states with a fixed number of particles which are therefore eigenstates of the number operator. Here states are sums of states of all numbers of objects — each a product state, compatible with the restrictions of equality of momenta and angular momenta for all states in the sum.

The idea of expanding a field in terms of harmonic oscillators is misleading. There is no reason why harmonic oscillators should be important. What is actually being done is to write the field as a Fourier series (or integral) over group representation states that are eigenstates of momentum operators. There are fundamental reasons for doing so, but there are also other representations. And Fourier series (or integrals), so harmonic oscillators, are not always relevant, as with spin.

II.5.a.iv *Justifying these expressions*

The reason for taking this view of fields is that these are the states seen in experiments. Thus experimentally there is a state of a single Δ resonance, and one of $N\pi$, with the Δ produced in a reaction becoming a sum of a Δ and $N\pi$. After a time (actually infinite), only the $N\pi$ state is found, experimentally. Likewise, experimentally, there are states of a single N, and of a single electron, of a single π, and so on. There are also states of an N and k pions, say produced in a reaction, as well as, for example, an $N\pi$ state with these particles moving toward each other, colliding giving a Δ, and also states of nN plus k pions, and so on. And when particles are not too close to each other their states can be reasonable regarded as essentially plane waves (or wavepackets).

These are the experimental states, and expressions for fields used here are those describing the states that are actually found.

II.5.b What do we mean by "single particle"?

It is often said that light (say in astronomical observations) is measured to a single photon. This statement carries a subconscious connotation of a photon as a little ball, which is clearly not correct. Yet the statement itself has meaning, certainly experimental meaning. What do we mean by a single photon?

This measurement is really one of energy. Energy is related to frequency by $E = f$, because of the kinematical meaning of frequency [Mirman (1995b), sec. 3.4.2, p. 53], and because energy is the eigenvalue of the time-translation generator, the Hamiltonian, so is conserved (which gives it its content and importance). Thus a single photon of frequency f has energy $E = f$, and this is what is meant by a single photon, an electromagnetic wave of this frequency, and this energy. A collection of n photons then has energy nf. This gives the value of n. There is no implication in this that the photon is a "particle", whatever that means, or is confined to a point, whatever that means.

The electromagnetic radiation may be in the form of a wavepacket, but that is not a particle. The energy then is the expectation value of the Hamiltonian between the states giving the wavepacket, integrated over its various frequencies. This has even less meaning as a single "particle", although it can be considered as a single object.

The number of photons, and similarly other elementary objects, is then the ratio of energy to frequency, E/f, and is an integer. Why an integer? The photon, like all (single) objects, is described, must be described (sec. I.8.a, p. 44), by a sum of Poincaré group representation basis vectors (with the sum over different translation representations). The basis vectors relate its energy and frequency. What we mean by n photons is an object, or better system, described by a product of n identical representations (or if they are moving in different directions, or have different frequencies, a product of different representations). But the number of representations is an integer, so that of photons, E/f, must be, by definition (of a photon as a basis state of a particular type of Poincaré representation). If frequency and energy have specific values (within a very narrow range) the number of photons E/f is determined.

II.5.b.i *How one photon differs from two*

What is the difference between a single photon of frequency $2f$, and two of frequency f? They both have the same energy. Of course, if they are moving in different directions, they belong to different representations of the translation groups. And for fermions, even if momenta are the same, their spin directions must be different, so they also belong to different basis vectors.

However it is possible to have two photons (or pions, or ω's, and so on) with identical momenta and spin. Two objects are given by the product of two basis vectors. Both the frequency and energy are measurable, thus their ratio

$$n = \frac{E}{f} \tag{II.5.b.i-1}$$

is an experimentally determinable number, and this is the number of the elementary objects, say photons.

Another reason the product of two states, and a single state with the same energy, are different is that they interact differently. The behavior of light in a diffraction grating, or slit, or going through a crystal, for example, depends on its frequency. Also if a photon of energy E strikes an atom with excited states of energy $E = f$, and $\frac{3}{2}E$, it can be absorbed. Thus a state of two photons can result in absorption, but a single photon of energy $2E$ (almost) cannot be absorbed.

Consider an atom at rest with ground state mass m. Initial states of the system consisting of the atom, $|a)$ plus the photon(s), $|y)$ are either

$$|system) = exp(imt)|a)2|y)exp(ift), \tag{II.5.b.i-2}$$

or

$$|system) = exp(imt)|a)|y')exp(i2ft). \tag{II.5.b.i-3}$$

Both have the same total energy, and photon energy. The final state for the second is the same as the initial state since it cannot interact, but for the first it is (with one photon absorbed)

$$|fs) = exp(i(m + f)t)|a)|y)exp(ift). \tag{II.5.b.i-4}$$

By Fourier analysis the exponentials of the initial and final states must be the same, which is true for absorption of one photon of frequency f, but cannot be with a photon with frequency $2f$. For the first

$$exp(i(m + f)t)exp(ift) = exp(imt)exp(ift)exp(ift), \tag{II.5.b.i-5}$$

which is correct. For the second

$$exp(imt)exp(i2ft) = exp(iMt), \tag{II.5.b.i-6}$$

which cannot hold since there is no such M;

$$m + \frac{3}{2}f \neq m + 2f. \tag{II.5.b.i-7}$$

This, Fourier analysis, explains why there is a difference between a state of energy $2f$, and two each of energy f, why one can interact, the other cannot. This shows the meaning of a multi-particle state.

The concept of a single particle really refers to a single basis vector. Can it have a classical meaning [Avron, Berg, Goldsmith and Gordon (1999)]? For example we can consider

$$hN = \frac{U}{\nu}, \qquad \text{(II.5.b.i-8)}$$

to give the number of photons N in the field, where U is the energy of an electromagnetic field, ν the frequency. However this has a problem, h depends on the units. Does the number of photons depend on the units? Actually U must be proportional to ν [Mirman (1995b), sec. 3.4.2, p. 53]. It is

$$U = hN\nu, \qquad \text{(II.5.b.i-9)}$$

so this equation is actually a statement of the requirement coming from group theory. This holds true both in classical physics and quantum mechanics, as it must because it is purely a group theoretical result, actually coming from geometry.

II.5.b.ii *Product states and single states are distinguishable*

We see then that product states can be distinguished from single ones experimentally. States in a product may be functions of different variables, as with particles moving in different directions. But even if identical, there is a relationship between frequency and energy, and this is different for a single state and a product — the ratio is the number of states in the product, and varies with the number of terms. For a massive object there is a relationship between energy, mass, and frequency, in every system. Frequency can be measured in an interference, or double-slit, or other diffraction experiment. Mass can be determined, independently, by behavior in a magnetic field, since the charge can be obtained with another independent experiment. In these experiments products of different number, and different types, of states behave differently.

Suppose that a photon is captured. That establishes the amplitude was 1, either it is captured, the atom went to a higher energy state, or it was not. This can be checked by studying the atom. And the number of photons captured, the actual experimental result, not the expectation value, is an integer. The matrix element of the Hamiltonian, is nonzero only between states of the same energy. Being a Poincaré momentum operator, it is invariant under space and time translations, so conserves energy. This is independent of the amplitude. Hence the energy of the final, minus the energy of the initial, states of the atom equals the energy of the photon, which is equal to an integer times the frequency (which can itself be measured).

II.5.b.iii *How functional forms of single and multiple Poincaré states differ*

What is the mathematical difference between a state of a single Poincaré representation, and the product of several? The clearest case, being the most extreme, is of two identical scalar objects, with plane wave states. One moves along z, the other in the xz plane, so the states are (always unnormalized)

$$|1) = exp(ipz - ip_4t), \quad |2) = exp(iqzcos(\theta) + iqxsin(\theta) - iq_4t), \tag{II.5.b.iii-1}$$

with

$$p^2 - p_4^2 = -m^2, \quad q^2 - q_4^2 = -m^2. \tag{II.5.b.iii-2}$$

How does this differ from

$$|c) = exp(ir_zz + ir_xx - ir_4t), \tag{II.5.b.iii-3}$$

$$r_z = p + qcos(\theta), \quad r_x = qsin(\theta), \quad r_4 = p_4 + q_4? \tag{II.5.b.iii-4}$$

For it

$$\begin{aligned} r^2 - r_4^2 &= p^2 + q^2 + 2pqcos(\theta) - p_4^2 - q_4^2 - 2p_4q_4 \\ &= 2(-m^2 + pqcos(\theta) - p_4q_4), \end{aligned} \tag{II.5.b.iii-5}$$

and is not a constant, but a function of momenta.

To see the effect of this consider a boost in the zt plane, giving a different coordinate system, but one which physically has the same status. Then

$$p = p'cosh(\eta) + p_4'sinh(\eta), \tag{II.5.b.iii-6}$$

$$qcos(\theta) = q'cos(\theta)cosh(\eta) + q_4'sinh(\eta), \quad qsin(\theta) = q'sin(\theta), \tag{II.5.b.iii-7}$$

$$p_4 = p_4'cosh(\eta) + p'sinh(\eta), \tag{II.5.b.iii-8}$$

$$q_4 = q_4'cosh(\eta) + q'cos(\theta)sinh(\eta), \tag{II.5.b.iii-9}$$

so

$$\begin{aligned} r^2 - r_4^2 = 2\{&-m^2 + (p'cosh(\eta) + p_4'sinh(\eta)) \\ &\times(q'cos(\theta)cosh(\eta) + q_4'sinh(\eta)) \\ &-(p_4'cosh(\eta) + p'sinh(\eta))(q_4'cosh(\eta) + q'cos(\theta)sinh(\eta))\} \\ = 2(&-m^2 + p'q'cos(\theta) - p_4'q_4'). \end{aligned} \tag{II.5.b.iii-10}$$

Although this is form-invariant, variables have different values in the primed and unprimed systems. Thus for a basis state that is the product of two basis states, the equivalent of mass squared depends on the system of the observer — it is not a constant. This is true for all values of θ. The Poincaré group then distinguishes between basis states of a single representation and those that are products from different (although identical) ones.

In this way also representations of inhomogeneous groups, like the Poincaré group, differ from semisimple-group representations. It is not possible, using only the rotation, or the Lorentz, group to specify whether a state is of a single representation, or a product reduced into a state of an irreducible one.

II.5.b.iv *Amplitudes, energy and particle number*

A statefunction has an amplitude. Can changing the amplitude of a state change its energy? The energy is actually $|A|^2 f$; why is it taken as f? All photons no matter what their frequencies have the same amplitude, since a photon of one frequency can be taken to have any frequency, but in a different frame. The only difference between photons of different frequency is the frame in which frequency is measured; there is a frame for which any photon has any chosen frequency, and all frames are equivalent. So photons of all frequencies are equivalent, and must have the same amplitude (which is 1).

While $|A|^2 f$ is the energy expectation value, it is not the energy.

A single basis vector must have a definite amplitude, 1, although in an expression for a total state, $|S)$, which may consist of a sum over different numbers of objects, the statefunction appears with a coefficient, $c(t)|y)$, where $|y)$ is the statefunction for the object (a photon say), the basis state for the representation describing the object, with the coefficient time dependent since photons can be created or annihilated. Thus the energy expectation value for a single object, say a photon, but not the entire state, is time dependent. A state itself is defined by the action of a creation operator on the vacuum (the state for which all eigenvalues of the group diagonal operators are zero).

States transformed into each other by group operators must have related amplitudes, $c(t)$. Amplitudes can have any value. They give the probability of finding a particle somewhere, for example. If, say, a particle is produced by decay, at any time the probability of it being found anywhere is less than 1, and time dependent. Integrating over all time an amplitude can be different from 1, if there are various decay modes, and can be greater, if some modes give more than one particle.

II.5.b.v *Wavepackets*

We have defined a single particle by the relationship of energy to frequency. Since fundamental objects are described by single representations of the group (usually taken as ones with the momentum operators diagonal), this is the correct definition (sec. I.8.b, p. 48). But actually every observed particle is a wavepacket (otherwise its probability of being in our galaxy would be quite small indeed (sec. II.3.c, p. 78)). What do we mean by a single particle if it is given by a wavepacket, that is its statefunction is a sum of basis vectors from different irreducible representations? This is a question to be considered for each experimental setup. But often a general answer can be given.

A wavepacket is taken to be greatly different from zero only in restricted regions in both position and momentum spaces. Usually it is also taken to be smoothly varying in some sense, so if, say, it is bimodal the meaning of single particle might have to be looked at more closely. The statefunction (in position space) is then an integral over momentum with the sum of the squares of the coefficients equal to 1. The energy of an object is the expectation value of the Hamiltonian for this wavepacket, and almost certainly the measured value is close to the expectation value. Of course there is a slight chance, since the statefunction is a wavepacket, that the object is produced with an energy much different from this, but experimentally such values are found only rarely.

How do we distinguish experimentally one object with energy E from two, each with energy $\frac{1}{2}E$, when their directions are the same (otherwise they would clearly be different)? If the levels of the detector are discrete, and are spaced about E apart, then in most cases the single object will excite the detector leading to its detection. But only extremely rarely will this be the case for two particles (in some very rare cases the production mechanism gives a highly energetic pair, as it must since the wavepacket has contributions from all momentum values). Thus again the particle number is found from the relationship between energy and frequency, or momentum (sec. II.5.b, p. 86), but for a wavepacket these are expectation values.

More relevant to the definition relating energy to frequency, for a fairly narrow, well-behaved (in a sense that we do not explore in general) wavepacket, almost all energy and frequency values differ only little from the most probable ones, which are (about) those given by expectation values of the Hamiltonian and momentum operators. The relationship between these values is different for one object than for more than one. The definition then of a single object works for a wavepacket, as well as a statefunction of a single frequency (thus energy and momentum).

With complicated wavepackets the question must be looked at, and it may be that the definition of a single object becomes vague.

II.5.b.vi *Why discreteness?*

Why are states discrete; there is one photon, or two, or ..., but never 1.5? Integral powers of basis vectors are basis vectors, thus can be statefunctions, nonintegral powers — like $(sin(\theta))^{\frac{1}{2}}$ — are not basis functions, thus cannot be statefunctions. Thus there are always an integral number of objects, which does not imply that they are point particles (whatever that is), only that the statefunctions of systems are composed of integral numbers of basis states. Nor does it imply localizability of an object, which simply means that its statefunction can be a narrow wavepacket, a sum of translation basis states.

Ultimately the Hamiltonian is — must be — in the form of a product of states, so is different from one in which there is only a single state. It acts differently on a product of states with total energy $2E$, than on a single one with this energy.

II.5.b.vii *Why is energy change equal to an integer times the frequency?*

If an object is absorbed, by conservation of energy and the kinematical meaning of frequency, the energy change of the detector is equal to the frequency of the object. However it can also be 2 times that. Why not 1.3 times the frequency? The Hamiltonian acts on — must act on — one state at a time, giving the change of energy of the detector equal to the frequency of the object. For a product of states, then (independent of amplitude) the Hamiltonian can be considered to annihilate one photon, raising the energy of the detector, then the second, raising the energy of the detector again, and so on. This, by the nature of the Hamiltonian (that it is, must be, a product of basis states), must be done in discrete steps, so each annihilation must give an energy change equal to that of the frequency — one photon is annihilated at a time. Hamiltonians cannot annihilate a partial photon. It thus is proportional to photon annihilation operators, perhaps to some power a^n, which is meaningless if n is not an integer. And it can have only this form. It is here that the requirement of integral values of the number of particles enters, in the requirement imposed on Hamiltonians by the necessity that they be integral products of basis vectors, which themselves must be integral products of basis vectors.

The reason thus that n must be an integer is that a statefunction of a physical object must be a representation basis state (sec. I.8.a, p. 44; sec. I.8.b, p. 48). These basis states are a complete set of states. If n were not an integer then a creation operator would create a state that is a basis state to some nonintegral power. While a basis state to an integer power, or a product of states, is a basis state, so acceptable, one to a nonintegral power is not, so cannot be a statefunction of a physical object. A Hamiltonian that produced a state that could not be a basis state cannot be an operator of a consistent theory, it is not an inhomogeneous operator (an Abelian translation generator) of an inhomogeneous group (one containing an invariant Abelian subgroup) like the Poincaré group.

That is why there can be 1 photon, 2 photons, ..., n photons, ..., but not 1.5 photons.

II.5.b.viii *Does a photon interfere with itself?*

One question related to the nature of single-particle states is whether, in say a two-slit interference experiment, photons interfere with each other, causing the pattern or whether each photon interferes only with itself [Dirac (1956), p. 9], and interference between two never occurs. The first question that must be asked is whether these statements have meaning? Can we tell the difference? Suppose that the light flux is so low that (almost certainly) only a single photon is present at any time, that is the experiment is, has to be, described using just a single representation basis state. In that case there will be no interference pattern, so the distinction between photons interfering with themselves, or with others, is nonexistent. If there are a large number, there is no way of telling what is interfering with what, so these statements are also without content.

Take the experiment to continue for a long time — there is a large number of photons — but the flux is so low that only one goes through at any time — the probability of finding more than one is extremely low. For this the statefunction is a wavepacket — it is confined to a small interval in both time and space (it must be confined in space otherwise we could not tell where it is, so could not say how many go through at once) — actually a sequence of such well-separated wavepackets. Such a wavepacket consists of many waves, so is in that sense not a one-particle state, although the total energy carried by the wavepacket can be so small that it is able to give but a single dot. Here different waves can be considered (but only picturesquely) to go through different slits. What interferes with what? Of course, we cannot tell. These statements and questions about whether or not a photon interferes with itself or with others make no sense.

II.5.b.ix *Double slits*

Double-slit experiments raise questions about, and illustrate, the crux of the concept of a single object beyond that just discussed concerning how a single photon can go through two slits (which is not a difficult question — being vacuous it is not a question). But how do opaque screens between slits act on a single photon, requiring it to go only through them?

Objects cannot go through parts of the screen between, and outside, slits, thus we do not consider them, except that we take statefunctions of objects zero there. However in order that objects not traverse these barriers, they must interact with the atoms in them. Can a single object interact with all the atoms and still go through the slits? We assume that the objects are not absorbed by the atoms, the usual case for we always regard the entire beam of particles as traversing the slits. The statefunction of each particle is spread out, being nonzero within the slits, and also at the barriers, and for some distance within them (say, decaying exponentially, as usual). Thus there is a nonzero probability that an atom in a barrier will interact with the particle. Since it is not absorbed its statefunction along the path from the atom into the barrier becomes a sum of the incoming one (at that position) plus that produced by the atom when it is (virtually) excited and then decays. A complete statefunction, of a single particle, therefore is the sum of an incoming statefunction plus those produced by all atoms, and this goes to zero inside the barrier.

The statefunction is then a wavepacket, along the plane of the slits, and perpendicular to it, as well as in time — this a sum of the incoming wavepacket plus those produced by the wall. The wavepacket is (almost) zero in the plane, except at the slits. Because it is a wavepacket, it cannot be considered as a single object. Rather it is a sum of single-object states, with each term having an amplitude such that the sum of the squares of the amplitudes over all terms equals 1, perhaps for a short period of time and if so 0 otherwise. This wavepacket is almost zero, except at the slits.

The expectation value of the energy is obtained by taking the expectation value for each term in the sum, in the usual way (and each has a value less than its frequency because their amplitudes are less than 1), and adding. It is the expectation value of the energy that determines how many dots each wavepacket can produce.

II.5.b.x *Linear polarizers*

Crystals that give linearly polarized light cannot produce single photons. Photons are circularly polarized — because the statefunctions of the Poincaré group describe these states [Mirman (1995c), sec. 4.3.2, p. 65]. How then are linearly polarized states obtained? Linearly polarized light is a sum of circularly polarized states, one clockwise, one counterclockwise. An atom thus emits circularly polarized photons (as is well-known; this is used to get selection rules for angular momentum).

For a beam traveling through a crystal to become linearly polarized it must interact with it, with its atoms. These absorb the circularly polarized photons, become excited (perhaps virtually) and decay, emitting circularly polarized photons. However in some cases the direction of polarization is reversed. Thus the resultant beam of photons has a state that is a sum of basis states, and this, with proper coefficients (forced by the properties of the crystal), describes a linearly polarized beam.

Individual photons, however, are circularly polarized. Linear polarization is really an interference effect. Two states of circular polarization interfere so that the sum giving linear polarization in the "wrong" direction is zero.

Models, which cannot be considered here, should clarify these effects by showing how interference produces states with zero amplitudes, especially how properties of crystals lead to extinction of waves with one polarization but not the orthogonal one, and how the point and space groups of crystals [Mirman (1999), sec. IX.6.c, p. 499] quantum-mechanically determine which polarizations are allowed.

II.5.c Gauge transformations

While gauge transformations are fundamental for electromagnetic interactions it is often not realized why this is so, why these are necessary, and how they affect electromagnetism and its interactions with objects. We review this next.

Minimal coupling of the electromagnetic potential is required by gauge invariance, which is a partial statement of Poincaré invariance for massless objects [Mirman (1995c), sec. 3.4, p. 43]. Thus it is required by Poincaré invariance (and would hold true even if this were broken, except, perhaps, for unusual or extreme violations); it is thus a consequence of, not invariance, but geometry and its having the Poincaré group as its transformation group. Gauge transformations are simply a subset of Poincaré transformations for massless particles, and these only.

Thus the electromagnetic gauge group is a subgroup of the Poincaré group. It might seem that it can be defined as that group leaving electromagnetic fields invariant. However this is neither relevant nor correct. Electromagnetic fields, not being physical, cannot impose conditions on physics, and they are not gauge invariant [Mirman (1995c), sec. 3.3, p. 37].

II.5.c.i *Gauge transformations act on space coordinates*

While Poincaré transformations are sometimes regarded as "external", acting on coordinates of space, gauge transformations seem to be "internal", not acting on coordinates. But actually gauge transformations are Poincaré transformations, so the distinction is misleading. The reason that the potential can undergo gauge transformations (and the gravitational connection undergoes general coordinate transformations — the helicity-2 analog of electromagnetic gauge transformations) is that phases of statefunctions of massive particles to which the photon is coupled are arbitrary, so the transformation is able to vary them to offset the change in the potential, which is thus unobservable. In the gravitational case, the term added to the connection by gauge transformations is canceled by the term resulting from the transformation of the statefunction under these corresponding changes; this term is a consequence of the statefunction being a tensor, requiring the covariant derivative, rather than the ordinary one, for its transformation [Mirman (1995c), sec. 1.2.2, p. 8].

II.5.c.ii *Gauge transformations are not space-dependent coordinate transformations*

It is incorrect to say that gauge transformations require that we can, or have to be able to, perform coordinate transformations at each point differently. Transformations are of observers, and each observer looks at all of space. Thus a transformation is of the same coordinate system over all space. To transform an object being looked at requires a physical action on it, and this changes it. Changing a field differently at each point gives a different field. Gauge transformations on mass-zero fields are not transformations that are space dependent, but are Poincaré transformations for these objects, only.

Transferring to an observer who has a differently-oriented coordinate system at each point is not a rotation and not a transformation like those used in physics (and at best the group of these transformations, if any, has to be defined, and shown to have geometrical relevance, if possible).

II.5.c.iii *Gauge invariance and charge conservation*

Gauge invariance has nothing to do with charge being conserved globally as well as locally. If it were conserved globally but not locally, that would mean that charge could be destroyed at one point, and instantaneously an equal amount of charge would be created elsewhere, perhaps very far away. It would require signals with infinite speed. No one has ever constructed a Hamiltonian, for example, that gives this. Of course, mass, baryon and lepton numbers, and so on, are conserved locally, but for these there is no gauge invariance.

Those who say that gauge invariance is necessary for charge to be conserved locally should explain why there is no corresponding invariance for angular momentum, or momentum, or energy, for example, even though these are conserved locally. If a system is in a momentum eigenstate then its position cannot be determined so it is meaningless to say that there is local conservation, since there is no way to measure the values of the various quantities in any region. However if it is a wavepacket then the Hamiltonian acting on it, moving it in time, commutes with the operators for charge, mass, momentum, and so on, thus does not change the values of these. The expectation values for them are (almost) zero except in a region of space — the statefunction is a wavepacket — so their values in that region remain unchanged and they are conserved locally, independent of whether there is gauge invariance, and this is the same for all these quantities: charge, mass, angular momentum, and so on. Physics is local because all terms in Hamiltonians, and the other momentum operators, are functions of the same point in space and time. This is what is meant by locality. Those who do not agree should define what they mean by it.

II.5.c.iv *What field is the electromagnetic field?*

There is a strange belief that the electromagnetic field is physical, measurable, and gauge invariant. Of course it is none of these [Mirman (1995c), sec. 3.3.1, p. 37]. That is not measurable is trivial. It is measured by placing a charged object in it, and observing the object's motion. However charged particles are governed by Dirac's equation, or the Poincaré invariants for other spins, and these contain the vector potential, not the electromagnetic field. Thus it is the vector potential that is measured, not the field. Perhaps most important, it is the potential, not the field, that is a Poincaré basis vector, and only these can have physical meaning.

II.5.c.v *Duality*

There is no wave-particle duality (sec. V.2, p. 201), nor is there duality of electric and magnetic fields. These are not physical objects, the potential is [Mirman (1995c), sec. 3.3, p. 37], as is obvious from Dirac's equation. That there are two fields is purely an artifact of formalism and definitions. Time and 3-space are treated differently, as are derivatives with respect to them; this makes it appear that there are two fields, rather than nonphysical functions of the actual physical fields.

It is believed that the field is measurable, while the potential is not: one can use a test charge and the Lorentz force law to (in principle) measure electric and magnetic fields. But actually the only way to measure is to use the Dirac equation, or the corresponding pair of equations for other spin particles [Mirman (1995b), sec. 6.3.2, p. 116], and these contain the potential, not fields. The Lorentz force law is inconsistent [Mirman (1995b), sec. 1.4.1, p. 9]. So it is the potential, not fields, that are measurable. That there is in the formalism two fields has no meaning, no physical implications, no physical significance; it is a result merely of the way that we write these (unphysical) fields, purely formalism, not physics.

II.5.d Misconceptions about quantum field theory

Just as is quantum mechanics, quantum field theory is plagued by misunderstandings, misinterpretations, and too often, just mere confusion. Here we mention just a few, but probably only a small fraction, of these.

II.5.d.i *Second quantitization*

The history of the development of quantum mechanics includes the guessing of operators of systems from their eigenvalues (the classical variables). This program of "quantitization" (that is guessing) then led to "second quantitization", a description of "fields" by "quantizing" the Schrödinger wavefunction. Actually it is incorrect to distinguish, at a fundamental level, a "field" (which presumably means a multi-particle state) from a single-particle one. They differ merely in the number of particles, whose meaning is clear (sec. I.6.d, p. 32). Thus "second quantitization" is merely a way of introducing a different formalism which is useful for states that need not contain only a single particle [Cao (1997), p. 158; Dirac (1956), p. 229, 233; Schiff (1955), p. 341; Schweber (1962), p. 148].

Creation and annihilation operators provide a useful formalism for writing these statefunctions, which can also be written as multi-particle states. But these are identical, being just slightly different notation. Thus there is no such thing as "second quantitization". It is just a grandiose name for a particular version of (guessing) formalism.

II.5.d.ii *Quantum field theory is not a result of relativity*

It is often believed, for some reason, that quantum mechanics combined with special relativity results in quantum field theory (sec. II.1, p. 54). However, it should be clear that quantum field theory does not depend on a transformation group, certainly not on the Lorentz group. For its general properties the Galilean group is just as good, or just as irrelevant, as the Lorentz group. That particles are created or annihilated and that they decay is not a consequence of the group, but of nonlinearities (interaction terms) in the governing equations. Fundamentally, quantum field theory is merely a description of multi-object states whose statefunctions obey nonlinear equations. Of course, these statefunctions are group representation basis states, an additional (but necessary) requirement.

II.5.e What are particles and what are forces?

The apparent distinction between particles and forces is merely that between fermions and bosons. It is misleading to say that bosons carry, or produce, forces between particles. A nucleon emits a pion which is absorbed by another nucleon, resulting in an interaction between the nucleons (this is a visualization which needs some care). But a pion can emit a pair of nucleons, necessarily an even number since these are fermions (actually a nucleon and an antinucleon), which is then absorbed, giving an interaction between pions. And nucleons can interact by transferring two or three pions. These differ only in that for interactions carried by bosons, the number of bosons can have any integral value, while for fermions, the number must be even, and fermion number must be conserved.

Nor does matter (electrons or nucleons, say) result from the transformation of energy (that is pions or photons) into mass. Pions and photons have mass, and it makes as much sense to say that a collision of a pion with a nucleus resulting in more nucleons is a conversion of energy to mass as to say that the collision of two nucleons resulting in pions is such a conversion, or even that the decay of a muon or lambda giving photons or pions converts mass to energy. All objects (before and after decay) have both mass and energy.

We can emphasize, as many like to do, that forces between fermions are carried by bosons. However we can also emphasize that fermions are merely those objects that emit and absorb bosons. Neither fermions nor bosons are in any meaningful sense more basic. The only difference between "particles" and "forces" is statistics.

II.5.e.i *Product states must be symmetrized, giving the real difference between bosons and fermions*

Statefunctions, for any number of particles, are products of representation basis vectors of the relevant transformation group (the rotation, Lorentz, and Poincaré groups, plus those on internal variables, like isospin, but in systems like crystals, perhaps subgroups of these [Mirman (1999)]). The number of basis vectors in this product is what is really meant by "the number of particles". There is no implication of any particular properties of these "particles" like localization. A basis state can be a momentum eigenstate, thus completely unlocalized.

However a product, for identical objects, is not a proper basis state — it must be symmetrized, being symmetric for integral spin, antisymmetric for half-odd integral spin. The latter objects are often called "matter" (or terms like that), while the former are referred to as "energy". But both fermions and bosons, including photons, have mass and energy. The only difference is the form of symmetrization and consequences of that. Thus statements like two particles annihilate and are converted into pure energy are rather silly. A proton and pion can "annihilate", "destroying" each other, being converted into pure something-or-other, such as a Λ. Or a proton and antiproton can annihilate going into "pure energy" such as pions (or into a Λ plus an anti-Λ). Conservation of angular momentum requires that the parity of the number of fermions (odd or even) be conserved in a reaction. Thus the total number of fermions (antiparticles giving negative values) is constant, while the number of bosons generally is not. But that really is the only difference.

Indistinguishability of identical particles in quantum field theory comes from the symmetry built into commutation or anticommutation relations of field operators. However this does not mean that indistinguishability or symmetry results from quantum field theory. Rather the formalism of quantum field theory is chosen to reflect these requirements, in particular those coming from the rotation group which relates the statistics of fermions and bosons to their spin [Mirman (1995b), sec. 8.1.1, p. 147].

II.5.e.ii *Fermions are more real only intuitively*

Intuitively fermions may seem more "real". But (macroscopic) objects that we deal with are collections of fermions, except for light whose emission and absorption — from fermions — is what is usually perceived. It is this experience that makes fermions seem more important. However this perception is merely a consequence of fuzzy thinking.

Likewise there is a belief that the electromagnetic field is a mental construct brought in (only) for the purpose of analyzing interactions between charges (these being regarded as "real"). This regards the electromagnetic field (light) traveling from a distant galaxy to our eyes as having reality only in terms of the interactions between the charges in that galaxy and those in our eyes. This light can be scattered along the way thus giving the interesting concept of a mental construct being scattered by dust particles.

And in what sense are charges more real than fields? Charges give interactions between objects and others they are in contact with. Are these charges more real than the objects? If an electron is "real" because it interacts with an electromagnetic field (potential), is the field really inferior since it also interacts with an electron? A gravitational field acts on photons, implying that photons create a gravitational field (which then acts on other photons). Should we then say that a gravitational field is a mental construct brought in (only) for the purpose of analyzing interactions between photons? But if a photon is merely a mental construct than a gravitational field is a mental construct used only to discuss the interaction of mental constructs. Also a pion, like a photon a boson, is needed to discuss interactions between nucleons. Is a charged pion merely a mental construct of the same kind? If it is, then the electromagnetic field is again a mental construct needed only to discuss interactions between mental constructs. If not, then how does it differ from a photon — merely by being charged, which would mean that a neutral pion is less "real" than a charged one? And is an object with nonzero rest mass more "real" than one with zero rest mass, but actually with energy, so mass?

This illustrates how our classical concepts are carried over leading to confusion and often nonsense.

II.5.e.iii *What can we see and feel?*

One reason that a distinction between matter and fields (sometimes loosely called energy) has arisen is that intuitively we can feel and see objects, so regard them as real, but intuitively we cannot see the electromagnetic field, so regard it as, if not fictitious, a mere construct.

But, as often happens, intuition is wrong and misleading. In fact we cannot see objects, we see electromagnetic fields (actually potentials), classically called light; in quantum mechanics the term photon is used. These are what reaches our eyes. Nor can we feel objects — there are forces between atoms, ones in objects and those in our bodies, and information about these forces is transmitted to our brains making them believe that they feel objects. These forces are electromagnetic, so it is fields that we perceive, not solid objects. Hence objects, chairs, tables, people, should be taken as mere constructs, electromagnetic potentials as real.

Fundamentally, since nature is quantum mechanical and is a collection of microscopic objects, elementary particles, both objects and fields are constructs, and in principal have the same status, equally valid constructs differing only in their statistics.

II.5.f Is field theory consistent?

It is often argued that field theory is inconsistent because of the infinities appearing in perturbation theory. However perturbation theory is unlikely to be an exact expansion, but only an asymptotic one. Problems in an approximation scheme are more likely due to that than to the underlying theory — we cannot conclude that a theory is inconsistent because there are problems in an approximation method of studying it. If problems arise in perturbation theory, that implies the approximation method is wrong, not that quantum electrodynamics is.

Computation of "self-energy" of an electron gives infinities, which prejudices many physicists. All that this shows is that "self-energy", or mass, cannot be computed using electromagnetic theory (sec. IV.2.m, p. 168). And this should be obvious. There are many charged particles, and they (with the same charge) have different masses. There is no "self-energy", and such computations are inherently meaningless and quite misleading. They imply nothing about quantum field theory (although perhaps they imply something about those who work with it).

II.5.f.i *Inconsistencies of classical physics are irrelevant*

There is a reason why physicists are prejudiced against. and confused about, quantum field theory, although the reason is wrong (sec. V.2.e.iii, p. 208).

It is clearly incorrect to attribute infinities to objects being point particles, a classical concept which has no meaning (sec. IV.2.e, p. 157) in quantum mechanics (and has, at best, questionable meaning in classical physics). In quantum mechanics objects are given by statefunctions which are spread over space.

That there are inconsistencies in classical mechanics is also irrelevant — classical physics is inherently inconsistent [Mirman (1995b), chap. 1, p. 1]. But inconsistencies of classical electrodynamics has again prejudiced many physicists, who regard problems with it as due to electrons being point particles. But the concept of point particles is vacuous, nor does it have anything to do with problems of classical physics — it has nothing to do with the inconsistencies of classical physics. Thus classical inconsistency implies nothing about the consistency of quantum electrodynamics. However this history, and the unthoughtful use of it, has been misleading and the source of great confusion.

The reason that classical field theory is inconsistent is illustrated by the equations from Newton's second law and from the Lorentz force law being inherently incompatible. This has nothing to do with point particles, and has no relevance to quantum field theory.

II.5.f.ii *How do we tell if a theory is consistent?*

What do we mean by saying that a theory is consistent? If we can find, for every set of initial conditions, for all systems to which it (alone) applies — for all governed only by the theory — a complete description of their behaviors, then we would regard it as consistent. Perhaps there is some subtle point that makes this argument not fully rigorous, but it is difficult to see what it might be. Certainly it is very close to being so.

In fact, we can do this for quantum electrodynamics. There is a completely well-defined procedure, with no ambiguities, no freedom, that gives (in principle) a full description of every system, governed only by quantum electrodynamics, under all conditions, with all predictions as precise as we wish. It is so well-defined that it can be computerized. There are steps of doubtful validity, perhaps, but this is due to approximation methods, not to the theory. The final answers, provided by using all steps of the procedure, are completely reasonable, finite, and in full agreement with experiment (to the extent now known). Any disagreement, no matter what, would show that the theory is physically wrong.

The argument that quantum electrodynamics is fully consistent appears very strong — it gives complete, reasonable, finite answers to all relevant questions. What more can we want?

Also, fields are Poincaré basis states, although of nonlinear representations. It is unlikely that the Poincaré group, or its representations are inconsistent. The reason that they are Poincaré basis states is that, since there are different observers, they must be transformable by the group of the geometry, and that is the Poincaré group (sec. I.7.b.iv, p. 39). There would be serious problems for physics and geometry (and other branches of mathematics) if these lead to inconsistencies.

Chapter III

Statefunctions and Probability, What and Why

III.1 Some fundamental concepts of quantum theory

Fundamental differences between classical physics and quantum mechanics and quantum field theory come from the role played in quantum physics by two constituents, statefunctions and probability. Thus to understand quantum physics we must see the meaning of these and the reasons they are necessary. Here we continue this study [Mirman (1995b)].

It is necessary in considering such topics to also try to eliminate some distracting misapprehensions that have arisen about these subjects. So a large part of these discussions (unfortunately) have to emphasize this.

Often analyses begin with primitive concepts, probability, observer, statefunction, and so forth, that cannot be further analyzed. But if they cannot be analyzed, how can they be related to physics? Then such words are vacuous with no connection to the real world and thus no way of helping us understand it. How can we tell if theories based on them are correct if they use words that have no meaning and thus lead to predictions that can have none? Mathematical systems can be based on symbols (which might be sequences of letters that are the same sequences as words) that are not specified beyond their relationships to other symbols. But physical theories based on meaningless symbols must therefore themselves be meaningless. This is often forgotten by those who construct rigorous systems (which, for some reason, they claim to be about physics) with sequences of letters the same as those of words that are used in physics. Clearly defining (physically) and analyzing all terms must be done, no matter how difficult that is — and it can be quite difficult. This then we can merely try to begin.

Essential for any complete, fundamental, physical theory is the specification of the sets of transformation operators. One in particular is emphasized, the momentum operators, including the time-translation momentum component, the Hamiltonian. The most familiar is the Dirac Hamiltonian, and this is usually the only one considered. But it is for a special, but very important, case that of spin-$\frac{1}{2}$ objects. It is so familiar it can be misleading. Thus we have to consider how to construct Hamiltonians in general.

Also we emphasize the roles of geometry, and interactions. For the gravitational interaction we take space as curved. It is worth reviewing the generic relationship between interactions and how we view geometry. This is only mentioned to indicate the existence of such topics, but discussed later [Mirman (2001)].

III.1.a Is quantum mechanics complete?

One purpose of the present discussion is to show that quantum mechanics is a fully correct, meaningful theory, entirely reasonable with no interpretative problems or difficulties (except emotionally). However many have questioned whether quantum mechanics is "complete" (sec. IV.2.w.v, p. 180)?

Quantum mechanics, given any physical situation, and all necessary information about it, does completely describe all experimental results (to the extent that they can be calculated, which is not relevant to its "completeness", but only to our calculational ability). What more can we want? What more is it possible to ask for?

Its description is often given by probability distributions. But this is required by experiment, as we see next, and geometry through group theory. Any "incompleteness" is not an indication of problems with quantum mechanics, but is a property of nature [Ballentine and Jarrett (1987)].

What seems missing from quantum mechanics is the emotional satisfaction that too many achieve by finding descriptions that are in accord with their (highly restricted) intuition, imagination or desires. This usually means that quantum mechanics cannot provide a classical explanation for the behavior of physical objects.

Quantum mechanics is thus complete in a physical sense, no matter how incomplete it may be in an emotional sense.

III.1.b Is quantum mechanics mysterious?

Many physicists like to emphasize their inability to deal with the physical world by loudly proclaiming how mysterious quantum mechanics is. A person (let us hope not a physicist) who believes that nature is Aristotelian would find classical physics quite mysterious. Someone who believes that trees and rocks have spirits that direct their behavior would find all of science extremely mysterious.

But no scientist would regard these views as anything but highly irrational. It is an interesting question in abnormal psychology why so many physicists are so extremely eager to flaunt their irrationality and incompetence by proclaiming the mysteriousness of quantum mechanics.

Those who think about quantum mechanics carefully, who accept the world as it is, not as we may wish it to be, actually find it impossible to regard quantum mechanics, and also the world, as in any way mysterious.

III.2 Necessity for probability

Quantum mechanics is probabilistic. Is there a way of avoiding this? Why must it be true [Caticha (1998a), p. 1578]? Are the reasons experimental, fundamentally geometrical [Mirman (1995b), sec. 3.3, p. 47], or simply a result of (necessary) ignorance? And what does it mean, since we often (seem to) consider only a single system? Of course requirements, and meanings, must be in accord with experiment. So we start with that.

III.2.a Experiment requires probability

The members of a set of atoms decay at different times. It is an experimental fact that decays are uncorrelated (there is no pattern). The time at which an atom decays is independent of its position, momentum, or relationship to atoms that have previously decayed (for example its position with respect to them). It is an additional experimental fact that the number of decays per unit time varies about an average, and purely randomly. There is no way of predicting from past decays (or anything else) how many, or which, will decay in the next unit of time. To predict that a particular atom will radiate in the next time interval it must be distinguished from others. Information must be obtained about it that could be used for prediction. But there is no way of obtaining such information, there is no way of distinguishing atoms (unless they are forced into different states, which is irrelevant since we wish to predict the behavior of a particular atom from a set all in the same state). All atoms are identical in terms of our information about them, so telling which decay next is impossible, and experimentally atoms do not all decay at the same time. Hence there is nothing that can be done except find probabilities.

Arguments for other types of systems and observations are similar, and the need for probability general.

Probability is required by geometry, but also by many different experiments. A theory that did not give probabilities, and only probabilities, could not agree with observation.

III.2.b Required knowledge does not exist

It is not that we do not know when an atom will decay in the future, but that we cannot, such knowledge is therefore nonexistent. This is not a limitation on our ability to understand nature, or a necessity for ignorance, but rather a fundamental limitation on meaningful knowledge about the universe. We are not ignorant, for there is nothing to be ignorant about.

Classical probability and quantum probability differ in that the former comes from ignorance, the latter is intrinsic. In principle, given a classical system, it is possible to acquire enough knowledge to make exact predictions for all future times (ignoring external influences). Thus we do not make exact predictions because we have not bothered to acquire all that knowledge. In quantum mechanics the required knowledge cannot, even in principle, be obtained — it does not exist.

III.2.c Geometrical reasons for probability

Experiment clearly requires that nature be probabilistic, so theories describing nature, necessarily based on quantum mechanics, must be. This is not an accident that we are forced to accept because that happens to be, for some unknown reason, the experimental results. The probabilistic character comes from the way that we (have to?) describe geometry and so the universe. Geometry requires probability through its transformation group [Mirman (1995b), sec. 3.3.1, p. 47].

Physical objects are described by statefunctions, these giving all possible information about them. And statefunctions are basis states of representations of the transformation group (the Poincaré group, or perhaps a larger one of which it is a subgroup). Basis states are functions of transformation parameters, for example, angles for rotation groups (the Lorentz group is a rotation group), distances for translation groups, these defining the space. But basis states also bear labels for the representation and state. A single eigenstate of group state-labeling operators is a function of group parameters, thus it cannot specify a value for these. To get a sum of states that is almost zero except for a small range of parameters requires many states, and many representations. In that case the state and representation are not well-determined. Transformation parameters and labeling operators are conjugate. Both sets cannot simultaneously be specified completely.

An example is the set of conjugate variables, position and momentum, the parameters giving transformations of translation subgroups, and those giving its representation. Similarly angle and angular momentum are conjugate. And there are cases, like different angular-momentum components, for which variables or labels cannot be measured simultaneously (with complete precision).

Thus it is, in principle, impossible to have complete information about a system (which therefore does not exist) because statefunctions must be transformation-group basis states. We are left then with being unable to make exact predictions, and have only probability.

III.2.d Quantum mechanics requires ensembles

Thus in quantum mechanics we have to, and can only, consider ensembles [Ballentine (1970)], say do an identical experiment many times, or pick an object from a set, and we can predict only the fraction giving each particular outcome — not the behavior of a particular member of an ensemble.

Part of the confusion about quantum mechanics arises because only ensembles appear — in the way that we use expressions for statefunctions — but this is often hidden. Thus for the microscope frequently used to illustrate the uncertainty principle [Schiff (1955), p. 9] it may seem from the usual discussion that only a single particle is being viewed (sec. III.2.h, p. 112), and there are uncertainties in its position and momentum (with a minimum value of their product). Thus instead of a sharp line on a screen indicating position, there is a band, so an uncertainty in position. But lines and bands require many photons, many measurements, thus ensembles. The phrase "uncertainty in its" and the word "band" are mutually exclusive — combined they form an oxymoron.

Usual analyses consider that only a single photon is scattered. Suppose the screen is replaced by a set of counters, which register arrivals of photons, these being very close to each other (so that they do not give additional uncertainty). If there is but one photon (assuming its energy is not too large) only one counter registers [Mirman (1995b), sec. 3.1.1, p. 38] — there is no band, there cannot be. But if we repeat the experiment many times the counter that is struck differs for the various trials, with counters in the middle of the band registering more often, those toward the wings less. Thus plotting the number of times a counter goes off as a function of position gives a band whose curve has a maximum at the middle and decreases away from it. Again it is an ensemble that is used. Standard deviations, giving uncertainties, refer to a large number of experimental trials, and statefunctions give probability distributions. We cannot discuss the uncertainty principle for the microscope if only a single electron is viewed, and only once.

Probability (not only) in quantum mechanics refers to ensembles, explicitly or implicitly [Farhi, Goldstone, Gutmann (1989)], so our expressions for statefunctions (not statefunctions, which are physical objects (sec. III.3, p. 117)) are based on ensembles. But it is still possible to often make predictions, without absolute certainty, about individual systems.

III.2.e How do we deal with individual systems?

It would seem that since we cannot tell when a particular decay will occur we have no information at all about a single atom. Yet intuitively we act as if we do. Can there be single systems or only ensembles (perhaps hidden)?

Consider a professor who is doing an experiment with a single atom that ends when it decays, and has to schedule a class — but not during the experiment. When should the class be scheduled? Since it is impossible to predict when the atom decays, the answer is never. We cannot make predictions about a single system, only an ensemble.

Suppose that the atom's half-life is 10^{-9}sec, or 10^{9}yr. When should the class be scheduled? Of course, anytime. But if its half-life is 1 hour, it would be best to wait a few days for the class. Why? There is only a single atom so we should not be able to make predictions, no matter what its half-life. But how is that atom obtained? If there is a container of identical atoms, and the professor plucks a single one and uses its decay to end the experiment, then an ensemble, that of the container, is being studied. For a 10^{-9}sec half-life, and 10^{2} atoms, it is almost certain that every atom in that ensemble will have decayed within 1 sec. Of course, to give this rigorous meaning, we would have to consider the experiment repeated many times, giving an ensemble of experiments. But it would take a huge number of experiments to find one in which even a single atom has not decayed within the first second. Thus it is reasonable to assume, but of course it is not absolutely certain, that the particular experiment done is not that anomalous one. This is the justification for deciding when to schedule the class, and illustrates the meaning of probability in quantum mechanics and how we use it.

An atom can be obtained in other ways, perhaps produced in a scattering experiment. What is the ensemble? Again this scattering can (but need not) be repeated many times, giving an ensemble. With a 10^{-9} sec half-life every atom decays within a second, for almost every experiment, with a probability of almost 1, since almost all in the ensemble do. It is again reasonable to presume that the particular experiment actually done is not an extremely anomalous one for which an atom does not so quickly decay. Even if it is produced from a decay of another one, there is an ensemble of those, and almost all produce in their decays atoms that themselves decay within a second of being produced.

Another reason for the necessity of ensembles is that there has to be a set of atoms, at least in history, in order to know the half-life.

Thus while we might think that we consider a single system, actually we can (in principle) repeat whatever we do many times, with systems and experiments obtained from ensembles. These need not actually exist, rather all that is necessary is that they can be defined, that in principle we be able to physically produce them. And this is true for any isolated system (even if properties of the universe, like its expansion, change), for the definition does not require members of a set of experiments to be at different times or places.

Someone who drives a car long enough will eventually get into an accident and be killed. And the driver has no idea when that will be, so should not get into a car. Yet most people do. However if there were a road on which everyone traveled at five times the speed limit, most drivers would probably avoid it. Why?

Is there a difference between these two cases, of decaying atoms and of travel?

III.2.f What is probability really?

Many people may be unhappy with this definition of probability in terms of ensembles. They do not feel that it answers their question "What is probability in quantum mechanics, really?". But the only way of answering such a question is for them to specify what (type of) information they want as an answer (sec. IV.1.c.i, p. 147). If they cannot do that, it means that there is no other answer. Then what they are really saying is "Such an answer makes me uncomfortable, and I want something that is emotionally more soothing. And even if I cannot state what that is, I hope that someday someone discovers something that leads to a statement that I feel more contented with, that provides some relief from the pain of quantum mechanics". However this is irrelevant; here we are considering physics, not psychiatry.

If there are no ensembles, then probability becomes meaningless — but that does not stop many people from using it anyway.

It is in this sense that probability is an "intrinsic objective feature of the physical world" [Mermin (1998), p. 754] — a phrase actually without meaning, thus without value, giving a question (In what sense is it ...?, or does it only seem to be ...?) lacking content until an answer is provided.

Here is an example of a statement — relating probability to ensembles — that conveys real information, although it answers a different question (What is probability in quantum mechanics?) that then can be used as a way of providing content to a question (Is probability an intrinsic objective feature of the physical world?) that had none before the answer is known (sec. IV.1.c, p. 146). Knowing the answer, we can formulate the meaningful question that we were trying to ask using empty phrases. This is not unusual, and many meaningless questions, not only in quantum mechanics of course, are actually (unconscious) confused renderings of meaningful ones. Lack of clarity and specification of questions, and their well-hidden meanings, lead to much confusion.

III.2.g Is there a wavefunction of the universe?

Interesting examples of muddled statements about probability come from discussions of the wavefunction of the universe [Barrow (1997), p. 166]. A wavefunction has no meaning unless we can measure it — even if only very indirectly (as we measure that of an electron in a distant star by observing radiation by the star and developing theories, in principle using wavefunctions, to predict that radiation). And it is vacuous unless it (at least to some extent, even very distantly) predicts something — otherwise it is simply a function that is useful for the production of scientific papers, but has nothing to do with nature. What a statefunction predicts is probability (eventually — it may for example give statefunctions of scattered particles which then give statefunctions of particles scattered from them, and so on, until a final statefunction is measured, giving a probability distribution).

Some people seem to believe that there is a statefunction that predicts the probability for the universe having the form that it does. Suppose however that it predicts that the form of the universe that is actually observed is extremely improbable. What would be the significance of that? Could we check experimentally that the wavefunction is actually correct, and that the universe that we observe is really very improbable? If not then what would it mean to say that it is very improbable, or even extremely probable? How could we tell? And if it would mean nothing, then what is the sense of the wavefunction, what content would it have? It would predict nothing about experiment, it would have no experimental significance, would not be related, even very distantly, to observation — it would be an expression used in a mathematical game that has nothing to do with reality, with no connection to physics, nature or the universe.

To say that it would predict that the probability is strongly concentrated around observed values is to say nothing at all — there is no experimental difference between this and that saying that probabilities are strongly concentrated around values greatly different from those observed.

Probability has meaning only in terms of ensembles — ones that can be determined experimentally. We can talk about the probability of a scattering leading to a particular result, say a set of angles, energies, outgoing particles, even if we do an experiment but once. This actually says something because we can do experiments many, many times (provided the funding is sufficient — something quantum mechanics says nothing about). We can thus define ensembles, even if we choose only a single member. But (unfortunately) we cannot define an ensemble of universes, experimentally the only way the concept makes sense — our universe is a singular system. To say that there may be many universes, but we can observe only ours, is just to put together a contentless string of words. How do we distinguish between "may be" and "certainly aren't"?

Those who talk about the wavefunction of the universe should, if they really believe they are saying something, describe experiments to find that wavefunction by producing ensembles of universes experimentally (sec. IV.2.f.ii, p. 160).

III.2.h Uncertainty principles

An interesting question is what is the meaning of uncertainty in uncertainty principles? When we discuss the uncertainty of a value, what uncertainty are we considering? Here we often use a term that seems to refer to a precise value — there is a mathematical expression for it — but can be hollow because it is not fully defined. There is a mathematical expression, but the numbers to substitute are not specified. This problem is quite common. Is a quantum mechanical measurement a preparation of the final state, or the determination of the initial one? Is the uncertainty of position that of position before measurement, or after?

The content of an uncertainty principle may depend on the particular experiment — people can use formulas in many different ways, with different (although somewhat related) meanings, without realizing it. Thus it should be considered for each experiment.

This emphasizes the importance of careful definition of terminology and symbols. Uncertainty may refer to different concepts and the uncertainty principle then states several different physical principles. This is something that deserves, and has received, careful discussion [Appleby (1998,1999)], but probably not enough. (The definition of "probably" in the last sentence should be clear.)

But there is an uncertainty principle that follows from properties of statefunctions, which is that most discussions (seem to) refer to, and the one we consider here.

III.2.h.i *Ensembles and the uncertainty principle*

There is a relationship between ensembles and uncertainty principles, which are essentially purely mathematical requirements with fundamental physical consequences. In essence they come from Fourier analysis [Mirman (1995b), sec. 3.4.1, p. 52], or similar expansions for other variables, and say that in a wavepacket there must be many terms — for the momentum-position uncertainty principle, many momentum eigenstates — with more (making large contributions to a wavepacket) required for a narrower packet. What does this assert physically? If we measure a statefunction (many times) we find that the uncertainty of position (the width of the packet in coordinate space) times that of momentum must be greater than 1 (or some other number, depending on the units). So if we make measurements do they tell us about statefunctions before or after these? A measurement is of the statefunction that exists when we make that measurement, and thus on what it was for a short time before (for a longer time it could have changed).

But suppose that we find it in an eigenstate of some operator. Then we know that if we make another measurement, right after the first (before external interactions can change the state), we will find it in that eigenstate. The reason is that the first measurement changes the statefunction of the system to a state (for which the uncertainty principle holds, but with different values of the uncertainties than before the measurement) which is, in this case, an eigenstate. Of course, we can never find the system in precisely an eigenstate, with zero error. But the resultant state can be very close to one, so another measurement done very quickly will almost certainly find the value very close to the first. Does this not predict the uncertainty after the measurement? It does but this is a different question than raised by the uncertainty principle — it asks for the effect on the statefunction (say the size of the wavepacket) of the measurement. And the answer does not come from the uncertainty principle, but must be obtained from an analysis of the physical processes that form the measurement.

The uncertainty principle places restrictions on a statefunction at the time it is measured, giving limits on wavepackets, such as their size and number of momentum eigenstates. But we cannot determine a statefunction with a single observation. To find it we need an ensemble, a large number of experiments. Results of these must be in accord with the uncertainty principle.

Thus we must be careful about what questions we ask, and be sure they are the ones we intend to ask. There can be confusion about what the uncertainty principle states, and often this results from mixing different questions.

III.2.h.ii *Uncertainty produced by a microscope*

Consider measuring the position and momentum of an electron with a microscope (sec. III.2.d, p. 108). Here the initial state is known — both the electron and photon have (at least approximately) definite momenta and completely unknown positions. It is assumed that the momentum uncertainty of the electron after the measurement can have any value from zero to the final momentum of the photon, which is known from its wavelength (and the design of the microscope). In a single measurement a point of light appears on the screen showing that the photon hit there. We know the microscope's resolving power, so if a photon hits the screen within the central diffraction peak, we can trace its path back to see where the electron was, with the uncertainty given by the resolving power. Thus this gives the uncertainty in position, the photon's momentum gives the uncertainty in the electron's momentum. The uncertainties are in the values after the measurement, they are for the final state. Of course this also provides a restriction on the preparation of the initial state for the next measurement.

What is the problem? The word "uncertainty" is being used somewhat differently from its use in the uncertainty principle. And it is the latter use which comes from properties of the electron, dictated by quantum mechanics — that statefunctions be representation basis vectors so have uncertainties in position and momentum [Mirman (1995b), sec. 3.3, p. 47]. But these refer to a large number of identical experiments, each giving different values for position and momentum, with distributions having widths obeying the uncertainty principle.

Such concepts as resolving power and central diffraction peak refer to ensembles so it is necessary to say how, if possible, they can be used if we study but a single electron, what uncertainty means here, if any interpretation of uncertainty can be applied to one event, and if this analysis makes sense. At least the entire experiment must be analyzed very carefully and in depth. We must be very careful in the use of this principle in discussions of single experiments.

III.2.h.iii *The uncertainty principle is a consequence of the expandability of statefunctions*

Uncertainty principles are derived from Fourier analysis (or other relevant expansions, depending on the variables). What these furnish is best understood by saying that if we measure two variables given by noncommuting operators the product of the spreads of the values obtained is greater than 1 (in the proper units). This requires ensembles. Thus if we prepare many identical copies of a state and measure two variables we get different values for each measurement. Unless the state is unusual each clusters around some specific value, the expectation value, and there is a spread, with the number of times any particular value is obtained decreasing as its absolute distance from the expectation value increases. So this gives a restriction on values that can be obtained, what can be learned (for a set of experiments) from statefunctions just before measurements. This is what the mathematics requires.

An elementary, and quite familiar, example is a statefunction of the form $exp(ipx)$. Where is x? Obviously, anywhere. A statefunction that gives a narrow range for x is of the form

$$|s) = \int a(p)exp(ipx)dp, \qquad \text{(III.2.h.iii-1)}$$

and the more precisely x is determined, the less precise is p (the greater the range of p for which the value of $a(p)$ is "large"), and conversely. The uncertainty principle

$$1 \leq \Delta x \Delta p, \qquad \text{(III.2.h.iii-2)}$$

then follows. Other uncertainty principles are similar.

III.2.h.iv *Uncertainty and individual measurements*

What about an individual measurement? If we measure the position and momentum of a single electron, once, are there restrictions on the precision of the values that we get? Answering this question requires a careful analysis of every particular experiment to see how the measurement is made, and how measurements of the two variables affect each other. This is rarely if ever done so there is much uncertainty about the meaning of the uncertainty principle for such cases. What is usually done is to study the procedure abstractly with a rigorous manipulation of symbols whose relationship to actual experiments is not clear. But this relationship is the central question in the analysis. Another thing that is often done is to use an analysis that requires ensembles and apply it to a single case, which is then self-contradictory.

For example in the microscope experiment the precision of the value for the position is based on the resolving power of the microscope which is a result of diffraction, and a diffraction pattern requires a large number of events. However the position of a single electron is determined by where a single dot is on a screen, and this can be known with infinite precision (ignoring the discreteness of the objects making up the screen which is irrelevant to this particular point, but raises different questions). Thus to find limitations on determinations of position and momentum of an electron it is necessary to trace back paths of single photons producing dots and find what the limits on the precision of the values are. This is something needed for each specific, detailed, design of a microscope. We must be very careful about doing it abstractly.

These questions are not analyzed further here, and for a reason. Mathematically the statefunction gives probability distributions and uncertainty principles relate their widths. Thus they apply to ensembles. However if it is claimed that they have different meanings, for example for single experiments, it is the obligation of those who make such claims to explain what such principles actually say in the contexts in which they are applied, and to show that they really do hold in such contexts. Otherwise it is not necessary — really impossible — to discuss what other implications they might have. Those who believe that it is for ensembles only have no further responsibility, those who believe differently do.

III.2.i How does probability become certainty?

Quantum mechanics gives the probability of different events occurring. But we see only a single one. Although it may seem that to understand this we must understand human consciousness, all we really need do is understand how behaviors of macroscopic objects are determined by microscopic ones.

A dot appearing on a screen illustrates this (sec. II.2.a.i, p. 65). The statefunction of an impinging object is spread over the entire screen. Yet a dot appears at only one small region, although a different region for each repetition of an object impinging. Why [Mirman (1995b), sec. 3.1, p. 38]?

The initial statefunction of the system has the object approaching and the molecules of the screen in their ground state, say spread (approximately) uniformly. The final state has the molecules of the screen unchanged, except in one small region, in which they have been excited and transformed into different compounds, coming together, that is forming a dot, and with the object absorbed.

What happens is that the initial statefunction evolved, since it obeys a nonlinear equation, with many interactions between the object and the molecule, among molecules themselves, and among constituents of each molecule. And the initial statefunction is a very large sum of a very large number of products of individual statefunctions. As a result it evolved into a statefunction describing the dot, with the terms describing transformed molecules very large in that small region, and very small outside it. It is the peaking of statefunctions of transformed molecules that gives a macroscopic object, the dot.

Thus if the screen is illuminated the probability is very high that statefunctions of the reflected light will be that given by a screen with a dot in the region where the statefunction of transformed molecules is peaked. And if we wish to bring in human observers, human consciousness, it is this macroscopic object, and light reflected from it, that humans react to.

Quantum mechanics gives the probability that a dot forms in a specific region, which, if the statefunction of the incoming object is not uniform, differs for different parts of the screen. How does this occur? Why does the statefunction peak at one point, not at another? Why does it pile up at one point at all? The answer to the last question gives that of the previous one. It is essentially irreversibility that leads to this. The initial state of the screen has very low (relative) entropy — all molecules are in the same ground state (although, of course, arguments would not really change if this were only approximately true).

Final states have much higher entropy. Statefunctions then must describe transition from states of low, to those of high, entropy.

It might seem that if we knew the complete state of the initial system we could predict that of the final system, that is where the peak occurs, for given a statefunction at any time, Schrödinger's equation gives the exact statefunction at any later time — it is completely predictable. Then we would find it with certainty, rather than probability. However we cannot know complete statefunctions for single events. All we can know are those for ensembles. To find the complete statefunction we must do very many repetitions of an experiment. And each gives a dot at a different place — the statefunction gives the probability distribution of the dots. The statement "if we knew the complete state of the initial system" has no meaning for a single experiment. We cannot know where the peak occurs if we do the experiment only once.

Thus it may seem strange that we can go from a statefunction giving probability to a dot at a particular place. But that is only because we are mixing two incompatible questions. If we are interested in a single event, we cannot know the statefunction. Thus we cannot tell how it evolves; that Schrödinger's equation fully determines its time evolution is irrelevant for the initial statefunction (to the extent that it has meaning) is unknown. If we know the statefunction, then we are considering an ensemble so we cannot predict a single event with certainty, only give probabilities.

Attempts to use objects that refer to ensembles for individual events is a major source of confusion about quantum mechanics (sec. V.3.b, p. 212). But this does not mean that quantum mechanics is confusing or strange, but merely that people (physicists!) are careless, thoughtless.

It is essential that we not confuse questions that require ensembles in order to be meaningful with ones about single systems.

III.3 What is a statefunction?

The concept of a statefunction is fundamental to this entire discussion because it is to quantum mechanics. What is it? Mathematically, as is emphasized many times here and previously [Mirman (1995b), sec. 1.4.3, p. 11; (1995c), sec. 6.2, p. 98], it is a representation basis state of the relevant transformation group. But what does it mean physically?

There have been many discussions of its definition and meaning. So to analyze this concept it is necessary to consider also what it is not, why other definitions are incomplete or incorrect, why they do not properly state what usage and physics determine as the sense of the term, the concepts to which it must refer.

It is easy to regard a statefunction as some sort of physical object, real in a way that we cannot define, but which we feel intuitively that we should be able to define, and then if we could it would be real in that intuitively-pleasing sense. But we cannot define "real", and taking this term too literally leads to many of the serious problems in the interpretation of quantum mechanics. One way of dealing with these problems is to go to the other extreme, and assume that it has no physical reality (at least in an intuitive sense) but is purely a description of what we observe, or how we observe. Could this be correct?

III.3.a Does a statefunction report about us, or about a system?

Some believe that the mathematical theory is about our knowledge rather than about an external world, or even that the laws of quantum mechanics require the concept of consciousness. Related to this is the belief that "quantum jumps" are due to a sudden change of our knowledge, that there is a discontinuous change in probability functions because our minds acquire information about systems (although, of course, there is a change of a system, not really instantaneous, when we interact with it, but this view is that it is the change of our knowledge that causes a change in the system, not the interaction). Other examples [Fuchs and Peres (2000); Peres (1984)] are statements that a statevector is not a property of a physical system, but rather represents a procedure for preparing or testing such systems, or is a mathematical expression for computing probabilities. Similarly there is a belief that a statefunction is merely a summary of our knowledge prior to a measurement. Such statements are quite common [Englert, Scully and Walther (1999), p. 328], but no less incorrect for being so widely held. Physics is not determined by majority vote.

The phrase "computational mechanism" referring to the statefunction has also been used. Are Newton's laws also not computational mechanisms? Are there any parts of physics that are not? Otherwise how do they differ?

These base definitions of statevector, a fundamental part of quantum mechanics, on human actions. If human beings had never evolved, would there be statevectors? And if there were no statevectors, would there be anything, would there be a universe? Human beings are a quite recent addition to the constituents of the universe. If statevectors are dependent on human beings, did the universe exist before we did? This is another of the many examples of the human view that laws of nature are determined by human beings (rather than conversely). Laws of nature are believed thus to describe not the universe, but our knowledge of it, leading necessarily to the view that the universe does not exist, only our awareness of it. This is perhaps not a description of physics, but of the absurdity of human hubris.

A better statement [Mandel (1999)] is that — in an experiment (a very important qualifier) — the state represents (certainly a better word than "is") not what is known, but what is knowable, in principle. Here it is emphasized that by focusing on what is knowable, in principle, and the irrelevancy of what is actually known, anthropocentrism and references to consciousness are completely avoided. It is unfortunate that such statements (although incomplete) are not far better known, and part of the accepted beliefs about quantum mechanics.

How reasonable are views that statefunctions represent (only) our knowledge of systems? No one would say (let us hope) that our knowledge of an electron determines the behavior of other objects (it interacts with). Likewise is the view that a statefunction is a procedure for preparing or testing physical systems. But a procedure (the statefunction of an electron in a star) is unlikely to determine the behavior of other objects.

Beyond that an electron in a distant star has a statefunction, as does one in a crystal or molecule, or .... Using statevectors we can calculate behaviors of stars, or crystals, or ..., and (potentially observable) radiation. But in no way does a statefunction describe a procedure that we carry out, or our knowledge of what is going on inside a star, nor even a crystal. Our knowledge does not determine the behavior of objects.

Of course those who do believe that it represents a procedure can undoubtedly explain how a statefunction describes a procedure that we carry out inside a star.

It may be argued that these statements are not completely incorrect because there is some ambiguity of language. Thus we can say that a use to which we put a formula for a statefunction is to represent our partial information. That would be much better. But even that is too restrictive. We use the formula for the statefunction of an electron in a crystal to calculate properties of the crystal. Thus we can (in principle, perhaps even in practice) use statefunctions of objects in a diamond to determine the diamond structure [Mirman (1999), sec. III.5, p. 159]. And here statefunctions do not give probability but exact structures. The reason of course is that the diamond structure is a theoretical construct. If we studied structures of many diamonds, we would find different ones, and can (in principle) obtain a probability distribution for them. But these all differ only somewhat from an ideal one, and that is fully determined, but not experimentally determinable.

These views show again that (too) many (not only) physicists base the universe upon themselves. They view laws of nature to be, and designed to be, properties of human beings (perhaps in more than one sense). Yet the universe exists, and is governed by these laws, whether we are here or not. Perhaps it really would not matter if we were not here — we would not know physical laws, and would not have to be concerned about what they mean. But the universe was here long before we were, and governed by these laws. And now we can see the effects laws had before we arrived. We were not needed then, we are not needed now. The universe did quite well without us.

III.3.b The completeness of the description given by a statefunction

Does a statefunction provide a "complete" description of an individual system? The word that causes problems is, of course, "complete". Statefunctions provide all the possible information we can have and which we can use to predict behaviors of systems — but we are limited in the possibilities. Thus for a collection of atoms, they give complete predictions of probabilities for decay. This does not completely describe the set of atoms for it does not tell when an individual one will decay. But such additional information cannot be provided, there is — in principle — no way to obtain it. Statefunctions then give as complete a description as is possible — as nature allows. This may makes some people unhappy, but human beings can only observe nature, not design it to fit our desires.

Suppose that the collection consists of a single atom, what can we say? If that were all, then we can say nothing, we could make no predictions about its decay. But if we can consider it, as we usually can, as a member of an ensemble, though we possess but a single member, then meaningful predictions about decay are possible. And in that sense it describes a single system, and gives a complete description — as complete as possible.

III.3.c What we really mean by statefunctions

Statefunctions are given mathematically, being mathematical functions. But they are not mathematical objects — the interactions of an electron determine its statefunction, and its statefunction is determinative of statefunctions of other physical objects, but interactions do not act on mathematical objects, nor do mathematical objects result in interactions. Of course, statefunctions are not physical objects in the sense we like to think of physical objects, ones we can see and touch and feel. We cannot touch statefunctions (at least in the way that we like to think of touching).

What is a statefunction then? The danger here is in the "what is". This is a question that expects too much of an answer. What is the statefunction of an electron in a star? It is a mathematical function, in one sense a purely mathematical object, that we can find from theory, and use to find other functions that eventually lead to ones describing our observations. It is thus a concept that we use, and need, in a chain of reasoning, perhaps an extremely long chain, to go from one (perhaps very large) set of observations, say those allowing us to construct a theory of electrons in stars, to another set of observations, say of radiation from a particular star that might enter our telescopes.

However we have to be careful about regarding it as merely a mathematical object; it is a physical object but with the unfortunate property that we can not define what this physical object is beyond treating it as a mathematical object used in, and determined by, physics. It may seem disappointing to say that a physical object is defined by, and only by, its mathematical properties. But how else can we define it, and how else can we give its physical properties, than by expressing these in mathematical terms? We can not see, or touch, or feel or taste it. But even for those objects that — we think — we can, it is really the mathematics that expresses the physics (sec. III.4, p. 129) — however for statefunctions, or electrons, this reality is starker.

III.3.c.i *Definitions are determined by use*

This is what statefunctions are — as can be seen from how the term is used. And the way that it is used defines what it is (but, of course, this definition should be translated into — substantive — words and sentences so it can be used, and it must be used in a way that gives a proper description of what we experience). Problems arise not from the nature of statefunctions, nor definitions, but from trying to give them properties that they cannot have — usually ones that are pleasing intuitively, that satisfy us, but have no meaning in contexts in which they appear. Such overdefinitions lead to confusion. And it can be seen from many discussions of the meaning of terms like statefunctions, that while often precise definitions (but ones that have no content) are given, they are not in any way applied, and have no relationship to ways in which we do use concepts these terms name (but do not really define).

There is a possible objection to the name "statefunction" because this implies that it represents the state of a system, rather than expectation values for quantities that are measured in studying the system. But this objection implies that there is something that is the state of a system, and that the statefunction is not it. Of course, we have to be very careful with names. But here, how do we distinguish between a state of a system, and its statefunction?

The problem with this question about the meaning of statefunction, as with so many others, is that those who ask it want an answer that is not correct, cannot be correct, is not possible, a classical, intuitive, picturesque answer.

These emphasize that definitions are dangerous, that much confusion arises (and not only in quantum mechanics) from them, often because they seem to say something but do not (or at least they do not say what they seem to say). Also confusion occurs because definitions are wrong, they define what we think a term or concept is, but not what it really is — that is determined by how it is used; often we ourselves do not understand how we use a term.

And it is essential to realize, and accept, that we must deal with nature as it is, not as we would like it to be. It is our responsibility, not merely as scientists but as cognizant, thinking beings to properly describe reality, not to find ways to shape it to our liking. We can see what reality is by the way that we actually treat it, not by the way we wish to treat it, or believe that we are treating it. Although views here are based on a very abstract conception of nature, something that many find highly objectionable, it is unlikely that those who disapprove can show how this model differs from our actual, but not our believed, treatment of the world in which we think that we exist.

III.3.c.ii *What is a physical object?*

Much of the confusion about quantum mechanics results from failure to distinguish between a physical object and its mathematical description. But is it possible to make this distinction?

What then is a physical object (sec. II.5.a.i, p. 84)? We have a clear (perhaps not fully correct) idea of what a macroscopic physical object is, we can see, feel, taste it (or at least we think we can). But we cannot see or taste statefunctions of electrons, our senses cannot react to them (although they can do so for statefunctions of macroscopic objects — those whose sizes approximate, or are greater than, ours — thus we can see and taste statefunctions of classical objects). However this difference between statefunctions of electrons and of food results from statefunctions of food acting directly on our senses, while electrons affect our senses by acting on other statefunctions, which act on further ones, and so on, and finally at the end of a chain are statefunctions that act directly on our senses. This difference between statefunctions of macroscopic and microscopic entities has lead to much misunderstanding about what physical objects are — for many (almost all?) people try to define them in terms of their classical intuition.

A physical object is one that (necessarily) appears in a theory about the physical world, that is part of a physical theory in such a way that it determines what other physical objects do (with the possibility in principle of) leading to statefunctions that affect our sensations. We can write (in principle) a formula, or algorithm, for statefunctions of electrons in stars. A statefunction is a physical object, it determines (probabilities of) behaviors of other objects (not only in a star), leading to (a possibility in principle of) an effect on our senses.

III.3.c.iii *Physical objects and their mathematical descriptions are not the same*

There is a difference between a physical object and the mathematics we use to describe it and its effects on the rest of the universe. It is meaningful to say that statefunctions determine the behavior of other electrons (not only) in a star. But we would not say that the formula for a statefunction determines the behavior of electrons (it allows us to calculate how the statefunction determines the behavior of other objects). The formula is the mathematical expression describing the physical object.

The statefunction of one object determines statefunctions (behaviors) of others because that is what appears in interaction terms in equations governing objects (determining statefunctions). This may seem like a mathematical statement. However the equations are the mathematical expressions of physics — equations do not determine statefunctions, they only provide a means for us to find how statefunctions are determined. They allow calculation of how one statefunction determines another.

There is an important distinction between a physical object and the mathematics, such as a formula, describing it, even though this mathematics provides the information, all information that we can possible have, about the object. The mathematics describes the object so all we know about it is the mathematics, including statefunctions and equations, that we use to determine and express our knowledge. But our complete knowledge about a physical object is not a physical object.

We say that a statefunction obeys a differential equation, or that it determines the behavior of other objects by appearing in nonlinear terms in their differential equations. However a statefunction, a physical object, cannot appear in a differential equation, or obey one, it knows nothing about these, never having taken courses in calculus. Correctly, a formula for a statefunction obeys a differential equation; we describe statefunctions (and systems) by means of (formulas that are solutions of) differential equations. However while we cannot distinguish between statefunctions and formulas for them, these telling everything that we can know about statefunctions and so systems whose statefunctions they are, there is a difference. Realizing this will help prevent statements that a statefunction tells about our knowledge of a system, or that it gives probabilities of finding specific experimental values. Our knowledge about an electron in a star cannot determine the behavior of other electrons; but the statefunction of the electron does. Our knowledge of the formula allows us to determine how the statefunction determines other statefunctions, but the statefunction does not allow us to do that.

III.3.c.iv *Physical objects must be statefunctions*

An electron, proton, and so on, are sets of statefunctions (sec. III.3.c.vii, p. 126). They can be nothing else. An electron does not appear in Dirac's equation for a proton, another proton does not, a pion does not, a photon does not, their statefunctions do.

Notice in Dirac's equation for the proton,

$$i\gamma_\mu \frac{d\psi_p}{dx_\mu} + m\psi_p + ig\phi_\pi\psi_p = 0, \qquad \text{(III.3.c.iv-1)}$$

the pion does not appear, (the formula for) its statefunction does. A pion cannot determine the behavior of a proton, its statefunction can (and does). We tend to think of a proton emitting a pion as if the pion springs out of the proton like a jack-in-the-box. Clearly this makes no sense. Such pictures of physical objects may provide an escape from mathematics but they are not really about physics.

Statefunctions are the physical objects. There can be nothing else, for nothing else can affect other physical objects, so cannot be observed, cannot be detected, thus cannot exist — physical objects can therefore have no meaning beyond being (classes of) statefunctions. Only (formulas for) statefunctions appear in equations. Only statefunctions can affect other objects (other statefunctions). Only statefunctions can be detected. Only statefunctions have meaning, can exist.

To avoid excess words we say that statefunctions obey differential equations, and so on. But it should be understood what words are being left out, and what statements like this really say.

It is essential to distinguish between physical objects and mathematical properties and objects, the mathematics, that we use to understand, describe — and manipulate — them.

III.3.c.v *When is a basis vector a statefunction?*

It has been emphasized here that physical objects are (given by) statefunctions and that these are basis vectors of representations of the transformation group of space, the Poincaré group (or perhaps a group of which it is a subgroup). However not all basis vectors are statefunctions. By basis vector we really mean a formula, which is of course not a physical object. Not all formulas that are basis vectors are statefunctions. Relevant ones give all possible information about the statefunctions with which they are correlated, including how these are seen by different observers — their transformation properties. Statefunctions, which we can discuss as formulas describing physical objects — providing all possible information about them — or as physical objects themselves, have physical sense because they are (correlated with) objects of the external world.

To have a meaningful, coherent, usable, useful organization, systemization, of our sensory inputs we have to picture these as coming from an external nature, this with objects that we can conceptualize in some useful way. The external world exists in this sense — and in the sense that such a coherent picture is possible. Certainly these are sufficient (and necessary) for its existence. To these external (physical) entities we associate mathematical functions so that we can manage them. These, basis vectors that are so associated, are statefunctions. But not all basis vectors can be so associated. Developing a physical theory from the mathematics requires, in part, finding which can be, which are.

Thus we know, and what we know, is our sensory inputs, determined by those basis vectors that are statefunctions — and the consistent, coherent, extraordinarily complex picture that they provide. And this is the only way that we can know, and all that we can know — statefunctions (sec. I.2, p. 5; sec. III.4, p. 129).

III.3.c.vi *Are we relevant?*

This formulation uses the possibility of an effect on our senses as a requirement for physical theories being meaningful. Does this not regard physical theories as being based on the existence of human beings, since whether a theory has meaning is determined by what it says about the possible effect on our sensations. It is true that our existence is necessary for physical theories to be meaningful, for it is we who construct and use them. We are also necessary for languages to have meaning, and a theory is a form of language. But that is very different from saying that we are necessary for physics to have meaning. We determine what language is, but sounds — oscillations in, say, air — exist whether we notice them or not. We should not describe sounds, oscillations, with reference to ourselves just because we make use of them in certain ways. We should not describe physics with reference to ourselves, but we must tell how we learn about it, and what concepts mean — how we define terms referring to these concepts — and it is here that our senses and our neural processes are relevant.

Definitions are necessary for us to communicate among ourselves, but they are not necessary for the universe to exist or for laws of physics — regularities in behaviors of objects in the universe.

It is sometimes said that science and faith are different ways of knowing, they are ways of knowing about different aspects of reality. This is really not true — one is a way of finding knowledge, the other a way of finding emotional comfort. Too much of physics is actually the latter (as seen, for example, by not distinguishing physics from the theories we use to learn about it).

III.3.c.vii *How an electron differs from its statefunction*

Physical objects come in various types. What is the difference between, for example, an electron and its statefunction? Perhaps it is most accurate to say that the type of object called an electron is a set of statefunctions with certain properties (such as mass and spin values that determine the mathematical form of the statefunctions, including the group representations to which they belong). An electron is a single statefunction of this set. The set of properties that define this set defines the electron. Physics does not deal with electrons, only their statefunctions. These (actually the formulas for them) are what we solve for using Dirac's equation, these are what appear in equations governing (statefunctions of) other objects. Electrons appear in no other way in physics; they are thus only their (set of) statefunctions.

We can consider then an electron as either (a member of) a class of statefunctions, or the properties specifying this class. What else can it be?

This is true in general, a physical object is identical to its statefunction (or can be taken to be a set of statefunctions, or the set of rules for determining, or defining, these). Anyone who disagrees can state explicitly differences between, say, an electron and its statefunction — and show experimental consequences of these differences.

III.3.c.viii *How do electrons, statefunctions and formulas differ?*

One source of confusion is that the word statefunction has two different, although closely related, definitions (where the definition is determined by use not by dictionaries). It is a physical object, but it is also a formula for that object. Thus it is the statefunction of an electron in a star, or crystal, the physical object, that affects other statefunctions, other physical objects. Statefunctions, physical objects, produce interference patterns in double-slit experiments. But we use the formulas for these physical objects to calculate such affects, patterns, probability distributions.

Thus an electron is a class of physical objects; when we speak of an electron scattering from a proton, we are referring to that particular statefunction in the form of a wavepacket moving on a line toward a particular proton (statefunction), interacting with it, and then moving away. But to study this scattering we use the formula for the object. Thus "electron", its "statefunction" and the formula for it are three different concepts, but so closely intertwined that they are often difficult to distinguish.

When we say that a statefunction appears in Dirac's equation we really mean that a solution of the equation is given by a formula and this we use to determine the behavior of the electron (how the statefunction varies in space and time) and how it determines the behavior of other physical objects, that is to compute their statefunctions. Since such statements are quite awkward we use fewer words as abbreviations. But it helps to realize that they are abbreviations.

III.3.d Multiparticle states

One of the objections to regarding statefunctions as — in some sense — "real" is that for a system consisting of more than one object (n objects) the statefunction is defined over a configuration space of dimension $3n$, which therefore is not the space we live in. Is this true? If it were each particle in our bodies would be in a different space (a different universe?). That would make it very difficult for us to exist.

A statefunction for a hydrogen atom, consisting of three objects, can be written (schematically)

$$|Hatom) = \psi_p(x_p)\psi_e(x_e)A(x_y), \quad \text{(III.3.d-1)}$$

a product of those of the proton, electron, and electromagnetic potential (without which atoms would not exist). It would be a strange kind of atom if the proton and electron were in different spaces. Further if we regarded x_p and x_e as completely independent, being orthogonal coordinates like x and y, they would have to refer to different coordinate systems, which would mean that positions of the two particles could not be correlated.

What does this statefunctions mean? It gives the probability of finding a proton at x_p and an electron at x_e (plus photons) — in the same space. Statefunctions are defined over a single space, real space, in which protons, electrons and we (and all particles in our bodies) exist simultaneously. But they can be at different points within this space. It is useful at times, say for calculational purposes, to regard space as having higher dimension, so particles exist in different ones. But of course that is merely a way of labeling and calculating. There would be nothing to calculate if they really were in different spaces.

These statefunctions are wavepackets with their centers at different points (of the same space). Measuring from their centers gives functions depending on different variables, but only because we take the origins as particle-dependent.

What are the conditions on this statefunction [Mirman (2001), chap. IV]? It must be a basis state of a representation of the Poincaré group, so being an eigenstate of its two invariants (we leave open the interesting question whether it can be created in a reducible representation so being a sum of eigenstates). Further there are state labeling operators, so that if it were in a momentum eigenstate, it would be labeled by its momentum eigenvalues. In general it is in a sum of such states (a product of wavepackets). And the objects that make up an atom are also described by Poincaré representations, irreducible ones, so their statefunctions are eigenstates of its invariants — but now with interactions. Also they are labeled by the state labeling operators. For electrons, protons, neutrons, the two equations given by the invariants can be replaced by Dirac's equations, with interactions [Mirman (1995b), sec. 6.3, p. 114].

Operators acting on a statefunction of an atom are sums of those acting on the individual objects, with interaction terms all being evaluated at the same point — interactions, so theories, are local — the electron interacts with the electromagnetic potential at the point at which it is, as does the proton, but at a different point, and the electromagnetic potential interacts with the electron and proton at the positions where they are. Of course this statement is wrong since neither the electron nor proton are at a single point, their statefunctions are spread over space. The correct statement is that in the equations governing the electron, and the separate set of ones governing the proton, there is only a single value of the coordinate which is the same for all statefunctions in each equation (including interaction terms) — the same in the derivative, in the statefunction of the electron (and similarly proton and potential) and in all statefunctions appearing in the interaction term.

Since operators for an atom are sums, the variables acted on by individual operators are different. Thus we have a set of coupled equations, three for the hydrogen atom, one for each object, with interaction terms that have different variables in each of the three equations. But these variables are all just different positions in the same space.

Variables are denoted differently in the different equations for the three objects. Solutions are then functions of different variables — different positions in the same space. However the variables in any one equation must all be the same. It is important that this distinction be clear, else variables can be misleading.

For the helium atom the statefunction is (schematically)

$$|H_e) = \psi_N(x_N)\psi_e(x_e)\psi_e(x'_e)'A(x_y), \tag{III.3.d-2}$$

writing just the statefunction for the nucleus, N, which is a product. But statefunctions for electrons must be antisymmetric so

$$|H_e) = \psi_N(x_N)A(x_y)\{\psi_e(x_e)\psi_e(x'_e)' - \psi_e(x'_e)'\psi_e(x_e)\}; \tag{III.3.d-3}$$

with ψ and ψ' distinct, but not different. This gives the probability of finding one electron at x_e, the other at x_e', with no way of distinguishing which electron is which. However there must be more, else antisymmetrization would have no content. In a reaction, such as one creating the atom, states must be antisymmetric [Mirman (1995b), sec. 8.1, p. 146]. For example, an incoming state of two electrons might have no symmetry (the electrons being uncorrelated) but the only possible resultant states are those for which statefunctions of the pair are antisymmetric — the Hamiltonian acting on an arbitrary state projects out the one with the proper symmetry.

All particles, no matter how many, exist in only one space, there is only one universe, and statefunctions of any system depend on coordinates of our single space, that of the universe that we, and all other objects, exist in.

III.4 Reality is mathematical

In quantum mechanics all objects, including ourselves, are described by statefunctions. And that really is all that can be said about them. Physical objects thus are replaced by mathematical functions. Is there anything different that can be done, that can be learned about physical objects, is physics nothing more than mathematics? Are we ourselves nothing more than mathematical functions — mere products of Poincaré representation basis states?

III.4.a We can be replaced by robots

Consider classical physics to see if there is a difference. Take a robot with sensors (which may include internal ones) able to, say, measure wavelengths and amplitudes of incoming light, or forces exerted by objects with which it is in contact. This robot sees the world as purely mathematical, its only knowledge of it, and itself, is given by numbers provided by its sensors.

Objects with which it interacts are then merely mathematical functions with properties of these giving the readings of the sensors. Information about objects is stored in the robot as a set of numbers. These predict what sensors will read for every possible thing that could be done to an object. Thus as far as the robot is concerned the world is given by sets of numbers, purely mathematical. Someone observing nature by means of the robot would have full knowledge of it (in principle), but would not have a subjective feel, knowing merely functions describing objects.

But we are no different than the robot. Our knowledge of the world comes from numbers provided by our sensors, values giving the nerve impulses from our eyes, ears, fingers, and so on. Objects of the world are then no more than functions that tell what the readings of our sensors are under given conditions. Subjectively we interpret these incoming numbers in terms of pictures and sensations, but actually all that we know are functions provided by our senses — readings carried by our neurons, and from these sets of numbers we find those functions describing objects which then predict what our sensors measure under different conditions.

Physical objects are stored in our memories as functions (sets of numbers, these giving say strengths of synaptic connections), though not digitally (analog is the wrong word, since there seems no simple analogy between properties of an object and synaptic strengths and other variables describing it in our memories). But this is irrelevant, however stored, as far as our memories, and thus our perceptions of the world, are concerned, objects are merely sets of numerical values.

Of course objects that we deal with are far more complicated than electrons, thus we can do far more with them, this perhaps hiding the reality that these are mere lists of numbers. And we cannot produce subjective images of electrons as we can of macroscopic objects. So that an electron is just a function, a statefunction, is clearer, and starker, than for a chair or computer.

But ultimately there is no real difference. Our perception is different for an electron than for a chair, our (purely) subjective view is different, but the world "out there" that determines perceptions is no more than a set of mathematical functions, whether considered quantum mechanically or classically.

III.4.b The apparent reality of the classical world

The reality of the classical "real world" comes from it being observable in many different ways, by sight, sound, touch, and so on, by different people and instruments, all simultaneously, and these all give the same description. Is this true of statefunctions, or are these just mathematical? The differences are that statefunctions have to be determined by a series of measurements, and also that a measurement affects the state (the statefunction) in a discernible manner, but for classical physics changes can be made so small as to be unnoticeable. Thus we can regard an object as fixed, and giving the same results for many measurements, using different senses, or measurements by different people, different instruments.

However it is possible to have a large quantum system, which needs a statefunction to describe it, and for which also this is reasonably true. In that sense classical physics and quantum mechanics are the same, and the universe is purely mathematical.

Consider then a nucleon. It can be prepared many times (perhaps by many people) in the same state, many times so its statefunction can be measured. It can be observed by different "senses", say photons and pions. These give different information, just as with human senses, but there are aspects on which they agree, say position, momentum, spin, isospin. Thus, as with a macroscopic object, we can say that "there is something there", and objectively so.

What is the difference between this and a classical object? First there are more steps between a measurement and the perception of information. Photons have to be amplified and energies changed, or they have to be converted to other objects, say tracks, or motions of pointers which are then read. But this does not make the statefunction more or less mathematical.

Also each repetition of an experiment gives different results, only a set is reproducible. And the set has ranges of values. This is true classically as well, but the ranges are so small as to be imperceptible — all members of a set of experiments seem to give the same results. But these results are still mathematical instructions to our neurons. Different senses, and different people, appear to see the same thing, get the same information, because ranges are so small. For a quantum system, a large number of experiments is required to obtain the same thing, the same results.

We accept, we have to accept, the existence of an external reality, because of the consistency of the different reports of it, not only from different senses, but indirectly, from different people, reporting on the same thing, from different instruments and so on (sec. I.1.b.i, p. 3). In this sense there is no difference between classical physics and quantum mechanics, although in quantum mechanics it is more difficult to get information about the external world, and in a way more artificial, more indirect, thus there are fewer consistency checks. But in all cases, information going to our brains, coming as it does from our senses, is nothing more than sets of numbers, encoded in spikes (and perhaps other forms) of neural activity.

III.4.b.i *Is there an external reality?*

Why do we believe in an external reality, not only for objects that we can see and feel and hear and taste, but electrons, atoms, stars, galaxies, and so on? Why do we believe in such concepts as nuclei or stellar interiors? We have (in many cases) direct observational evidence, through our senses, and much indirect evidence. This includes observations of objects that affect others, stars, say, act on other stars, in different ways, and we can observe stars and other objects that they affect. Thus we have a vast web of interconnected observations, these consistent with all others, with the requirement of consistency highly complex, as there are many observations. The only way that we can have a coherent, meaningful picture, a way of dealing with all this information, is by replacing it by a model of an external reality, and such concepts that provide this information (sec. I.2.f, p. 8). But that is what we mean by saying that there is an external reality, what we mean by such (extremely general) types of concepts.

There is in this way no fundamental difference between the reality of our own bodies, those of others, of chairs and tables, of stars, exterior and interior, atoms, electrons, pions. The types of information may differ somewhat in different cases, but the concept of an external world is the same, and in terms of what reaches our brains all these are just sets of numbers coherently related.

The reality that we experience is purely mathematical. Of course this disagrees with our intuition. But our intuition is based on the way our neural system interprets — deals with — the mathematical input coming to it. Ultimately, reality is what we make of it.

III.4.b.ii *Why the world seems as it does*

While we may logically argue that nature is purely mathematical, that is not how we understand it (and intuitively not how we can understand it). Why does the world seem as it does?

If we have, say a table, it does not appear to be a set of statefunctions — we can see it, see the graininess of the wood, feel its hardness, when near a flame feel the heat, if newly made smell the wood, perhaps fold or bend it, none of these mathematical.

For an electron, no matter how we picture it, we treat it as a mathematical object, and the way we treat it is the way we define it. An electron is a, or a set of, solutions of Dirac's equation, with certain parameters, usually in the form of a wavepacket, localized in space so almost zero outside a small region. If in an atom it might be spherically symmetric, and almost zero much beyond the radius of a sphere. An atom is a product of functions, wavepackets, these correlated so the regions in which the various states are much different from zero overlap or are close to each other. A molecule is then a similar product of such statefunctions of atoms, and so on. All these are purely mathematical, not in our pictures or our minds, but in our treatment of them — which gives our definitions of objects.

A table is a set of these statefunctions, of atoms, molecules, lined up in three dimensions. We look at it, what do we see? Of course, we never see a table, even classically, what we see is light that has been scattered by it. The set of lined up statefunctions — wavepackets — that is a table has the property that it scatters light in such a way that our minds picture it as a hard, solid grainy object. But scattering of light is purely mathematical. We can, in principle, take statefunctions of atoms of a table and of light, purely mathematical objects, and compute outgoing light to find its properties. We can find, mathematically, that an outgoing beam has spectral properties equivalently to those (we interpret as being) of wood, and that a scattering object, a region of wavepackets, is grainy — something that can be defined and determined purely mathematically. When we touch it, particles of our fingers, wavepackets making up our fingers, scatter from it, or are in potential fields produced by it, and the scattering is that produced, in a way that is defined and can be determined mathematically, by hard objects. That is particles of our fingers scatter from the outer surface, moving that very little, and do not (appreciably) penetrate beyond a very small distance.

Thus an object, say a table, that volume in which statefunctions of objects of which it consists differ from zero by a significant amount, has a consistent set of properties that produce scatterings, of light or of objects making up fingers, and so on, that allow us to construct a consistent picture. And this is the same picture as that of others who view the table construct (to the extent that we can communicate our pictures). Also it is consistent with properties that we have found elsewhere, and so pictures we have constructed of objects, like wood, of which a table consists.

We have then a consistent, coherent, necessary, and very useful set of mental images (patterns of neural activity and structure), but ones that are purely phenomenological. If however we analyze what we are actually doing, we find we are really taking mathematical objects, wavepackets governed by equations which include as nonlinear terms other statefunctions, and that these determine statefunctions of parts of our bodies, which in turn determine those of our neurons, whose patterns of activity are our pictures of reality.

III.4.c Artificial intelligence is completely artificial

We considered a robot to clarify these points. To prevent confusion and misleading implications, it is necessary to emphasize that this does not mean in any way that these can think. It does not imply that a robot can have the intelligence of a human, or even a fish, ant or worm. Since there has been so much nonsense written about this, it is necessary to discuss the subject, and artificial intelligence (a subject that desperately needs a lot less artificial intelligence and much more real intelligence) in a little depth, even though it is not otherwise relevant. (artificial intelligence is technically defined as a very minor computer program written by a very conceited computer programmer.)

III.4.c.i *The immense complexity of brains*

Human brains are estimated to contain (at least) about a 100 billion neurons (plus more than that of glia), and each has about an average of about 15,000 connections to other neurons. But their complexity is far greater. Each synapse can have different strengths, probably continuously variable, distributions of each chemical within, and between, neurons, sizes and shapes, and so on, of neurons, can all vary, almost, if not exactly, continuously. In addition it is likely that the synaptic cytoskeleton (the internal skeleton of the cell) affects, at least, synaptic efficiency [Weng, Bhalla and Iyengar (1999)] giving another way of storing data and thus increasing the amount of data the brain can hold. For the worm *Caenorhabditis elegans*, there are 80 different types of potassium-selective ion channels, 90 ligand-gated receptors, and about

1000 *G* protein-linked receptors [Koch and Laurent (1999)]. The numbers of each type can vary producing an immense storage capacity. And this is for a worm; it would be disturbing if humans were less intelligent.

There are also indications that learning can produce greater thickness in certain brain layers, differences in amounts of neurotransmitters, more connections between neurons, and greater branchings of neuronal projections [Kempermann and Gage (1999)]. This increases the possible amount of stored data even more, making it vastly greater than the vast amount allowed by other ways.

While small differences are not likely to be important, the number of states, and likely the number of relevant states, of a brain is enormously greater than that of any conceivable computer. But it is more than that. The program, the data set — for a brain the data set is the program — is enormously larger. Not only is wiring, from before birth, extremely complex, but the amount of information starting before birth, from all external and internal sensors, reaching the brain is immense, every second of our lives. For example, it is estimated that there are about 10^6 nerves from an eye to a brain. View a rapidly changing scene to get an idea of how short a time is required for each frame to be viewed, and each contains this number of bits (perhaps more if each nerve encodes its information in several bits). This gives an impression of the amount of information carried from an eye to a brain in a second — for each second — and throughout life. But sight is only one sense.

Some believe that a brain is a million times more complex than a computer. We can see that a complexity of $10^{million}$ would be a more reasonable estimate (or likely a far larger one).

It is clear to anyone who has ever done any thinking that the brain is very much not a huge database, with thinking merely combining of data items. This is especially obvious when thinking about novel situations, which most situations in life are.

The total information stored, even if most is lost, is (to put it very mildly) huge. There is no way that anywhere near even a minute fraction of all this information can be put into a computer. It is not a matter of some brilliant algorithm or extremely fast computer: there is no possibility of a computer being as intelligent as, certainly a human, probably even an ant, or worm. The complexity, the quantity of information, is beyond the powers of any artificial system. It is this difference in the magnitude of information that makes computers, or robots, and neural systems so totally incomparable. Comparing a brain with a computer is like comparing a galaxy with a pebble — although the difference between a brain and a computer is vastly greater than that between a galaxy and a pebble. True the latter are both forms of matter, the former both information storing and processing systems, but members of each pair are so different as to be incomparable. Perhaps a better analogy is to say that it makes as much sense to compare a computer with a brain as it does to compare either with a galaxy. They are completely different types of concepts.

Thus while considering robots helps us understand our own neural processes, and in a sense while underlying methods of learning about the world have similarities, the vastness of the difference emphasizes the immense distinction between living brains and artificial ones. While artificial intelligence is a name given by some conceited computer programmers to their programs, that intelligence is completely artificial and has nothing to do with the term used in referring to animals, certainly not to higher ones.

III.4.c.ii *Intelligence, lack of intelligence and language*

Many developments in many fields, unfortunately including physics, are based on a fundamental technique of modern intellectual life, the method of false analogy (sec. IV.2.m.i, p. 168). To use a single word for thoroughly different concepts may make it seem as if they are analogous, when in fact they have nothing to do with each other.

Language can be dangerous (chap. IV, p. 142). Using the same words for concepts so vastly different, so qualitatively different, as neuronal operations of animals, certainly higher ones, and especially those processes that provide the vague idea of intelligence, and computer programs and databases, leads to muddles, confusion, error and hinders discovery and development. Another such term applied to vastly different concepts is "think", as in can computers "think" (sec. IV.4.a.iii, p. 188)? No computers don't think, and people who say such things as "computers think" don't think either.

III.5 The Hamiltonian, its construction and necessity

A Hamiltonian is needed for a theory based on the Poincaré group (as fundamental theories must be), being one of the required momentum operators, the generator of time-translation, so is always necessary. If states are chosen to be eigenstates of the Hamiltonian, energy conservation is automatic (actually even if the states were sums, wavepackets, the expectation values would still be constant in space and time). Only if this were not possible could the conservation law be violated.

It seems to be generally assumed that a Hamiltonian appears in some mysterious way and that it is a physical law that it equals the time derivative, and that its eigenvalue equals the energy, itself a mysterious quantity appearing for some unknown reason. However the concept of a Hamiltonian independent of its definition as a momentum operator is nonsense. It is defined by its exponentiation being the time-translation operator, so it equals the time derivative (when acting on a statefunction in quantum mechanics, or giving the change of classical variables).

One difficulty is use of undefined terms, like energy, that may have intuitive meaning, but are not carefully given in a theory, resulting in circular reasoning. Energy is correctly defined as a Hamiltonian eigenvalue (with an i since time translation is given by an exponential; any constant merely comes from different units for the two sides of the equation stating that energy is a Hamiltonian eigenvalue, and would not appear with proper choice of them). To assume that energy is defined otherwise, and then take its equaling this eigenvalue as a physical law is not only circular, but absurd (as is also true when the time-dependent Schrödinger's equation, which just states this eigenvalue relationship, is regarded as a physical law). Heisenberg's formulation of the time derivative as a commutator (in classical physics, a Poisson bracket) of the Hamiltonian with an operator is equivalent, but just transfers the time variation to operators. The Feynman path integral method is the same, using the time-translation operator, rather than its derivative.

Commutation relations also follow from the formalism, from the form of the operators which are Lie algebra generators, and are not some (mysterious) independent assumption. They merely relate the action of operators and their eigenvalues (as with the momentum operator, the space translation generator, and its eigenvalue, momentum). The form of operators (with interactions) may be an assumption, but (four)-momentum is a translation operator, so its properties, including its commutation relations with its eigenvalues, four-momentum, are determined by that. Likewise angular momentum is a translation generator, in an arc, giving the form of its realizations by derivatives, and commutation relations, between its components, and also with its eigenvalues.

III.5.a Why the energy is the Hamiltonian eigenvalue

There is a vague feeling that energy is some special characteristic — independent of any definition — that objects have (like people have souls) and that its relationship to the Hamiltonian is a physical law. But energy is meaningless unless defined. That the energy is equal to the Hamiltonian eigenvalue is not a law of physics — energy is defined as the expectation value of the Hamiltonian (the eigenvalue for states for which the Hamiltonian is diagonal) — thus their equality cannot be a physical law.

One reason that it is so defined is that the (exponentiated) Hamiltonian, being the time-translation operator is, necessarily, an operator whose eigenvalues and expectation values, for fixed states, are constant in time, thus energy is conserved. And this conservation is not a physical law but a consequence of a definition, although it is a physical law that there can be such Hamiltonian operators — this following from the invariance of space under the Poincaré group, one of whose elements is obtained by the exponentiation of the Hamiltonian, and also such fixed states. Because energy, and momentum for the other momentum operators, and likewise angular momentum, are conserved, they are of great importance, so are given names and special attention. This importance, and that they have names, make it appear (for some not-understood psychological reasons) that they in some way exist independently of their definitions — expectation values. But they are not some abstract qualities (like the soul, for example), although they are often thought of that way.

III.5.a.i *Schrödinger's time-dependent equation*

The time-dependent Schrodinger's equation is not a dynamical law, and it does not come from symmetry, or consistency, but rather is the definition of the Hamiltonian as the time-translation generator; the time-independent equation actually is no more than the nonrelativistic approximation of Dirac's equation. There is no physics in the time-dependent equation. Physics enters in the form of the Hamiltonian.

Schrödinger's time-dependent equation cannot be wrong, it cannot disagree with experiment — it is a definition. If there is a disagreement with experiment that does not show the equation is wrong, but that the Hamiltonian is incorrect. A quantum mechanical Hamiltonian is not determined by a classical Hamiltonian, but they both have to be determined in some way. However a quantum mechanical Hamiltonian is fundamental, a classical Hamiltonian is simply the limit of that (since classical mechanics is inherently inconsistent [Mirman (1995b)]) and (many) quantum mechanical Hamiltonians come from basic properties of space or physical systems.

III.5.a.ii *Lagrangians and action principles need not exist*

There is no reason to assume a Lagrangian, or to believe that one is possible (that it works for collections of spin-$\frac{1}{2}$ particles seems an accident), but a Hamiltonian is necessary. A Lagrangian is a purely formal construct used, in certain cases, to set up a problem, to, say, get equations of motion. But we cannot guarantee that a system has a Lagrangian, or that equations obtained from one that we might guess are correct. Likewise an action principle cannot be assumed [Mirman (1995b), sec. 6.3.3, p. 118]. A Hamiltonian is meaningful, and required. Thus if we find the right form for it, the system is correctly described.

III.5.b Construction of a Hamiltonian

How is a Hamiltonian constructed? The use of Dirac's equation is misleading, this being a special case. Linearization, which Dirac used to guess the equation, is wrong. It happens to work for spin-$\frac{1}{2}$ objects, but does not for other spins [Mirman (1995b), sec. 6.3.2, p. 116]. Matrices are not necessary, nor is a Clifford algebra [Jancewicz (1988), p. 30]. Physicists like to generalize from a single case — so are often mislead.

III.5.b.i *Relativistic wave equations*

Indeed much of the interest in constructing relativistic wave equations, or equations of motion, comes from a belief that statefunctions of all objects obey them, even though this is known to be true (and is true) for but a single case, that of spin-$\frac{1}{2}$ objects. Basis functions for other spins do not obey only one equation.

This provides another example of one of the fundamental methods of modern science, the method of jumping to conclusions. This method is widely used, not only in physics, but in all of science. Indeed much of paleontology, for example, would not be possible without the invention of this fundamental technique.

III.5.b.ii *Using creation and annihilation operators for Hamiltonians*

To construct a Hamiltonian in general we use creation and annihilation operators, $a^*_{p,s,m}$ and $a_{p,s,m}$, these creating and annihilating objects of spin s, and z-component m, with four-momentum p [Mirman (2001), chap. IV]. They transform according to the spin state. This satisfies one condition given by the Poincaré invariants [Mirman (1995b), sec. 6.3.1, p. 115]. The other is that

$$p^2 = m^2, \tag{III.5.b.ii-1}$$

for an object of mass m. Thus the Hamiltonian (without interactions) is

$$H = \int a^*_{p,sm} a_{p,s,m} \delta(p^2 - m^2) d^4p. \tag{III.5.b.ii-2}$$

The state created by these operators satisfies both conditions given by the Poincaré invariants. So this Hamiltonian acts like the time-translation operator. Putting in interactions is done the same way.

What variables appear in a Hamiltonian? It is an operator, so must be a function of operators. These might be, for example, position q, and momentum $i\frac{d}{dq}$ (or in the dual space, momentum p and position operator $-i\frac{d}{dp}$. However we cannot use something like $i\frac{dq}{dt}$, which has no meaning, being the derivative of an operator. This is really a symbol for the time-derivative of the expectation value, $i\frac{d(|q|)}{dt}$, between states ($|$ and $|$).

III.6 Topology and geometry

Poincaré group transformations define the space [Mirman (1995c), sec. 10.1.1, p. 166]. Point x is given by the statefunction $\psi(x)$, or $|x)$, which is obtained from the statefunction at an arbitrary point 0, $|0)$ by a translation:

$$|x) = exp(ip_\mu x_\mu)|0); \qquad \text{(III.6-1)}$$

p_μ is the momentum operator in the x_μ direction, $\mu = 1, \ldots, 4$. Momentum can be nonlinear, dependent on statefunctions.

Suppose that momentum operators (like the Hamiltonian) are nonlinear? What effect does that have on the geometry [Mirman (1995c), sec. 10.1.3, p. 169]?

Representations and states are labeled by a set of numbers, including ones we denote as (eigenvalues of) p_μ. The basis vectors that we regard as relevant are functions of momenta, in particular they are eigenfunctions of momentum operators, $-i\frac{d}{dx_\mu}$ + nonlinear terms. Operators $i\frac{d}{dp_\mu}$ (or the corresponding covariant derivatives) thus also exist; we take these as position operators; their eigenvalues, x_μ, define a space. The resultant basis vectors are functions of position. Thus we have two sets of functions dual to each other. An eigenvalue of operator $i\frac{d}{dp_\mu} + \ldots$, is position, that of $-i\frac{d}{dx_\mu} + \ldots$, momentum.

That we not only can, but must, realize momenta (translation operators) as $-i\frac{d}{dx_\mu} + \ldots$ follows from the group being the transformation group over a space, in particular one allowing translations. This realization leads to the commutation relations between position and momentum, which are regarded as being part of the foundation of quantum mechanics. These are the result of this realization, coming from the geometry, not an extra assumption.

Because of this duality, from group representations we can define a space, and with it a geometry. It is important to notice that this is a property of this type of group, an inhomogeneous group (sec. I.7.b.iii, p. 39). It is possible because momenta commute. It cannot be done for the rotation group, for example. This has two angles, but only one commuting generator, so only one eigenvalue. And while eigenvalues for the commuting case are continuous, as is required for coordinates, those of semisimple groups are discrete. Coordinates arise from Abelian, or inhomogeneous, groups.

The rotation group is a subgroup of an inhomogeneous group so angles can be obtained by defining first linear coordinates, and then angles in terms of them. Or we can realize the transformation operators as derivatives with respect to variables, which we call angles. But these we do not obtain from the representation labels, say by taking a dual space; they have to be defined independently. They are not the only variables we can use for the rotation group; operators can also be realized as derivatives in one or two variables [Miller (1968), p. 45, 49; Mirman (1995a), pb. X.7.c-7, p. 297], which have no clear relationship to angles, to the structure of the group or the meaning of the operators, but there could be a relationship, something we do not consider here. This is similar to aspects of the realization that gives the conformal group [Mirman (2001), chap. III].

This starts to illustrate how properties of the group are fundamental in determining geometry, and physics, and of course conversely.

Under the Poincaré group, p^2 is invariant, and so is x^2. Corresponding to the invariants in one space we have invariants in the dual space. For nonlinear representations momentum is no longer given by $-i\frac{d}{dx_\mu}$; there are terms added that depend on basis vectors. Likewise coordinates are no longer given directly by $i\frac{d}{dp_\mu}$. In principle given the momenta we can transform to the dual space and find the coordinates. This, however, is part of the problem to be solved, finding functions of space that are eigenfunctions of momentum operators. While these operators are obvious for linear representations, determining them is a large part of the challenge of finding nonlinear representations. There are conditions on them: they are basis vectors of the homogeneous subgroup, of the defining (or conceivably perhaps adjoint) representation, and they commute. The difficulty is that momenta depend on basis functions explicitly, so that we cannot realize operators of the Lie algebra of the group independently of the representations, as for linear representations [Mirman (1995a), chap. XIII, p. 367], rather finding realizations is part of the problem of finding representations [Mirman (1995b), sec. 6.1, p. 109], something equivalent to what is understood, although in different terms, in general relativity [Stephani (1990), p. 88]. At this point we merely try to see how much we can understand without having explicit expressions for momentum operators.

We thus have coordinates x, dual to the nonlinear momenta p. We can also define (perhaps only locally) variables z such that the distance squared is

$$ds^2 = dz_\nu dz_\nu = g_{\rho\sigma} dx_\rho dx_\sigma, \qquad \text{(III.6-2)}$$

where $g(x) = g(z)$ is the metric, necessary since the x's define a curved space. This introduction of a set of "linear" coordinates is possible as long as all functions and coordinates are well-behaved, but since space need not be globally flat, it is possible only locally. The two sets of coordinates are related by this equation.

We can now consider momenta q dual to the z's. The position of a point is then given by operator $exp(iqz)$, which is, by definition of the z's, and thus the q's, the same point as that given by $exp(ipx)$. We can (we assume) define a complete set of functions of the z's, $exp(iqz)$, which then give a complete set of functions of the x's, $exp(iqz(x))$, since the z's are known (and hopefully single-valued) functions of the x's, and the x's single-valued functions of the z's.

For gravitation we have momentum operators; a specific one, the Hamiltonian is [Mirman (1995c), sec. 8.2.1, p. 147]

$$H\Gamma^{\lambda}_{\mu\nu} = \Gamma^{\lambda}_{\mu\nu;t} = \Gamma^{\lambda}_{\mu\nu,t} + \Gamma^{\lambda}_{\phi t}\Gamma^{\phi}_{\mu\nu} - \Gamma^{\phi}_{\mu t}\Gamma^{\lambda}_{\phi\nu} - \Gamma^{\phi}_{\nu t}\Gamma^{\lambda}_{\mu\phi}, \qquad \text{(III.6-3)}$$

where the ordinary derivative is with respect to coordinates z, which are known in terms of the coordinates x defined by the momenta of the field.

Since we can transform from an arbitrary origin to both any value of x and any value of z, statefunctions expressed in either set of variables must be basis states of the operators taking them from the origin to these points. In the z's, by their definition in terms of a constant metric, momentum operators are linear, and their basis states are of the form $exp(ikz)$. Basis states of operators expressed in terms of the x's are complicated since these operators are nonlinear. Each set of basis states can be written in terms of the other, as there is (locally) a one-to-one relationship between points expressed in terms of one set of variables and those expressed in terms of the other — the points are the same — even if values of their coordinates are not. Thus for consistency each set of basis states must be complete in the sense that all states of the other set can be expressed in terms of it.

So we then have, in principle, the definition of space in terms of the group, and of a curved space in terms of the nonlinear representations defined over a flat one.

In this view, or perhaps language would be better, objects like the metric, connection, curvature, are not so much properties of space, but rather functions defined over it (sec. I.6.a, p. 28).

Chapter IV

Language and its dangers

IV.1 Misuse of Language

A large part of the distress resulting from interpretations of quantum mechanics (and much else) comes from use of meaningless or misleading language, of phrases that seem to say something but do not, of ones that assert something different from what we think they do, or which have connotations to which we respond rather than to their denotations. Since this is so fundamental it is discussed here in depth, with a large number of examples. Misuse of language is so extensive that many examples occur elsewhere throughout as part of discussions of specific topics. By considering these, often much too common, cases it is possible to develop habits of analyzing what we read — and what we say, thus helping to notice, avoid and eliminate such errors. This problem is not unknown [Englert, Scully and Walther (1999), p. 328; Omnes (1992), p. 371], although certainly not fully appreciated.

Since the impediments arising from language are so central, it is useful to survey its role first since other misunderstandings (not only of quantum mechanics), even if not explicitly caused by it, often come about because of linguistic inaccuracies, errors, fallacies.

It is an interesting psychological question why quantum mechanics, especially it seems more than other subjects, impels so many physicists, and science writers, to say all the absurd things that come across so glaringly in reading discussions of it. Perhaps some of these samples can stimulate thinking and help clarify this strange psychology.

Illustrations of meaningless and misused language run through may discussions, unfortunately not only, of quantum mechanics (sec. V.7, p. 251), and form such essential parts of these that it is difficult to treat language separately. Thus there are many examples throughout this book. It is useful however to give some explicitly to illuminate and emphasize these points. That is the purpose of this chapter.

These will also help guide the reader in finding (undoubtedly many other) cases of meaningless language in the present discussions.

IV.1.a The name's to blame

Problems start with the name, quantum mechanics. But quantitization is neither universal nor necessary, it is not central to quantum mechanics or laws of nature, as free particles, ones moving through barriers or in bands in crystals, as well as many other such cases emphasize. Discreteness really results from boundary or regularity conditions, and that statefunctions are group basis vectors; it is not fundamentally a property of these physical laws. Yet the name has misdirected far too many beliefs and discussions, and caused much misunderstanding. (A better name might have been functional mechanics, or even transformational mechanics, since an essential aspect is that the objects are functions of space and time, that is basis vectors of the transformation group.) Unfortunately quantum mechanics is now so entrenched that it is impossible to change.

The term "wavefunction", misleading because, among other reasons, quantum mechanical objects do not wave in a classical sense, and because in barriers, for example, there is not even oscillation, is somewhat less strongly established. Here "statefunction" or sometimes "statevector" is preferred. This is a function that gives the state of the system — that completely describes it. Nothing else, neither waviness nor quantitization, for example, is implied. It is thus, at least, less misleading. But occasionally wavefunction is used, for variety, or because it is appropriate in a particular context, such as when discussing statements in which it appears.

Of course names often mislead. For example, if the TCP theorem were not called a theorem would its triviality [Mirman (1995c), sec. 4.2.6, p. 60] have been more easily recognized? Too often people base their thinking on labels, names, not content.

The name "general relativity" has been criticized as being misleading (sec. I.7.b.vi, p. 41). But it is the accepted name, so is used here.

Capacitance, despite its explicit and well-known definition, is a name that has mislead too many people to think that it is the amount of charge that a capacitor can hold.

An amusing (now) example of a name (actually a sequence of letters) being misleading comes from the Schiaparelli's observation of linear features on Mars, which he called, in Italian, canali. But of course the word "canal" has a different implication in English leading to the fantastical belief in intelligent creatures on Mars who constructed artificial canals. Intelligent creatures however are not mislead by sequences of letters.

IV.1.a.i *Classical physics can be quantized, quantum mechanics not*

It is interesting that classical computation is done with discrete bits, while quantum mechanical analogs use qubits given by linear combinations of states with continuous coefficients. Thus despite the name quantum mechanics, it is classical theory that is actually quantized.

Also a classical vibrating string, fixed at both ends, has only discrete wavelengths and frequencies, so energies — it is quantized — while in the quantum case, unless the string exists forever, the energy is continuous, but perhaps strongly peaked.

Quantitization in classical physics is ubiquitous. Many (almost all?) musical instruments are based on allowed wavelengths forming a discrete set. Electromagnetic waves in waveguides are quantized and lead to many other such examples. Classical quantitization is of great technological and economic importance.

Perhaps classical physics should really be called quantum physics.

IV.1.a.ii *Statefunction is a better term than wavefunction; why?*

A name that has lead to much confusion is wavefunction — implying waves. But, for example, wavefunctions of particles (another bad name) within a potential barrier do not "wave", but exponentially decrease. This name also leads to questions like "what waves?" (apparently some people still like the ether). However if there were something that "waves", then a wavefunction would be needed for that something [Mirman (1995b), sec. 1.4.4, p. 13], and there would be a question of "what waves" for that wavefunction, and so on, giving an infinite regression. Some might like this material that waves to consist of little balls (with springs?) but unless we are prepared to consider finite balls, likely infinitely hard, with no way of knowing about their interiors, hardly a scientific viewpoint, that will not work. And besides such a view would not lead anywhere [Mirman (1995b), sec. 1.5, p. 14].

We shall not wave(r), but take the term as statefunction.

IV.1.a.iii *Defining statefunction by use*

As is usual with terms and concepts, statefunctions are defined not by what they are (they're statefunctions), but by how they are determined — as representation basis vectors, with equations giving them (sec. I.6.d, p. 32) — and how they are used. Definitions are pragmatic (sec. III.3.c, p. 120). But genuine understanding of the practical use of them, so their definitions, may be difficult. While concepts of a representation basis state, and a particular one, can be well-defined mathematically, physical meanings of group transformations may be subtle. And seeing how they are actually used can often be quite tricky. General, but empty, definitions do not clarify, but hide the lack of correct ones — given by the use of concepts.

Statefunctions are broadly used, appearing in many vastly diverse circumstances and for multitudinous entities, free objects, ones inside others — atoms, molecules, crystals, nuclei (both atomic and cellular), planets, stars, for systems, simple and complex, including objects and observers, and various combinations. It is important not to assume definitions that restrict their use, for that would eliminate much of physics, and our ability to understand it.

Thus, except occasionally for variety or to use the same terminology as a quotation, we use statefunction, giving the state of the system (a nice circular definition, since there is no way of knowing the state except from the statefunction, which therefore defines the state). There is no implication of any properties attached to this term, and a large part of analyses consists of finding its sense in particular (but often very general) cases, and how to determine and use it.

While the term wavefunction is, experimentally, misleading, statefunction, implying knowledge of the state of a system (although we have no way, in principle to distinguish between a state and a statefunction of a system), is less misleading. We write the term as statefunction, a single word, to emphasize that it should not be misinterpreted. An artificial term, or perhaps a term like systemfunction, might be better. But these are unfamiliar, while statefunction sounds less so. Thus we use that, but with the understanding that it should not be broken into parts which are then incorrectly interpreted.

IV.1.a.iv *Is a force an interaction?*

Terminology is often carelessly used as a result of not distinguishing between forces and interactions. In quantum mechanics the correct term is interaction — there are no such things as forces. Behaviors of objects are determined by equations governing their statefunctions (like Dirac's equation); nonlinear parts, those which (in general) contain statefunctions of other objects, are called interactions. One reason that it is important to make this distinction is that for electromagnetic interactions it is the potential that appears in interaction terms, not the electromagnetic field (which classically gives a force). And it is (not only for this reason) the potential, not the field, that is physical and measurable [Mirman (1995c), sec. 3.3.1, p. 37].

Also, despite statements that occasionally appear, an interaction does not involve particle exchange (especially since neither particle nor exchange have clear meaning here), although it is often useful to consider such a picture when more than one fermion is involved (although it can also refer to bosons). Emission of a photon by an electron (in a scattering process), or decay of a neutron (into three fermions) are examples where there are no pictures of exchanges. In general "particle exchange" is only a part of a phenomenological picture, if that.

IV.1.b Meaningless statements often seem meaningful, so confuse

A preeminent problem in the misuse of language is construction of sentences that seem meaningful, but really are not. An analogy should make this clear: "The analogy is frantically chartreuse.". This sentence is grammatically correct, with a quite familiar structure, all its words are fully meaningful, but the sentence is absurd, it does not say anything — the words do not go together. Modifiers cannot modify what they seem to be modifying. Many of the riddles, and beliefs, of quantum mechanics, physics, life, are based on sentences like this. Here it is clear that words do not, can not, go together, so the string of them is nonsense. But in many other cases it is not obvious, so we do not realize that we seem to be saying something but are not.

This is not an oxymoron — a statement containing opposite, conflicting, terms, canceling each other. Here the statement contains terms attached to each other, but which have no sense as such. They cannot be so attached.

Words are defined, in part, by the contexts in which they can appear. But they are often used in association with other words that provide contexts in conflict with their definitions. Thus a string of such words is self-contradictory or simply void. And such strings are quite common.

IV.1.c Questions meaningful and meaningless

The process of science is, to a large extent, that of providing answers to questions. This is particularly evident in quantum mechanics for which many discussions involve answering specific questions about nature, physics, quantum mechanics itself.

One major cause of confusion — not only in quantum mechanics, not only in physics, not only in science — is meaningless questions, attempts to answer them, and discussions and analyses that are based on them. This has long been known [Bridgman (1960), p. 28]. It is unfortunate that this particular discussion is so little read. Careful study of it, and its sample questions, would prevent much folly, misunderstandings, blundering. Other people [Wallace (1996), p. viii] also recognize this, but they are far too few. In general much of the language used (and, of course, not only in physics) consists of terms and concepts not fully (if ever) defined or analyzed despite attempts (too little studied) to do so [Bridgman (1960)]. Although some such attempts are so out-of-date as to be ancient (indicating how fast science has been moving) they still provide models of analyses often much needed, but too rarely undertaken. Physics would have been much clearer, better understood, better known (by physicists, and all others) — and more of it would be known — if such endeavors were studied and built upon.

It is thus a fundamental question whether questions have meaning. Often these deep, philosophical arguments about quantum mechanics (about nature, about everything else) are no more than attempts to answer questions that are merely sets of words, which in the sentences that the questions are, have no sense — mere strings of letters that together are nonsense. This is a major cause of much of the confusion, and not only, about quantum mechanics.

IV.1.c.i *What questions actually ask something?*

A question has meaning if, but only if, a type of meaningful statement that answers it can be given. What is the mass of the proton, to 100 significant figures? It is possible to give a statement that answers this question, it has content, even should it be always impossible to actually find an answer.

What is a question? It is a request for a statement that provides information. But if it does not specify what (type of) information is being requested, it is not asking for anything. And then it is not a question — it is vacuous, an empty string of words. Since the information requested has not been specified, it cannot be given — the question cannot be answered. Then it is no more than a set of words without content, though that may be hidden by the fact that the words have meaning in other contexts.

IV.1.c.ii *Examples of questions lacking content*

Questions like "what is the actual physical situation described by the statefunction?" or "what is consciousness (beyond aspects of neural activity)?" cannot be answered for there is no information that answers them — they are thus meaningless. They express an emotional feeling that the physical descriptions are unpleasant, but there is no way of providing statements that answer such questions — there are no such statements. Answers — except for ones relating to concepts like statefunctions, or neural activity — that would be acceptable, not because they told something about the physical world but because they were emotionally satisfying, are mere strings of words providing psychological comfort, but not information about the external universe. These do not give a possible basis for scientific work. Considering such questions cannot lead anywhere.

So "Does quantum mechanics explain, or merely predict; do we understand it?". How are these questions distinguished? Also "Is it complete?, What notion of reality does it imply?". What is meant by complete? How can we tell? And what is reality, beyond what quantum mechanics predicts? Do these words actually ask anything? Such questions are regarded as mind-teasing. Could they be mind-teasing because they are mere sets of words without content, gobbledygook?

"What is a system physically like, when not being measured?" Many people regard it as evasive to not answer such questions. Or they feel that a statement like "an electron has a statefunction that gives probabilities of measurement outcomes" does not tell anything about a system when it is not being observed. But such questions cannot be answered, they are vacuous — there is nothing that can be said, except for information given by measured or calculated statefunctions, that answers them, however emotionally unsatisfying that may be. There is no combination of words that provides an answer — a physical answer, although there might be many combinations that provide emotional relief.

"When, for a single electron going through a double slit, does a dot appear on the screen? Is it there only after someone looks or before?" Of course, the formation of a dot is independent of human actions, of whether, if ever, a human being looks at it. And this we can check experimentally. Thus the formation of the dot can be registered in many ways, by cameras, by devices that record current produced by its formation, and so on. Does this mean that the wavefunction is not the complete description of the electron and screen before the human being looks? In fact, the wavefunction does give the probability of a dot appearing as a function of time — something that can be checked, although human beings need not for it to have meaning.

What more can be said, thus what more can be asked for? Those who are unhappy with this, who desire that the wavefunction, or something else, provide more information, must state what they want, and how to get it. Otherwise they are merely saying that they are unhappy with the way nature is and wish that it would be something else. However human beings do not decide what nature is, only observe it, no matter how unhappy we may be with observations.

"What is time? What is space?" Such questions excite because they seem so fundamental. Well what is time, what is space? What type of answers can such questions have? What kind of information do they request? And if they have no answers, if they request no information, can their allure be due not to the understanding of nature that they provide, not to the guidance they give in seeking such understanding, but solely due to the excitement that they provoke?

It is also essential that the correct question be asked. Thus [Mermin (1998), p. 755] the question "Is your sensation of blue the same as my sensation?" is a completely reasonable question with a definite answer "yes" or "no", except that it leads to another question (else it misleads): "How do you know?". To answer the first, it is necessary to answer the second.

IV.1.c.iii *How meaningful, and meaningless, questions fundamentally differ*

There is an essential difference between meaningful, and meaningless, questions. Meaningful ones can be answered (at least in principle) by observation of the world of which we are part, and they guide us in finding required observations. Meaningless ones cannot be, nor do they even hint at what observations are needed — because they are not requesting information about nature, but merely desires for subjective, psychological succor. They confuse, not guide.

IV.1.c.iv *Unhappiness with inability to answer meaningless questions*

Unhappily though it may be, nature does not allow us to answer many types of questions — there is no way of giving (what we regard, from our classical intuition as) a complete description of everything — one that tells us things about our experiences, past, present or future, that we would not know without having that description. And that is the requirement that a description have physical meaning; physical descriptions tell us things that we would not know without them, but they cannot tell us everything. An answer to a question that does not have meaning in that sense, it does not tell us something that we would not know without the answer, has no meaning in any sense. And if a question does not have (at least one) such answer, if we cannot even imagine what type of statements might provide such an answer, then it asks nothing. This does not require that an answer provide a description of our experiences. It might be a statement somewhere in a long chain of reasoning — perhaps linked to other chains. But such an aggregate must give information about our experiences that we would not have without it. And the answer must be necessary for it to do so.

There are undoubtedly large numbers of people who are unhappy with these analyses, with the definition of probability, or of consciousness. They do not feel that it answers their questions such as "What is probability in quantum mechanics, really?" (sec. III.2.f, p. 110), "What is consciousness, really?". But the only way of answering such a question is for them to specify what (type of) information they want as an answer. If they cannot do that it means that there is no other answer. Then what they are really saying is "Such an answer makes me uncomfortable, unhappy, and I do not like it". However this is irrelevant; the purpose of physics is to describe nature, not to provide happiness.

What is wrong with such a desire? If we ask for a physical definition of probability, then it is possible to provide a set of experiments (in principle) whose results give an answer — probability then is defined by the types of sets of such experiments, and the types of their results. Likewise for consciousness we can specify a set of procedures that lead to knowledge of neurophysical activities that differentiate between conscious and nonconscious states. Of course finding these may be technically impossible, and may never be completely possible (which is certainly true of consciousness). But they provide a means of at least starting to attack problems, of providing some knowledge, some information. But it is impossible to even begin to study, physically, the problem of finding emotionally comfortable answers. Of course, different people find different pictures comfortable, and uncomfortable. But more fundamentally, all we can do is find what nature is like, and if what it is like makes people uncomfortable, there is nothing that we can do to change nature. We are subject to the desires of nature, it is not, and cannot be, subject to ours.

IV.1.c.v *Questions that do not seem to have meaning but may*

It is easy, much too easy, to ask meaningless questions, and even to develop detailed theories based on answers to them — theories that are completely without content. But it is not always easy to know whether a question really asks anything. Too many seem to but do not — which is why they are so misleading. Can the reverse be true, can a question seem senseless but actually imply ones that are not only meaningful, but physically important?

For example: "Is an atom a clock [Bridgman (1960), p. 72]"? Since clock is undefined there can be no answer to this. But since clocks are theoretically important, and in practice defined in terms of properties of atoms, might this be a hint of meaningful and important questions? We want objects that we use as clocks to have certain properties: reproducibility, allowance of comparison of different ones, both those at (almost) the same point, and ones (very, very) far away, say. Then questions of the properties of atoms, and whether they can lead to such useful objects are quite sensible, and consequential.

It is helpful to keep such examples in mind, for in discussions (of interpretations of quantum mechanics) there are many sentences that seem to be real questions, but are not. However in some cases, certainly not all, these indicate real ones that are struggling, often unsuccessfully, to get out.

Is there a particle with mass twice that of an electron, and half the charge? This is semi-meaningful. It is possible to specify, in principle, a way of finding such a particle. But it is not possible (at least within current knowledge) to find a way of showing that there is no such object.

Quite relevant at present is the question of whether objects have certain properties, like spin direction, before they are measured? This is meaningful for certain properties. It makes sense to say that an electron has a particular mass before it is measured because the mass value is part of the definition of an electron. And it is meaningful to say that in a collision the object emitted by an atom is an electron since properties of atoms are quite well-known — it would affect a huge part of physics if some atoms were found to have particles with different properties in their outer shells.

However can we say that if a pion at rest decays into two fermions the spin direction of the one moving to the right is along the positive (arbitrarily chosen) z axis before we measure the direction? In fact we cannot because it makes no difference whether it is true or false. This use of empty statements has mislead so many about so much regarding (not only) such topics as EPR experiments (sec. V.3, p. 208).

However the situation is more subtle. The statement has no content in quantum mechanics, but it does in classical physics. Thus in many analyses it is important to distinguish between questions and statements that refer to objects for which quantum mechanics must be used and those that we can treat phenomenologically with classical physics. And it is essential that statements and questions really do have meaning. The strange, even absurd, conclusions that (too) many people (physicists!) draw from many of these discussions show why care about the content of sentences, and questions, and the realm in which they are actually being used, is so important.

IV.1.d Can physics be replaced by logic?

It might be thought that difficulties, including those with language, can be avoided by replacing physical statements with logical ones (perhaps logical symbols only). Thus some [Omnes (1992), p. 377] believe that statements in language can be replaced in principle by some mathematical objects, and reasoning can be proved correct like an equation. Then the logical structure of quantum mechanics can be completely formulated in terms of mathematical entities, reducing its use to mathematical equations. Clearly if that were true there would be no problems with interpretation of quantum mechanics.

But this statement illustrates why it cannot be true; it contains (at least one) meaningless word "replaced". In fact it is just this word, or some equivalent one, that is so fundamental to so many problems in understanding (not only quantum mechanics) and to so many meaningless papers whose vacuity is hidden by their being filled with (meaningless and irrelevant) equations.

How are physical statements replaced by logical or mathematical ones, or words replaced by symbols? It is this replacement process that involves physical understanding, assumptions and interpretation. Mathematical and logical symbols, by themselves, have no significance. Giving them physical content is the real problem, and the foundation of physical theories. Too often they are given content by using words, phrases, symbols and sentences that themselves have no content. But that only makes things worse by hiding their lack of substance.

Logic and mathematics are only relevant to physics if their symbols (words are symbols) have actual (preferably clear) physical content.

Suppose that experiment disagrees with a theory: how should the search for reasons, perhaps misunderstandings, and if necessary a new theory, be started? If physical foundations of a theory are clear they can be studied and weaknesses as well as possible reasonable changes discovered. But if the foundations are purely logical, and their physical relevance uncertain, it may be almost impossible to know how to attack problems. This is one reason why physical theories should be based on physics, not logic, not mathematics.

IV.1.d.i *Validity of self-validation*

A related idea [Benioff (1999)] is that if quantum mechanics is universally applicable, then it should describe its own validation. However a physical theory is validated by comparison with experiment. This requires that concepts be given experimental content, and that can only be done by intelligent objects (if there actually are any in the universe). Intelligent objects must be extremely complex (a hydrogen atom cannot be intelligent). For a theory to describe its own validation would necessitate that it also describe intelligent objects, and how they interact with their external surroundings, and each other, and the physical processes that result in language and its content, the physical processes that relates words to their experimental meaning, and that decide such relationships. But intelligent objects are all different, and mutually interact. And the number of such potential objects is infinite. There is an infinite number of ways of constructing them. Such a theory would have to describe these objects taking this infinity into account.

Perhaps then "describe its own validation" is a quite interesting phrase, but one that has no validity, "language" that does not say anything.

IV.1.d.ii *Providing physical sense is quite difficult*

An example of an attempt to replace physics by logic is the suggestion that a basis for quantum mechanics is an orthocomplemented, orthomodular, separable lattice. This might provide a firm, rigorous foundation for quantum mechanics — as a purely mathematical theory. But unless these concepts are related to what occurs in the physical universe it does not provide a physical theory. How do words like "orthocomplemented", "orthomodular", "separable", "lattice" apply to physics? What is their experimental meaning? How can that be determined? Unless these can be answered (in a way that makes sense physically), with answers having well-defined physical content rather than being merely sets of meaningless (including physically meaningless) words that replace other sets of meaningless words, the rigorous theory is not a theory about anything having to do with the actual world.

Esoteric words and terminology can be quite seductive. They give us an image of knowing subjects mere mortals are incapable of grasping, that we are priests of science and mathematics privy to the great mysteries of the universe, that we are superior even to others in our own fields. Their emptiness, their nothingness, especially in the contexts in which they are used, their irrelevancy, are irrelevant (or perhaps that is part of their seductive appeal). The emotional rewards that they provide are more powerful than any desire to really understand nature.

Giving (relevant) physical sense to mathematical terms is a very difficult and subtle problem. The rigor of the mathematics, instead of aiding the process can hinder it, and hide its nonexistence, so cover up the fact that the rigorous mathematical theory has nothing to do with physics (sec. I.2.f.v, p. 11).

Physicists are often almost captivated by equations. After all they are more precise than words, which often appear ambiguous, imprecise, fuzzy, aren't they? This emphasizes again how mathematical criteria can deceive when applied to physics. Of course, in a certain sense they are precise, but not physically. While relationships among symbols may be given very carefully and precisely, these symbols have to be related to physical phenomena otherwise the equations have nothing to do with physics. And defining symbols — physically — is itself ambiguous, imprecise, fuzzy. Thus the precision of equations is superficial, hiding the imprecision beneath, the imprecision arising from the process of giving symbols and relations physical content. Precision, by hiding the fundamental imprecision, then misleads and confuses, leading to greater fuzziness, and more likely error.

Unless words, symbols, equations, logic, can be properly and intelligibly correlated with physical objects and processes, they, and arguments based on them, are worse than nonsense, for they make it far more difficult to formulate (potentially) correct and sensible views of nature.

This emphasizes how often it is so difficult to be sure that words and sentences that we do use actually form physically, mathematically, logically, meaningful, reasonable, statements.

IV.1.d.iii *Is mathematics better than physics?*

But beyond this is a vague feeling that mathematics is, in some way, better than physics, that it is an advance to replace physics by mathematics, even if that mathematics has nothing to do with physics. And it is also impressive to read (or perhaps just look at, for they are often unreadable) so many papers filled with equations but with no attempt made to connect them to physical concepts. Such papers do not lead to physical predictions, to explanations of physical phenomena, but merely to sets of equations going nowhere, that are apparently not intended to go anywhere. They are aimed not at understanding physics, but rather are designed to give physicists the nice, warm feeling of seeing pages and pages filled with Greek letters and mathematical symbols. They provide a fascination that it seems many physicists do not find in physics. And in the midst of the difficulties, and at times frustrations, of trying to understand nature the sight of all these formulas is so comforting.

Of course this allure of equations extends even into the "educational" activities of physics teachers, who often do not seem to realize that they are supposed to teach physics, but instead work very hard in teaching students to substitute numbers in equations. But however fascinating numbers, Greek letters, mathematical symbols, equations. formulas, logic, may be, to a real physicist physics itself is far more fascinating.

IV.2 Examples of statements meaningless, misleading, and often nonsense

The number of strings of letters and strings of words having no substance, of statements whose connotation differs from their denotation, of linguistic errors ranging from the most glaring to the most subtle is, unfortunately, quite large. Ones given here are not necessarily the worst or the most representative, but are at least enough to stimulate readers to notice others — and avoid them. That so many examples are given indicates how deep the problem is, but providing this list is helpful in preventing at least these errors from being repeated.

IV.2.a Can a system be in all states at once?

An example of the type of statement that often appears, and which seems substantive if not thought about, is that because a statefunction is a superposition of states a system is in all basis states at once. It is an interesting exercise to explain what this means, or at least, what the author is trying to say.

Actually since a statefunction gives the probability of a particular experimental value an ensemble is needed for it to be intelligible in this sense. If it is a sum of basis states, then it says that different repetitions of a measurement find different basis states with frequencies given by the coefficients in the sum. However to say that it is in all these states at once means that a single experiment finds it in all basis states. This obviously is not true (unless the sum is a single basis state, as a sum of states of spin components along z is a single basis state for the component along z'). But if anyone disagrees they can give an example of an experiment that shows a system in all states at once.

Statements about a system being in all basis states at once are mindless as stated but are trying to say something reasonable, and having difficulty doing so. Perhaps those who make such statements do not know what they want to say.

Another related statement is that not only properties of things but things themselves are subject to quantum superposition. The statement about superposition of properties can perhaps be understood as referring to superposition of statefunctions giving the properties (like spin component). But can, say, electrons (certainly an example of "things" that quantum mechanics deals with) be superposed? What are superposed electrons like — experimentally? Or could this be an example of the superposition of words into phrases that, if taken seriously, advocate the most bizarre notions, the most bizarre delusions?

IV.2.b Can an object exist in many quantum states simultaneously?

The word "exist" does exist in many sentences. Undoubtedly the reader should have no difficulty in understanding what it says when it appears in them. Undoubtedly writers in whose works the word exists should have no difficulty in explaining what actually exists in their sentences and what the word refers to. There are many examples (sec. V.4.b.iii, p. 226).

Thus it is believed [Seife (2000)] that an object may exist in many quantum states. And bizarrely (a word much beloved by physicists and journalists) it may inhabit several quantum states at once! And if an atom is bumped (?) its waveform can "collapse" (another much beloved word) by forcing the atom to commit to one of its possible states.

Obviously an eigenstate is not in the least bizarre but a superposition of eigenstates is quite bizarre. Thus if the spin of an object is up along some axis, a very nonbizarre state being an eigenstate, it is also in a superposition of spin up and down along any other axis, dreadfully bizarre indeed. And if it is in a (nonbizarre) momentum eigenstate, it is in a (quite bizarre) superposition of angular momentum eigenstates — so is indeed bizarre, or if it is an angular momentum eigenstate — which is nonbizarre, it is in a superposition of plane wave states, momentum eigenstates — very bizarre [Mirman (2001), appendix]. Not only that but if it is in a (nonbizarre) position eigenstate it is in a (exceedingly bizarre) superposition of momentum eigenstates, and if in a momentum eigenstate, it is in a (very bizarre) superposition of position eigenstates.

Thus not only may an object inhabit several states at once, and be in several different places at once, but it may be both bizarre and nonbizarre at once! This shows that the laws of quantum mechanics, or at least those who write about them, are terribly bizarre indeed.

In a many-worlds interpretation (sec. V.6.b, p. 249) a system in a (coherent only?) superposition has one state in one universe, another in another universe. Thus we see that a system can be simultaneously in one and many universes! This must of course also be true of physicists and science writers who believe in this view, so it is not surprising that much of what they say is decidedly quite bizarre.

Not only that, but a classical light wave that is (nonbizarrely) linearly polarized is in a bizarre superposition of states of clockwise and anti-clockwise circular polarization, while a wave that is circularly polarized is in a bizarre superposition of states of horizontal and vertical linear polarization (as well as a continuum of other polarizations). Thus a classical light wave can also be both bizarre and nonbizarre at once!

Of course, a statefunction of an object can never be exactly an eigen-state of any operator, to infinite precision. This is true of both objects small and large. So all objects, including physicists and journalists, are in states that are certainly exceedingly bizarre — as is often emphasized by their writings.

IV.2.c Is the brain in a bizarre superposition?

Although comments about bizarre superpositions seem to have been meant as general, states that many refer to are quantum states of the brain considered as a quantum computer, in order to explain conscious-ness — an apparently undefined concept, whose lack of definition is likely necessary to allow this view of the brain! This view of conscious-ness was apparently proposed by a physicist and an anesthesiologist who were collaborating (but probably not closely enough; it might have been better if the anesthesia were administered before the idea was published). This is one of the all too common examples of physicists looking for the wildest explanation of extreme implausibility, totally un-related to what is known, and often of complete emptiness, rather than simple, obvious reasonable ones, close to what is already known, which are vastly more likely to be correct, or at least to lead to correct expla-nations and greater understanding (sec. V.4.i.ii, p. 239). Of course these models are (usually?) not based on computations, which undoubtedly would show their (to put it mildly) implausibility [Tegmark (2000)].

The quantum states of such brains do seem remarkably bizarre, and not only because they are in superpositions (of what we do not con-sider).

IV.2.d Do superpositions cling?

Superposition has other fascinating properties [Johnson (2000b)] such that all of a particle's possible states (its position and momentum for example) cling together in a condition known as quantum superposition. It would be fascinating to find how a particle's states (whatever these are) manage to cling together, perhaps like mountain climbers. However this statement does introduce an important new idea, that position and momentum are states of a particle. Beyond that presumably someone has actually seen position cling to momentum. It makes one wonder what such clinging looks like. Not only does superposition allow clinging but in it all possible states (are there states that are not possible?) stick together (an important discovery that quantum states are sticky), and if disturbed become unglued. What kind of glue is used to stick quantum states together?

IV.2.e Is there any point to point particles?

A phrase that has caused much confusion is "point particle". What does this mean? How, experimentally, can we tell if an object is a point, even classically (sec. IV.2.x, p. 182)? And if we cannot tell, then what do the words say? Anyone who mentions point particle should be required to describe an experiment — even a classical one — that shows an object is actually a point.

There is a belief that classical electrodynamics is inconsistent because it deals with "point particles". In fact it is inconsistent, trivially so, but the reasons have nothing to do with particles (objects really) being points. It would still be inconsistent, for the same reasons, if electrons were little balls [Mirman (1995b), sec. 1.5.1, p. 15].

We use the term particle here, not in this sense, but only to refer to a well-defined object, one that, for example, can be localized well enough to that it can be studied while other objects are ignored. Also generally, a word like "object" or "entity", carrying no implication of any properties, is better. But for variety, we sometimes use particle. And because the name is standard, objects like electrons and protons are labeled elementary particles, this referring to that set of objects, so we use particle for a member of the set, again with no implications attached to "particle". The term "particle" can (and definitely does) cause confusion, but there often seems no other good terms.

If we cannot believe in point particles because they have no meaning, because they are pointless, why can we believe in electrons or wavefunctions? These have well-defined properties, some holding true for all experiments, always, like the mass of an electron, others that can be varied in a well-defined, completely controllable, way, such as velocity or phase. These concepts are necessary in order to predict results of experiments from results of previous ones. Point particle is neither needed nor useful in any way, and cannot be defined.

An alternative theory to quantum mechanics (one aimed at fixing its nonexistent problems) attributes to all particles of the universe a definite position at all times. But what does this say? What does the string of words "a particle has a definite position" say if there is no possibility that we could find the position? And if we cannot define the position of a particle in the laboratory, can we define it if the object is in another galaxy? In what way do the sentences "A particle has a definite position." and "A particle does not have a definite position." differ if experimental results they describe cannot differ? One may be more emotionally comforting than the other, but this has nothing to do with physics.

Somewhat related to the concept of particle, but actually having content, is that of localization. But this seems to lead to difficulties with Lorentz invariance in that if one observer is certain to find an object in a definite finite volume (which means that an experiment is sure to get a positive result for a measurement in that volume and a negative one outside it), then another observer, moving with respect to the first, has a nonzero probability of finding it anywhere in space. Here the problem, but perhaps not the only one, is that words "definite" and "finite" contradict a fundamental property of quantum mechanics: statevectors are analytic functions of space. Thus it is impossible to find a statevector, therefore a probability, that is nonzero in a finite volume, but exactly zero outside it. This illustrates that words can be perfectly meaningful, but can say things that are wrong — for example they may assume properties that physics makes impossible.

IV.2.f The meaning of the wavefunction

The meaning of the statefunction has lead to many philosophical arguments. And it is for good reason that these arguments are philosophical. Consider, for example, the disagreement between those who believe, and who do not, that a statefunction is a complete description, so that its change that takes place on measurement is a real (?) change in the universe, rather than merely a change in our knowledge of the universe. It would be an interesting experiment that distinguishes these possibilities — real (?) change, and change of our knowledge, even if it were merely a gedanken experiment. But if there were no such experiments possible, no way of distinguishing these deeply-held positions, can there be real disagreement — beyond, of course, a disagreement in taste for the types of statements that people enjoy?

Does a wavefunction "exist"? Certainly as a function that we can write, as for example $exp(ikx)$. It exists in the sense that we can find it experimentally, calculate its value at other times and places, and use it to predict experimental results. "What actually 'exists' in nature, what is 'out there'?", does not ask anything. There is no way of answering this, no way of giving "exists" sense. It can be "answered" by a string of words that forms a grammatically and syntactically correct sentence, but which actually says nothing. It may alleviate the emotional discomfort resulting from not being able to give the question meaning, and help hide its meaninglessness, but that is all such strings of words do.

A very important question is "What is the statefunction (a better term than wavefunction, another example of a word that misleads)?" However the word "is" is undefined. The question has no answer, thus no substance. What people want is (which is defined for this sentence) some classical meaning for statefunctions, which, of course, cannot be. Thus they are quite unhappy. It is better to say that "the statefunction (actually our knowledge of it) is determined, given the experimental conditions, according to the following rules ... ", and "it gives the probability according to the following rules ... ". This avoids vacuous statements and should decrease confusion (but not unhappiness).

IV.2.f.i *How real is probability?*

Likewise the statement that a probability wave is a "real thing" and more than a tool for mathematical calculations, rests on the meaning of "real thing". But what meaning, besides an emotional, intuitive feel, does this have? And what more can it be then a tool (whose properties we can determine and influence) — and if it is more how can we tell?

If we know the statefunction of a system, and this consists of terms describing different physical situations (say spin up and spin down), is it correct to say [Everett (1973), p. 3] that an observation that finds the system in one state causes it to be in that state, and that perhaps the system had no objective existence until it is observed? Actually it is such statements that have no objective existence. Words, such as "causes", "objective", "existence" are here undefined — sequences of letters, but in this context not words. An interaction of an observer and a system can result in the system being forced into a particular state, but this is not what people who used the word "cause" in this context mean (although they themselves are unlikely to know what they are trying to say). And if we mean that the system can have an "objective existence" before the observation because it affected other objects, that is it was observed by them, the question is merely transferred to an earlier time.

That a statefunction is such a sum of up and down indicates that there is a nonzero probability of each result being obtained. It certainly does not imply that our observation causes the result. And there is no implication that the observed state existed, or did not exist, before observation, or was created by it — unless there is a way (in principle at least) of experimentally justifying such statements. There cannot be such an implication because these statements are empty.

IV.2.f.ii *What is the probability that the universe has a wavefunction?*

Then there is the wavefunction of the universe (sec. III.2.g, p. 111). But a wavefunction gives probabilities of outcomes of a measurement. What measurement are we performing on the universe? How is the wavefunction of the universe determined? How is the universe prepared in a state with a definite wavefunction? Undoubtedly those who use a term like this know how to prepare the universe in a definite state.

For the wavefunction of an object we can give a formula, and also tell how it (really the wavefunction not the object, which is actually its wavefunction) affects other objects (sec. III.3.c.v, p. 124), and its relationship to probability. That is what the wavefunction is; what more can it be? But how does the wavefunction of the universe affect other objects? Or could it be, again, that this phrase is merely a string of letters that only appears, when together, to form words?

IV.2.g Can statements be in two places at the same time?

Absurd statements like "particles such as electrons and photons are able to be in two places at the same time" appear in far more than two places at the same time (sec. V.4.a, p. 224). What does "be" mean in this context? Presumably exist. Ghosts exist in people's minds, but hardly in reality (although people who think objects can be in two places simultaneously may disagree). For something to exist physically requires that be experimentally observable, perhaps indirectly. Has there ever been an experiment in which entities like electrons and photons are found to be in two places at the same time? Has anyone even thought of such an experiment, or of a way of giving this sentence experimental meaning? This exemplifies how words are strung together to make a sentence that is grammatically correct, but which is worse than meaningless. It is total nonsense. The sentence does have meaning, but not physically. It says that there are experiments that show an object to be in different places at once. And this is absurd.

Thus some have reported the belief [Johnson (2000b)] that a bottle should be able to exist in a superposition of two (why only two?) locations, here and there (where?). And if it exists then we must be able to find it, and in both places simultaneously. This belief of course raises the question what is in the bottle that causes such ideas?

Besides particles, some believe that wavefunctions can also be in various places simultaneously. Of course this is almost true, but implications drawn from it need not be. Many think, in say an EPR type experiment (sec. V.3, p. 208), that there must be a collapse of the wavefunction because one counter must be made to know that a second one has been hit. This is based on the idea that originally the wavefunction is nonzero in both places and (the relevant part) suddenly goes to zero at one counter when the second is struck. However how do we know what the wavefunction is unless we measure it in both places simultaneously, and find it nonzero. But if that is done, the rest of the EPR experiment, the excitation of the counters (or of only one) cannot be done. Of course, we can write a function that gives a probability distribution. But we cannot say that for a single event the statefunction is, or is not, zero before being measured. Thus, contrary to conclusions people often draw, there need not, and cannot, be any communication between the counters.

A similar view is shown by the statement if a wavefunction is spread throughout a (say, one-dimensional) interval and then the object is detected at one end of the interval the other half of the wavefunction is forced to move to the detector in zero time. However if we know that the wavefunction is actually spread over the interval then we must have detected it at many different points, so the belief in the detection at only one is wrong, thus part of the wavefunction could not have reached the detector in zero time — the statement is self-contradictory. The phrase "spread throughout" is no more than a string of symbols unable to form a phrase unless we can measure the statefunction to be so spread out. In an ensemble of measurements, some give one detector finding the particle, some the other, so the ensemble is properly described by a spread-out wavefunction. But for an ensemble, the concepts of wavefunction collapse, and infinite-speed travel, do not occur.

What collapses here, as in too many other places, is not the wavefunction, but the foundation of the phrases, consequential definitions, and common sense.

IV.2.h Is inaccessible information really information?

Many seem to believe that there is a huge amount of information stored in a state of a quantum system that is inaccessible. But if it is inaccessible, how can it be information? What is inaccessible information? Is it nothing more than a contradiction in terms, an oxymoron? How is information defined? And how is it defined so that the phrase is not self-contradictory? How is the amount determined if it is inaccessible? Perhaps it is only a very small amount of information (measured and defined in the same way as infinite). And if it can't be determined, how can we tell that it is huge? And how is it stored? Is "stored" in this phrase actually a word? How do we know that it is really there?

IV.2.i How fundamental is complementarity?

Another quite popular question is "Is complementarity more fundamental then the uncertainty principle?". Ignoring the lack of definition of fundamental, and that the uncertainty principle is a direct consequence of the (necessary) properties of quantum mechanical statefunctions, this presents a remarkable view of quantum mechanics. Complementarity refers to particle and wave aspects, neither of which is relevant to quantum mechanics, but are purely classical concepts (at best). Particle and wave properties are for quantum systems worthless terms. Thus this fascination with complementarity bases the nature and properties of quantum mechanics on the irrelevance of classical concepts for it, so on vacuous words. That is like basing electromagnetic theory on the irrelevance for it of genes and earthquakes which is quite unlikely to lead to understanding of electromagnetic theory (or genes or earthquakes), and use of complementarity in quantum mechanics, as expected, has lead to many (nonexistent) dilemmas, but certainly not to understanding.

But at least the words "genes" and "earthquakes" do have meaning so basing electromagnetic theory on them is more reasonable than basing quantum mechanics on waves, particles and complementarity.

For example it has been said [Kim, Yu, Kulik, Shih and Scully (2000)] that complementarity is perhaps the most basic principle of quantum mechanics, and distinguishes the world of quantum phenomena from classical physics. Thus quantum mechanics is regarded as being based on the incompatibility of two classical concepts, waves and particles, or perhaps on two inapplicable properties both holding simultaneously — a self-contradictory statement. How can a theory be based on the inapplicability to it of concepts of another theory (and an inconsistent one)?

To regard complementarity as fundamental to quantum mechanics is to assume that a consistent framework, quantum mechanics, is grounded on its relationship to an inconsistent one, classical physics — a strange idea indeed, and one very uncomplimentary to those who say such things.

IV.2.j Time

Something that many physicists are working to determine is the nature of time. Well, what is the nature of time? How did time begin? (Let us hope that no one asks when did time begin?) Is there supposed to be a mechanism for creating time that would answer the question raised by "how"? What kind of answers can these have? And if they can have none, can they really be questions? Or are they merely rows of words that together are timeless in their vacuity?

IV.2.j.i *Can time have a history?*

The American Physical Society seems to believe in another property of time, as given in one of its advertisements [APS News (1999)]. It claims that today physicists have begun to unravel the history of time and the universe. The word "and" here is ambiguous (as happens quite often). Is this history the intersection of the history of time and the history of the universe, or perhaps the history of the intersection of time and the universe (?), or maybe something even more interesting. Or it might just be logorrhea (the excessive use of words), a too common disease, not only among those who study quantum mechanics, or physicists, but among the too many people who know lots of words, but little of their meaning.

However the history of time is a fascinating concept so worth further study. History refers to a chronological record of events, a story of them. Chronological means arranged in order of time of occurrence. Thus we can conclude that the history of time refers to a time-ordering of time. Or perhaps time is a set of time-ordered events (a nice circular definition), which leaves out only the definition of the events of which time is. Presumably those physicists who are unraveling the history of time (while getting their definitions and sentences quite raveled) know what these events are, and how to time-order the events making up time. And this must clearly be true since it is the official position of the main organization of physicists.

IV.2.j.ii *Is the universe bored?*

Time is fascinating in other ways [Wilford (1999)], as shown by the report about the rate at which space and time are expanding. Of course, when we are bored, time expands, when we are doing something pleasurable, or in a hurry, it contracts. It is an intriguing question whether this is also true for stars and galaxies, and the universe as a whole. Does the universe ever get bored? The distance between galaxies, and the distance to the edge of the observable universe, change with time. However it is nice to learn that time also changes with time (or could it be with space?).

IV.2.j.iii *Going backwards in time, and backwards in thinking*

Another interesting peculiarity of time is shown by the belief that antiparticles are particles that go backwards in time. This undoubtedly means (for what else could it mean?) that it is possible to measure the position of a particle now, and then measure what the position was at some past time — that is what this statement says (unless it is a denial of causality). Of course a simpler way of regarding antiparticles is that

$$exp[-iEt] = exp[i(-E)t] = exp[iE(-t)], \qquad \text{(IV.2.j.iii-1)}$$

although putting a minus sign in a particular position is much less exciting than going back in time and measuring the position of a particle then, after doing so at the present time. It is much less exciting, but it makes much more sense (which is why it is so unexciting).

IV.2.j.iv *Pointless properties of points*

Certainly time is not the only thing (?) that has fascinating characteristics as shown by the idea that space-time points can have properties [Teller (1997), p. 95]. It is indeed enchanting to consider what properties a point has. If a point has properties, does that imply that different points can have different properties? And it must obviously be possible to experimentally determine the properties of a point. Of course, it is true (fortunately) that there are functions of space and time whose values thus are different at different points. And these functions do have properties but that does not mean that the points themselves are, or can be, distinguishable, or have any attributes beyond location. But those who are more aware than most of us about the properties of points should enlighten the ignorant people that we are by telling us what these properties are, and how to discern them.

IV.2.j.v *Does time flow?*

Time often seems to be on people's minds as the statement, referring to what is called Newton's absolute time, that time of itself, and from its own nature, flows equally without relation to anything external. What does "time of itself" mean? How can itself modify time? And what is the nature of time? How can time flow? To flow means to move, that is there is a change of something (like position) with respect to something else (like time). Does time change with respect to something else? What? Does time change with respect to time? The word "equally" refers to the comparison to two (or more) things. What different things is the flow of time (whatever this means) being measured with respect to (time?)? This is another example of a sentence that gives the impression of saying something, but whose words do not go together thus is a mere sequence of words, expressing nothing.

While it is not clear that time can flow, it is unfortunately all too clear that words can flow.

IV.2.k Uncertainty, long may it wave

"The newborn universe, ..., would have been crisscrossed by waves of quantum uncertainty" [Glanz (1998a), p. 1248]. This does make one curious as to how uncertainty waves? Does it wave in time or in space, or both? How can we tell? What is the experimental difference between wavy and nonwavy uncertainty? Might quantum uncertainty go up and down? Does classical uncertainty also wave? If not, does it stay flat?

Why should there be a difference? If classical uncertainty does wave, why qualify the uncertainty with the word quantum? If uncertainty crisscrosses does it interfere with itself? If there were waves it would imply that quantum uncertainty has both positive and negative values. So what does a negative value of quantum uncertainty mean? Did uncertainty wave only in the newborn universe, or might it still be waving? And if not, when did it stop? If it is not waving now, is it flat?

IV.2.1 How excited are electrons?

"Electrons, for example, can be regarded as excitations of an electron field" [Wilczek (1999a), p. 13], so two electrons are identical because both are excitations of the same electron field [Kleppner and Jackiw (2000); Wilczek (1999b)]. What is an "electron field", how is it distinguished from an "electron"? And why should it (an electron?) be regarded as a "small-amplitude excitation"? "Why do electrons anywhere in the universe have precisely the same properties"? "Because they are all surface manifestations of ... the electron field". What are "surface manifestations", and what nonsurface manifestations of the "electron field" are there? What is an "electron field" that is not excited like? Also water, ropes, and many other things can be excited but their excitations need not be identical. It would be very unusual to find two waves in the ocean that were exactly identical. Why should electrons then be all the same? And if electrons had different properties, would we regard them as the same (type of) object?

It has also been claimed [von Baeyer (2000)] that particles are interpreted as ripples in an infinite, invisible continuum called a field. However if the field is invisible how do we know that it is infinite? And how do we know that it can ripple? These ripples are presumably up-and-down motions. What moves up and down (or is it rippling sideways)? This "explains" why the "profoundly quantum mechanical notion" of the identity of two objects had to await the development of quantum field theory. How does it explain (anything)?

Do these statements say anything, or do they merely camouflage our inability to give answers, or perhaps meaning, to questions (unless they merely camouflage empty thoughts)? It is interesting to wonder whether the statement that electrons are excitations of an electron field is a tautology. And if it isn't then how are the two parts of the sentence different? It is difficult to admit that we do not know something. Thus we often give very clear definitions that are completely circular, so do not say anything, but are very helpful in hiding our ignorance.

IV.2.l.i *Why are all electrons the same?*

However there is a real question of why electrons in different parts of space are identical. It is both more trivial and more profound than these "answers" imply. What is an electron, what properties define it? First is spin. The electron has spin 0.5. Why does an electron elsewhere not have spin of 0.5001? The trivial answer we know: the rotation group allows only spin 0.5, or other half-integer values and no others. Then there is charge. We know that allowed values of charge are discrete, multiples of a minimum one. Why we do not know. But with this rule we can tell that an electron elsewhere does not have a charge of 0.9997. The remaining property is mass. We know that only certain mass values are allowed, but have no idea of the reasons why, or why these. And mass, spin and charge are correlated; given the mass of a particle its spin and charge are unique — but the reasons are completely unknown.

However given the discreteness and correlation of these values, all electrons must be the same.

IV.2.l.ii *Babbling*

This is a particularly bad example of (lack of) reasoning. A question — "Why are all electrons the same?" — whose answer is completely obvious with the slightest thought, is instead answered with verbiage that is no more than useless strings of words. It shows not merely sloppiness, but laziness, in (lack of) thinking (in several ways), not only (even mostly) about physics. Thinking is replaced by babbling. And this is all too common, as can be seen from the number of people who say such things, and they are not merely the ones indicated here.

IV.2.l.iii *Photons also are overexcited*

Of course electrons are not the only objects that are excited. An example [Yurke and Stoler (1995)] is the (unexciting) statement that photons are excitations of modes of the electromagnetic field. An atom can be in its ground state, or in an excited state. But if an electromagnetic field is not excited, does it exist?

This raises the question what are photons? Actually they are basis states of the electromagnetic potential (not field [Mirman (1995c), sec. 3.3, p. 37]). The difference between a photon and the potential is that "photons" imply discrete numbers of basis states, while the potential does not — for it the number can be so large that discreteness, and the exact number, are irrelevant. In Dirac's equation, for example, the potential has a somewhat different meaning, that of an operator which acting on a state of photons (perhaps the vacuum) changes the number.

IV.2.1.iv *Electron fields illustrate the harm of meaningless strings of words*

These illustrate why meaningless questions and answers are harmful, and dangers of language without content. They fool us into believing that we have asked actual questions, answered questions, that we understand something about nature. And they distract us from trying to actually ask and answer meaningful ones and hide important physical questions, something that could have been done if only a list of properties defining an electron were made before a sequence of words without content was thought of. How is it possible to try to answer the question of why two electrons are identical if we do not first define electron?

Questions of why there only discrete mass ratios, and why these are correlated with spin and charge are perhaps the most fundamental of physics. Meaningless statements about meaningless ripples hide them. Even those who fool themselves into believing that they say something cannot even suggest that they answer such questions as why is there no particle with mass 3.5248 times the electron mass?

If we could answer this question, find reasons why masses are discrete, why there is correlation, then we would know why electrons everywhere must have the same properties, and why these are the properties that they can have.

Of course there is an important assumption here, which experimentally is true, fortunately: the laws of physics are the same everywhere.

Electrons are not like faces; if the face of a person here were exactly the same as that of a person there, it would be surprising and demand explanation. But only the slightest bit of thought makes it clear that the properties of an electron are few in number, discrete, and must be continuous over space, thus must be the same everywhere. That the properties of all electrons are the same does not require explanation by meaningless verbiage, which smothers, so discourages thought, which as statements like these make clear, is already woefully lacking.

IV.2.1.v *Are fields permanent?*

There is a belief that only fields, not the individual objects that they create and destroy (how do they differ?), are permanent [Wilczek (1999b)]. Aside from the possibility that a photon can scatter and become an electron-positron pair (does the electromagnetic field remain after the photon is converted?) this assumes that fields create electrons. Thus we must be able to measure an electron field, show that it is present, even though there are no electrons present. Otherwise if a field is not present, how can it possibly create electrons? And if we cannot show that it is present, how can we assert that it is? Besides if we cannot show that it always is, how do we know that it is permanent? But certainly it would be an interesting experiment if someone could detect an electron field, and simultaneously show that there are no electrons. How does this electron field, whatever it is, actually create electrons? What creates the field? Presumably those who hold this belief in immaculate creation know of experimental arrangements that (can, at least, in principle) demonstrate its correctness.

IV.2.m Self-energy

It is a popular belief that because a system with one charged object near another has a different energy than if they were far apart — one exerts a force on the other changing its energy — then a single object must also have some energy because it is in its own field. And the idea of self-energy does seem reasonable — at first. An atom consisting of charged particles does have a different energy then the total energy of the particles by themselves.

Does an electron have a self-energy? The energy of an atom, or of one particle in the field of another, can be varied by moving particles. Energy can be converted from one form to another. Is this possible for an electron? Can we separate the parts (?) of an electron? If not, how can we measure the self-energy of one? And if we cannot measure it, and cannot change it, then how do we know that it exists? And if we do not know that it exists, then what is it?

The mass of an electron, say, is a parameter in Dirac's equation. Is there any way of separating this number into components that we can assign to self-energy and to whatever else mass consists of?

Or could it be that self-energy should join the long list of terms that seems to mean something, that we can almost visualize, but that fade into ghostly apparitions, illusions, upon examination?

IV.2.m.i *Are mass differences due to the electromagnetic interaction?*

There are those who believe that differences in masses between objects (like the proton and neutron) are due to their differing electromagnetic interactions. Undoubtedly such people are able to turn off the interaction and check that the masses become the same. There is a difference in energy (so mass) between up and down states of a spinning object in a magnetic field. This can be verified (for example) by turning off the field.

However what does it mean to say that the mass difference between a proton and neutron is due to their electromagnetic interactions? Not only can this not be examined experimentally, but how do we know that they are related at all? One of the most important techniques of modern science is the method of false analogy (sec. III.4.c.ii, p. 135). It provides the very foundation for many of the statements about science, unfortunately. But here not only is the analogy false, it is silly.

IV.2.m.ii *What does experiment show?*

Are mass differences "due to" electromagnetic interactions? What does "due to" mean in this respect? Are they related? What does "related" mean?

But there is one experimental procedure that can be used to get a hint about these questions. There are many sets of particles that have the same differences in charge. Do these all have the same differences in mass (or even in ratios of mass differences to mass, or other simple patterns)? Also the Σ hyperon, for example, has states with opposite but equal charges and different masses.

Perhaps a study of lists of elementary particles could prove enlightening [Mirman (2001), appendix].

If mass difference is not related to interactions, perhaps the analogy is related to the lack of interaction between thought and language.

IV.2.n Uncertainties with analogies

The danger of overreliance on analogies is shown by the notion that the electromagnetic field can be decomposed into an infinite number of harmonic oscillators [Wilczek (1999b)]. Then it is argued that because the ground state of a harmonic oscillator (a nonzero mass in a particular nonzero potential) has nonzero energy, due to the uncertainty principle, so should the electromagnetic field, producing an infinite gravitational energy. But electromagnetic fields (or potentials) cannot be decomposed into harmonic oscillators, certainly not like masses on springs. They can be Fourier-analyzed, and each Fourier component has similarities to harmonic oscillator statefunctions. But this does not mean that each Fourier component has energy (sec. V.7.d.i, p. 257), or that there is a ground state. It is important to remember, because apparently many people do not, that mathematical expressions, like Fourier components, cannot have energy. Nor do they have ground states.

The proper way to discuss interactions for electromagnetic potentials and gravitational fields (connections) is to put interaction terms in the equations governing them [Mirman (1995c), sec. 7.2.5, p. 129]. Not only is there then no reason to believe in either infinite energy, or infinite gravitational attraction, but the usefulness of Fourier analysis of these nonlinear equations becomes quite doubtful, so therefore the analogy to harmonic oscillators becomes even more so.

The analogy of the electromagnetic field to an infinite set of harmonic oscillators holds for the free field (that is it can be Fourier-analyzed), so we can get expectation values of the momenta (including energy) which is, for Fourier component k_μ,

$$< k|p_\mu|k >=< exp(ikx)|i\frac{d}{dx_\mu}|exp(-ikx) >= k_\mu, \qquad \text{(IV.2.n-1)}$$

taking into account, in the well-known way the nonnormalizability of the statefunction. We can also consider more than one component, a wavepacket. But this, unless a δ function which is really unphysical, has amplitude that varies with k, and becomes (almost) zero for very large momentum. Thus there is no indications of unphysical infinite energy or infinite values of other momentum components. The analogy with a set of harmonic oscillators really just confuses.

We see again that a major tool of modern science is the method of false analogy. It is clear that it is fundamental to much of the discussions, not only, of quantum mechanics.

IV.2.o Is there any value in having values?

There are other problems (with language, not quantum mechanics). The "orthodox view" [Whitaker (1998), p. 31] is that individual spins do not have values of their z components (until measured). That might be orthodox, but what does "have values" mean, if they are not measured? And if they are, do they then not have values?

Similarly there is the statement that the uncertainty principle means that if a particle has a precise value of position, then its momentum cannot have a value (which means that it is impossible to measure the momentum, since that would give a value, or could it mean that there are measurements that do not give values, presumably giving something else?). This statement is supposed to be different from a "particle may have (?) such values, but we do not or cannot know them". This may be different, but not much different, since "may have" is without content if we cannot know. Such statements seem to lead to the view that a measurement gives a precise value, even if a precise value did not exist beforehand. Changing "have" to "exist" does not help; the statement is still mindless. How can we find out if a precise value existed before measurement? And if precision does (or did?) not exist will an experiment give several values?

"This suggests that there might be no real world in the absence of an observer". This statement can be left as an exercise for the reader (who hopefully is not in an unreal part of the world).

IV.2.p The quantum of action

A belief [Bohr (1935)] exists (a word that unfortunately has meaning in this sentence) that the finite interaction between object and measuring agencies, as well as many, if not all, of the properties of quantum mechanics, come from the existence of the (finite) quantum of action. Is there such a thing? Can the quantum of action be zero, or infinite? We can make reasonable guesses about what statements about the quantum of action are trying to say. Thus we can guess reasons that there is confusion.

Essentially action has (at least in the situations considered) a minimum value because

$$[x,p] = ih, \quad \text{or better} \quad [x,p] = i, \tag{IV.2.p-1}$$

[Mirman (1995b), eq. 3.4.2-2, p. 53]. Can it be any other way? The second form of the commutation relation shows that it would be very difficult to have it any other way — momentum operators (exponentiated) are displacement operators. From the realization that we (are forced to?) use it is clear that these commutation relations must hold [Mirman (1995b), sec. 3.4.1, p. 52]. But more basically, they follow from the definition of displacement, and the group structure.

There are additional reasons for the "quantum of action", that is for discreteness. These are boundary or regularity conditions. Thus the harmonic oscillator has discrete energy states, it has a "quantum of action", because only these give statefunctions that are well-behaved everywhere. Motion in two-dimensions has discrete angular momentum because the statefunction has to be single-valued. And these reasons lead to discreteness in many, many situations, as is (or should be) well-known.

To say that the quantum of action is finite is correct, but the statement carries an implication that it need not be, and that is not correct. The reason is not physical, but mathematical — coming from the mathematics that we (are forced to?) use to describe physics.

Notice how the word quantum misleads, certainly subconsciously.

IV.2.q Nonlocality

An idea that has fascinated many physicists is that in some way quantum mechanics is nonlocal. Reasons for the confusion, much of which is linguistic, have to be treated in depth (sec. V.3, p. 208) so here we just add a few remarks about language. Nonlocality is often based on a view that quantum mechanics exhibits correlations between spacelike-separated measurements that are inconsistent with any common cause explanation. Of course since "common cause" is undefined quantum mechanics can be taken as inconsistent with anything we wish (and too many wish to take it as inconsistent with something). It is only necessary to define "common cause" in a way that suits our purpose, no matter how absurd, and quantum mechanics becomes inconsistent with it. This illustrates not merely how language can be used carelessly to confuse, but how vagueness can allow any interpretation that someone finds profitable (for example allowing scientific (?) papers that are not only vacuous, but that allow scientists (?) to flaunt their superiority to everyone else because of their deep knowledge of the — nonexistent — weirdness of nature).

IV.2.r Can electrons scatter other electrons?

It is of course well-understood — and easily visualized — that an electron can be scattered by another, and how that scattering takes place — until words and sentences are analyzed. We tend to picture electrons (and other objects) as little balls, and fully understand how these little balls can scatter from each other, just like billiard balls. And if these balls do not come into contact we have clear pictures of charged — macroscopic — objects scattering from one another. Thus language, and pictures, developed and clear when used in a different domain, classical physics, mislead.

When we say that an electron is scattered we really mean that its statefunction is changed in a certain way, and while pictures of little balls may be helpful at times, they are not correct descriptions of actual physical events. What causes electron statefunctions to change? These are governed by Dirac's equation, and the change results from the interaction term in it. But this nonlinear term contains, not another electron, but the statefunction of the electromagnetic potential. And the equation governing that has an interaction term that includes, obviously not the other electron (it would be a strange mathematical expression indeed that includes a little ball), but the statefunction of the first.

Thus the statement that an electron is scattered by another is meaningful and correct if it is interpreted as a shorthand way of saying that the statefunction of one electron determines (through perhaps a series of equations) that of another. Otherwise it is mere words — "electron" does not refer to anything. What is an electron? Of course, its statefunction (sec. III.3, p. 117). What else could it be?

Language that seems to have perfectly reasonable content may actually be complete nonsense if words do not refer to the proper, actual, physical objects. And these are often very different from the way that we picture them.

IV.2.s The language of nature

Since this discussion is about language it is reasonable to ask what the mother tongue of nature is? So we can ask whether quantum mechanics is the language of nature, or whether it is purely something invented by human beings? Such a question, in some form or other, is regarded as raising a fundamental problem in understanding of the world. As in so many other deep philosophical questions, it provokes controversy, and has no simple answer (or even a complicated one) for a trivial reason: it is totally empty, a collection of words that seem to say something but actually say nothing. What does "language of nature" mean? Can a definition be given that is not merely an equally meaningless collection of words? Before anyone asks such deep questions, they should say what the collection of words expresses. Those who want to use the term "language of nature" should specify a way of determining whether, or not, mathematics or quantum mechanics is the "language of nature".

Of course this can be taken as purely metaphorical and there are no objections to this poetic sense. But people who raise questions like this usually do not intend them to be metaphorical.

IV.2.t Does quantum teleportation teleport denotation or connotation?

Quantum teleportation seems to have become quite popular, in part at least, because it appears to give credence to science fiction (as do too many discussions about quantum mechanics). This indicates a general problem (not only in physics): words and phrases both denote and connote, and for the latter they often bestow implications which are wrong, but are taken correct because the words are correct (their denotations are correct). Quantum teleportation refers to the transmission of information, not of mass (energy), except for the very small amount needed to send the information. Yet it is taken to mean the (instantaneous — which is not possible even with information) transmission, not only of mass, but of mass with the arrangement of atoms unchanged — thus not only of mass, but massive (extraordinarily massive) amounts of information.

So there are statements such as that quantum teleportation can in theory (this stated without the qualifier that it applies only to nonsensical theories) move people and things from one place to another without taking them through intervening points, and can do so at the speed of light [Rennie (2000)]. That this is obvious nonsense, and at the very, very least violates relativity (massive objects cannot move at the speed of light), conservation of mass and of energy (which disappear and then later appear someplace else) and the uncertainty principle, does not deter people from believing in an absurd connotation because they misunderstand the meaning of a phrase, its denotation — so reading into it meaning that it does not have, and cannot possibly have.

Of course it is well-known, and well-understood, that actual objects cannot be teleported [Zeilinger (2000)]. Yet there are those who still discuss it as if it were possible. This is not only misleading, and highly irresponsible, but amounts to lying to readers. However this emphasizes how seductive names are. The name overwhelms not only knowledge of physics and the understanding of what physics allows, but knowledge of the actual meaning of the term, what its explicit definition is. Connotation overwhelms denotation. Those who introduced the term to refer to something else acted irresponsibly by picking a name that they should have known would be uncontrollably seductive.

IV.2.u How many universes are there?

It is difficult to understand the universe. But some are not satisfied with merely one extremely difficult problem; they want many universes: "Is there a multitude of universes?" [Miller (1998)]. This contains two undefined terms: "Is there a multitude" (which can be broken down further) and "universes". The question seems perfectly reasonable, until analyzed. Of course, we know what a universe is, do we not? However this asks whether there is (a meaningless phrase in this context) or might be (equally meaningless) other (?) universes. What tells us that the other universes (?) are actually other? Or could this be an oxymoron (sec. IV.2.x, p. 182)?

It would be interesting to find if anyone could give an intelligible, relevant definition of "universe" (and one which can be modified by "other") — a set of words that says something, and is not merely an equally meaningless set, but perhaps with its emptiness less evident.

It might seem that the concept of many universes is (to put it mildly) either an indication of a serious inability to use language, or perhaps something even far more serious. Yet their existence seems to now be a scientific fact [Trimble (2000)]. Many are undoubtedly quite curious about observations that have demonstrated this.

IV.2.u.i *The anthropic cosmological principle*

This is related to the concept of the anthropic cosmological principle [Barrow and Tipler (1988)], that there are many universes, and that physical constants have (about) the values that they do because these are the only ones allowing intelligent beings able to notice the universe, and find the values of the constants. Of course the second part of this statement is true, trivially true. But what difference is there between this statement and the one that we do not know why constants have their values, we will never know, so let us just give up? Is there a real, not merely emotional, difference?

Does this statement of the anthropic cosmological principle differ from "physical constants have values that they do because God wants them to have these values"?

IV.2.u.ii *Do these questions exist?*

As for the question of whether other universes exist, the answer is simple. What does "exist" mean in this context? How do we know, how can we find out if these other universes are not part of our universe? It would be a fascinating experiment that would allow us to study universes that are not part of our universe. And if we could study them, why would we not regard them as part of ours?

Is space a "fundamental entity"? What does this term say; what is the difference between space being fundamental and it not being fundamental? And what kind of entity is it, fundamental or not?

Such questions are of course senseless, sentences of words that have meaning, but none when combined like this. Because they otherwise have meaning, they distract from the absurdity of putting them together in this particular way. Indeed this is true of some of the most important questions of nature. They are not questions about nature, but simply manifestations of misunderstanding and misuse of language. Many beliefs about quantum mechanics (as discussed throughout) are due to no more than such misuse.

Likewise there is the question does space "exist"? Spacetime is certainly needed in differential equations, and it also makes sense, and is necessary, to talk about the distance between two events. What more does anyone want of spacetime? This has generated a lot of philosophical arguments. The reason is of course "exist" has no real existence when used this way. Thus it allows much discussion — since these discussions consist of nothing more than worthless verbiage there is no way to reach any conclusion from them, and they go on forever.

IV.2.v Are physicists conscious of the meaning of consciousness?

Although consciousness is not properly part of physics, it often appears in discussions about quantum mechanics (perhaps unconsciously). A brief mention of it is therefore necessary, although it is really irrelevant to physics (but not everyone seems to realize this).

Some even believe that consciousness is necessary for the "reduction of the wavefunction". Thus human beings are needed for there to be a universe (presumably these people do not believe that an electron is conscious — although that is a better observer than a person, certainly than a physicist [Mirman (1995b), sec. 5.1.3, p. 88]). There are religious people who believe that the universe was created for the (sole?) purpose of having humans around. There are physicists however who believe that they created the universe, although for what purpose is not clear.

There is another problem with this view: it bases a physical theory on a term — consciousness — that is not defined, and that cannot be defined in the way that those who hold the view would like (which unfortunately is not unusual). It might be possible to define it in terms of neural activity, such as electric circuits and chemical reactions. That would mean that — any — electric circuits and chemical reactions would give "reduction of the wavefunction", which is in a sense quite true. Or do these people believe that only a special set of these do? And if so, what — physically — causes this class to be special?

Consider an external observer able to see details of our brain's operation (of course impossible, but for practical reasons which is not here relevant). It (he or she should not be used for observers [Mirman (1995b), sec. 5.1.3, p. 88]) would see light impinging upon our eyes, causing signals through nerves resulting in neural circuits becoming activated. This is a purely physical process, completely observable and understandable (only in principle of course). Done this way (as it should), there are no questions of consciousness, nor incentives to think of phrases like the "unfathomable consciousness perceptions" — mere sets of words with only emotional) content.

What is consciousness? This is often regarded as a hard question, more difficult than ones asking how neurons work. However this question is not hard, it is meaningless. A question has meaning only if it is possible to specify a type of statement that would answer it (sec. IV.1.c.i, p. 147). In principle an answer can be given (although certainly not details, which will never be completely known). It is a set of neuronal circuits, activation of neurons, perhaps including particular specific signals, coordination of such activity and so on. When we perceive something our brains react in a certain way, which can, in principle, be experimentally defined, observed and tested.

Considered this way the question does ask something. But those who ask it do not like this answer, it does not tell them what they wish to know. What they are really saying is that they find it emotionally unsatisfying. But there is no way of answering the question, and no way of knowing how to look for an answer, since the question is not about physics (or biology) but rather how to satisfy them emotionally. Different people need different types of satisfaction. And it is highly unlikely that any answer would ever satisfy everyone, or even anyone, emotionally.

Consciousness (of the sort that people seem to want to know about) is not something that physics is unable to come to grips with, rather it is not something. There is nothing to come to grips with.

Thus it is with blueness. Is everyone's sensation of blueness the same? Is "your sensation and my sensation" the same (sec. IV.1.c.ii, p. 147)? A person's sensation is a set of nerve impulses to certain cells of the visual cortex (some of which are known) and their excitation of other cells. What more do people want? What more can be given? What they want to know is whether others have the same feeling when seeing a blue object that they do. Activities of neurons can be compared, but not subjective feelings.

These questions are examples of sentences that give the impression of asking, but do not. They emphasize that language is dangerous. Concepts and words have no content when used in this fashion.

IV.2.w Physical reality; does it really exist?

A fundamental concept of physics, perhaps of life, certainly of the discussions of the meaning of quantum mechanics, is that of physical reality. Is there such a thing? What is it? And how do we know, how can we know?

Another phrase (or type of reality?) often used is "objective reality", although it is not clear how it differs from nonobjective reality. This belief (or terminology) then leads to a fundamental requirement on physical theories, which must take into account objective reality which is independent of any theory and the physical concepts with which the theory operates (but presumably we can still determine, in some way, the objective reality independently of theory). Concepts of the theory are supposed to correspond to objective reality, and which we use to picture this reality to ourselves.

This, of course, assumes that there is a distinction (besides mere words) between objective reality and concepts that correspond to it. But then there must be a way, an objective, experimental way, of making the distinction — this must have an objective reality. If we cannot tell the difference between reality and concepts that correspond to it, then what is the difference? Or might these differences just provide sentences that indeed seem to have a great deal of objective reality and which place strong restrictions on physical theories, but which turn out upon examination to be totally unreal, without substance, merely sentences consisting of words that in these combinations say nothing. Perhaps they have a great deal of significance in telling about the misuse of language.

Another statement of this problem [Mermin (2000)] is that the heart of the puzzlement induced by quantum mechanics lies a tension between reality and knowledge, between facts (whatever these are) and information. Do quantum states, or their wavefunctions, correspond to something in the real world (whatever this might be — especially experimentally)? What happens if they do not? It is suggested that quantum computation might be a way of reconsidering the knowledge-vs-reality muddle we have been been thrashing about in. Indeed undefined terms do lead to quite a muddle of puzzlement. Perhaps there would be less puzzlement if thrashing were replaced by thinking.

While sentences about reality are in reality quite often (always?) quite unreal, they are usually statements that are positive, asserting something about "reality": experiment shows that reality . . ., or some such thing. However in reality negative statements can be meaningless also. Thus a quite reasonable statement [Fuchs and Peres (2000)], at least until analyzed, is something of the form that quantum theory does not describe physical reality. This is based on a willingness to accept the possibility that we might never identify a reality independent of our experimental activity — implying that we might however be able to actually identify it. But how?

IV.2.w.i *Must experiments be published?*

Of course an experiment is something that physicists do in a laboratory. But if we look at, feel, touch, something, and gain information about it, isn't that also an experiment? Does it have to be publishable to be an experiment? If an insect looks at an object and decides whether it is food or a predator, isn't that an experiment? Is there really any fundamental difference in reality between an experiment that a physicist does in a laboratory and an experiment that a grasshopper does in the grass?

Aren't experiments going on all the time [Bell (1990)], whether humans do them or not? Are they not going on all over the universe? What else is life besides experiment? What else is physics?

If we have to accept the possibility that we can never identify a reality independent of our experimental activity this implies that there is a possibility that we can identify such an independent reality. But how else can we identify it without experiment? And if we cannot identify it can there be a reality independent of what we can identify? If quantum theory does not describe physical reality, what does? But if it cannot be described, then how does it exist, how can it have significance, what can it possibly be, how can it be?

And those who disagree, and believe that the concepts and the reality are actually different, should state — in intelligible sentences — what these distinctions are, and how we can tell.

IV.2.w.ii *Coherence, consistency and reality*

However we can say actual things about reality, such as that if an observer is present when, say, a detector clicks, it will acknowledge the objective occurrence of the click [Fuchs and Peres (2000)]. But what does the observer really acknowledge? Essentially that if it is in similar circumstances at other times or places, it obtains a similar perception. And if other observers are in similar circumstances their descriptions of their perceptions match the stated descriptions of each other (although the word "match" might be treacherous here, but it is still meaningful enough to be useful — the key requirement). It is coherence, consistency and consensus that define reality (sec. I.4.c.ii, p. 18).

What else can?

IV.2.w.iii *Local realism*

Another term often used in this connection is "local realism". It would be interesting to know what the users of this term regard as its antonym: "local fantasy", "faraway realism", "distant realism", "distant fantasy"? If a term has no antonyms does that not raise questions about whether it is really defined, says anything, or is even realistic? Another version is "local realistic description of nature". How do those who use this interpret "realistic" and "local"? Could there be "local fantastic descriptions of nature" (something that many discussions seem to give)?

Undoubtedly most people would agree with the statement that realism is a philosophical view according to which external reality is assumed to exist and have definite properties, whether or not they are observed by someone [Clauser and Shimony (1978)], unless of course they have to give a definition of "exist" which is necessary for the statement to have external reality, that is say something (although it may have internal, that is emotional, reality, especially if the words have no definitions). If anyone can do that, they can then define "reality" and "external" in sentences that are more than empty verbiage.

Actually then it is impossible to either agree or disagree with this view (although far too many people do both, many perhaps at the same time) because there is no such view.

A related statement [Yurke and Stoler (1997)] is that local realism involves a factual reality that exists independent of whether or not it is or can be observed. This suggests that there are other types of reality besides a factual one. Of course it would be interesting to know what these are and how they differ. This however hints at something else: a subconscious feeling that "reality" lacks content, and an attempt to give it some by modifying it with an adjective that, in this context, also lacks content. But the sum of 0 plus 0 is still 0.

It would also be nice to know how reality (of any type) can "exist" if it cannot be observed. How can realities that can and that cannot be observed possibly differ? And suppose that reality did not exist, what would be the consequences, what would be experimental differences from a reality that did exist?

Perhaps this existence is really being explained by language that itself does not exist — strings of letters that are not more than strings of letters that are not part of language.

IV.2.w.iv *Are there elements of reality?*

One of the problems raised by EPR experiments (sec. V.3, p. 208) is whether objects have, before measurement, an element of reality. This is a real problem, because obviously "element of reality" is not defined, and likely not definable (sec. IV.4.a, p. 187). It is therefore impossible to say whether objects do have such. That is one reason that there is so much controversy about EPR experiments.

An example [Einstein, Podolsky and Rosen (1935)] of the views about various forms of reality, and one that has greatly influenced discussions of quantum mechanics, uses definitions of the form (roughly): if we can predict with certainty results of a measurement of an observable, without disturbing the system, then there is an element of reality (to a physical quantity) which is, say, the result of the measurement.

This circular statement can be shortened to "if we can predict with certainty results of a measurement of an observable, without disturbing the system, then there are predictable results of the measurement".

The problem with this is not what it says, but what it suggests. Of course there can be no objection to the definition of the phrases "reality to a physical quantity" or "element of reality" — provided they are taken as only a summary of the sentences that define them (if the sentences actually do). If terms were merely such abbreviations they would be meaningful in that sense.

But this indicates a subtle, and unfortunately far too common, problem. Words have denotations and connotations (sec. IV.2.t, p. 173). But connotations are often not noticed, but highly influential in the way that we interpret words, phrases and sentences — thus can be grossly misleading, unnoticeably suggestive, perhaps both to the writer and reader, of ideas that are wrong, even though their denotations may be completely correct. Words like reality and physical are fraught with meaning, with significance, so can carry along suggestions that lead to much confusion.

We have some vague idea of reality and these seem to imply that such predictions show that this "reality" really does exist for such situations. That reality however has no content, it has no reality. It is merely a vague, intuitive feeling that does not correspond to anything in the external world. Thus the phrases do not tell us anything (especially about nature), despite our impressions of them, but are just abbreviations, and misleading because we do not think of them that way.

Also we must be careful that the words in a definition are meaningful, in their context, and unambiguous [Bohr (1935)]. Using a phrase as a substitute for longer statements (that is defining the phrase) would seem merely a way of saying something more compactly — but it could also hide problems with statements that might be more noticeable if it were explicit.

IV.2.w.v *Is the description given by a statefunction complete?*

Another word that appears in this context is "complete", as in the belief that the description of reality given by the wavefunction is not complete (sec. III.1.a, p. 105). But that means that if we take the — fully complete, correct — wavefunction of an entire system, we can in some ways predict experimental results that cannot be found from it. Certainly there is no reason to believe that, nor any experimental indication that it is true, or could be true.

Thus the question "whether a statefunction provides a complete description of reality (of a system?)?" is an example of a question that is so deep that many people get swallowed up by it. It consists of a set of words most of which say nothing when appearing in these phrases. What is reality? How is it described? How does it differ from what the statefunction of the system provides? What is the difference between a complete description of it and an incomplete one? If the statefunction does not provide a complete description (of whatever it is capable of describing), how can we distinguish between the information it provides and that which it does not? How can we say that there is further information if it can never be obtained nor is it possible for anyone to even specify what it is or what it might tell about a system (or is it about "reality"?)? It might seem that the word "provide" at least has content. But the question asks if the statefunction can provide something that is not specified. In that case how is this unspecified something provided?

These illustrate how easy it is to combine words that have content in other contexts but which do not in the combinations in which they appear, so form linear sets of letters with spaces that confuse because subsets of those on the line can be, elsewhere, more than just sets of letters.

IV.2.w.vi *Are fields real?*

An intriguing statement [Mermin (1998)] is that fields in empty space have physical reality but the medium that supports them does not. Ignoring the questions of whether fields in nonempty space have physical reality, or how space can be empty if there are fields in it, on first reading this seems to imply that it makes no sense to talk about fields as mechanical oscillations in some object, the ether. This is so obviously true that perhaps something else is meant. That is what makes the statement so interesting — it is a form of a Rorschach test. What is this mysterious medium, and how can it support (in what sense?) fields if it has no reality, physical or otherwise? Can fields really be supported by something that does not exist? In what way is physical reality different from ordinary reality? And how do we tell whether fields have reality, and what kind, and that the medium supporting them does not? How does the reality of fields differ from that of the (fantasy of the?) medium that supports them? Thus there is much room for titillating interpretations and speculations.

IV.2.x Oxymorons

Although much of the misuse of language is not due to use of oxymorons, there is still too much that is. In many cases it is difficult to tell because with empty statements it is not clear what the writer is trying to say (almost certainly not to the writer either). Many examples discussed here are, at least in part, oxymorons.

Thus many people talk about things like definite physical state of a system that is totally independent of measurement or observation, completely separate from them. But if it can not be measured or observed, then what is it, how can we possibly know of it, how can it affect what we do measure and observe? And if we cannot know of it, if it can have no effect, then the phrase referring to it is no more than a string of words that individually have (potential) meaning, but which strung together in this way are completely empty. The phrases "definite physical" and "totally independent of measurement" appearing together make the sentence an oxymoron — and such sentences are examples of a far too large genera.

The meaning of "multitude of universes" (sec. IV.2.u, p. 174) that is probably intended is such as to make it meaningless (it is, of course, difficult to know what people who use such collections of words really believe they think they are saying). However it is in reality an oxymoron. If we could observationally find other "universes" then we would have to regard them as parts of our, single, universe — the preface "uni" obviously means one. Of course if we could not observe them then how can we count them — as "multitude" requires us to do? Hence "multitude" and "universes" contradict each other.

Consider [Dürr, Nonn and Rempe (1998)]: A quantum object, however, reveals its wave character in interference experiments in which the object seems to move from one place to another along several different paths simultaneously. Ignoring the undefined "wave character", this means that we can observe the object moving along several different paths simultaneously, otherwise how do we know that it does so, and if we cannot know, isn't the statement vacuous? But add to this: any attempt to observe which way the object actually took unavoidably destroys the interference pattern. While neither sentence is an oxymoron, the combination of the two essentially is.

An example from classical physics, which verges on being an oxymoron are sentences discussing collisions by point particles (sec. V.2, p. 201). Of course, point particles can never collide — the chance of a collision is of measure zero. And since point particles can never collide, they cannot be observed so they do not exist. If they can collide, they are not point particles.

A nice oxymoron (at least let us hope that it is an oxymoron) is the statement [Weiss (2000)] that physicists suspect that there are many types of nothingness. It seems pointless to have to ask the obvious question: How do different types of nothingness differ? Apparently vacuums exist not only in vacuum. However it is more than suspicion that there are indeed different types of nothingness provided by strings of letters.

How much of modern physics is based on oxymorons?

IV.2.y Inflammatory language

A reason for so many errors in discussions of quantum mechanics is that physicists like to use inflammatory language about it. Consider "the quantum entanglement... is an extremely peculiar feature of quantum mechanics". The statement is not only peculiar, but it is modified in the most extreme way possible. Of course entanglement of the same type holds in classical electromagnetism, so it also is extremely peculiar. For some reason physicists enjoy insulting classical physics much less than they do insulting quantum mechanics. As a result, they believe that they understand classical physics (but are proud that they do not understand quantum mechanics). However very few physicists seem aware of what is wrong with classical physics [Mirman (1995b), sec. 1.5, p. 14]. If they did, they might insult it more.

There are far too many statements of this type (as seen, for example, in the present discussions) which the reader should have no difficulty noticing in far too many places.

Of course in a correct scientific paper inflammatory language or insults would never appear.

IV.3 Do objects have properties if these are not measured?

Can we talk about what objects are doing whether or not we are examining them? If we do not make measurements, what is there to talk about? But the word "measurement" is very broad. We can talk about what an atom is doing inside a star in another galaxy, though we cannot study it directly. But we can study consequences of its behavior — light emitted by the star. Also we have to ascribe to it properties and behaviors, even though we do not know, and can never know, of the existence of the star (and of course of the existence of particular atoms).

We construct theories of what the behavior is in order to understand, to predict, observations. To be able to deal with the universe coherently and intelligibly (sec. I.2.a, p. 5), we must regard atoms in stars (say in a particular class) as behaving in a certain way, though we cannot observe them, or even the stars that they are in. The behavior has meaning in this sense, it is described by pictures we construct to explain our observations, and which we then, for coherency, use for situations that we have not studied, but which we might, in principle at least to some extent, as for example we study, by the radiation emitted, the interior of stars.

Does the phrase "the actual physical state" refer to anything? Of course knowing the state is impossible unless we measure it, and then it means, and means only, the state that we have measured; the statement must be regarded as an abbreviation for this, else it is circular. However we have to be careful of the word "measure". We can take a large number of identical systems (ignoring the question how we know that they are identical, or how we can prepare them to be identical) and perform (identical?) measurements on them. From results of these measurements, we can say that after we make a measurement the statefunction is a particular sum of basis vectors — we can give the statefunction of the system. That is what "the actual physical state" means. This does not imply that the world we measure depends on us as observers, or on it being observed, or that quantum mechanics describes a world that is completely observer-dependent. This phrase can have content only in the sense that it refers to what we find by observation (although that can be very indirect). What else can it refer to?

There is really no basic difference here between quantum mechanics and classical physics. What does "actual physical state" of a roomful of people mean, except in these senses?

IV.3.a Are these the answers we want?

Answers like these are not (emotionally) satisfying to many people. They want to know, not only the probability of finding an electron at a particular position, but what the electron is "really" doing between measurements. But, of course, it is not "really" doing anything since there is no way of giving any more information about it. This is like asking why neutrons scatter from protons? The answer is that the Hamiltonians for these objects contain interaction terms, and applying the (exponentiated) Hamiltonian of the system to a state describing objects headed toward each other gives a state describing objects headed away, and with different probabilities of moving in various directions. Many people may not regard this as a complete answer. They want to know "the reason" for objects scattering, what "actually happens". One "answer" is that there are little green men standing on protons with baseball bats, and when a neutron comes along they hit it and it flies off. Since people have a better intuitive idea about baseball bats than they do of Hamiltonians, this is emotionally more acceptable. But it provides no further information. Such an answer may seem silly, but it is actually no more silly than many other questions and answers (not only) about quantum mechanics. These merely hide the silliness better.

IV.3.b Do we have to look for things to be there?

Is the Moon is there only when we look at it [Mermin (1985)]? The belief that it is provides an example of human egomania. The universe does not exist because we are observing it. Whether we look or not, the Moon has various effects, tides for example [Mohrhoff (2000)]. It is not only possible to find its effects now, it is possible to show the existence of the Moon and where it was millions of years ago by using ancient rocks. Also, "we" may not be looking at it, but other people are, and they can report to us that they have seen it. This however expands the referent of "we", which is not what those who ask such questions really take as a satisfying answer.

There is a more basic reason for objects to "be there", one that applies to things that neither we nor others can ever see (directly). Simple laws govern the Moon and say that it is always moving, whether observed or not. In order to assume that it is not there when we do not look it would be necessary for the laws to give a Moon that disappears (in some way) when our gaze is averted, and then reappears — at the proper place — when we return to look. Obviously such laws would, not merely not be simple, they would not be possible. Is the Moon there always? Do "yes" and "no" really differ as answers? If objects of nature can be described and predicted, if the universe and physics makes sense, the Moon must be there.

Besides what does "there" mean? In this context is there any "there" there?

While the Moon has many effects which obscure problems with the question, we can consider an electron, whose positions we measure at two different times, but take no notice between measurements. Is the electron "there" when we are not looking, does it "exist"? Suppose that we say that the answer is no, "The electron is not there, it does not exist, when we are not looking.", what then? What does this statement mean, what does it say, what experimental consequences does it have? If we say the opposite, then we find that there are laws, say those giving that the electron travels on a straight line with constant momentum. These laws have predictable consequences (although in quantum mechanics they are probabilistic) and they allow us, especially since they are part of a large set of laws, to deal with the world in which we exist. Thus to say that the electron exists when we are not looking means something — just that. It is not merely useful to regard it as existing, it is necessary in order for us to exist in the same world as it exists in. But to say that it does not is not merely without content, but more denies the very existence of laws, of the possibility of a coherent view of nature, so of the possibility of our own existence.

Of course we can say that it does not exist but it goes out of existence when we stop looking and comes back when we start again, at the same position and with the same momentum as if it did exist.

Aside from making the properties of objects dependent on us, this really says the same thing but in a very convoluted (and misleading) way. Questions of whether electrons exist when we are not looking are not about deep philosophical problems, but merely worthless verbiage (as so many deep philosophical problems are).

Does an atom in another galaxy exist, does it have properties, can we consider it in an excited state? It can have effects that require that we assume these, for example emitting light toward us. But even if the light is sent in a different direction, even if we cannot ever find any effects, we must assume them, though they be unmeasurable. In order to have an intelligible description of nature it is necessary to introduce unobservable, and experimentally meaningless, concepts. Try to discuss the behavior of a galaxy, or even any object available, without these concepts. But it is important that they be needed for coherence, that they (can) lead to consequences that are observable, otherwise the words for them may be just empty.

IV.3.c Where is it?

Does an electron have a position before it is measured? This is experimentally vacuous, and "position" is misleading. What is needed is a statefunction, and an expectation value for position determined by it, and these cannot be found until measured of course.

Also "position" is a classical concept, and one of the problems that people create in quantum mechanics is that they try to force objects to have meaningless properties. Electrons cannot have classical properties. It is like deciding that an electron should have a soul. If a theory is based on that, it would really cause paradoxes. An electron does not have a soul, it does not have a position. It has a statefunction which gives the probability that a position measurement will have a particular value. The statefunction can be such that the probability is almost zero except very close to some value. But that is all we can say, that is all we need say.

Those who are very careful about what they do, and what they require, find no paradoxes. Those who try to force nature to fit requirements that they find comfortable, like classical ones for quantum particles, or requiring physical objects to have souls, or think or feel pain, find all sorts of strange paradoxes, strange universes.

IV.4 Empty words are everywhere

While absurd verbiage is far too common in quantum mechanics, it is also far too common. A brief review of other examples can make readers more aware of this, so helping them avoid discussions of (not only) physics that do not discuss anything. And greater familiarity with emptiness makes it easier to notice the hollow jargon that so often passes for arguments about quantum mechanics.

IV.4.a Qualities floating in the air

We have developed a habit of regarding certain adjectives as denoting concrete, physical qualities (like mass or color), perhaps floating in the air, that other objects possess rather than as descriptions of them.

Meaningless terms are everywhere. They form the foundation for many philosophical discussions, and also discussions about physics, and about almost (?) everything else. One such term is "free will". It would be interesting to find an experimental definition of "free will", or even some other definition that does not consist only of hollow terminology.

Essentially these are terms without referents, they do not refer to anything, no matter how much they seem to.

IV.4.a.i *Is reality really real?*

An example prevalent in discussions of quantum mechanics is "element of reality" (sec. IV.2.w.iv, p. 179). In some cases those who use this term (or similar ones) give a definition so that it would seem that using the term is unobjectionable (providing the definition makes sense, which, at best happens rarely). This would be true if it were used in a computer program. But human languages are not computer languages and human beings are not computers (nor are computers potential human beings, despite what some of their designers think (sec. III.4.c, p. 133)). The term, no matter how defined, has subjective implications, not always recognized by those who read (or use) it. Thus attributes, or concepts, are taken (unconsciously) to have "reality", some quality beyond what a definition might state, since we have a subjective feeling about reality — it is a word that we have a psychological reaction to. But this (extra) reality has no real meaning — it is nothing more than a term that arouses a vague feeling due to a suggestion that the quality is comparable to that given by the same word used in other contexts (as when we apply it to a book — for which it is also much fuzzier than we might realize without careful thought). There seems to be a quality called "real", but what it is we really cannot say precisely (with sentences that themselves have say something, rather than merely seeming to). And because terms like these are so prevalent, they lead to much confusion about quantum mechanics.

Related terms like "physical reality" and "objective features of the physical world" are similar. How does the physical world differ from the unphysical one? How can we tell? And how does physical reality differ from unphysical reality? What is the difference between an objective feature and (presumably) a subjective (?) feature? These terms convey suggestions of content which upon examination turns out to be nonexistent. They are sets of words that are used in other contexts with meanings that are not possible for quantum mechanical contexts, but which thus carry over implications of knowledge not possible in contexts in which they are used — leading to strange paradoxes. These imply actual qualities that objects can have, perhaps ones floating in air until captured by an object, but ones that are only apparent and in no way actual, being totally unreal.

IV.4.a.ii *Why are atoms small and galaxies big?*

Consider the fundamental question of whether we know why atoms are small and galaxies big [Overbye (1999)]? This regards "small" and "big" as some sort of qualities independent of what they describe, like oxygen, for example: "Do we know why water contains oxygen?". (What is the definition of water?) We cannot ask why atoms are small; to another atom they can be quite large. The real question is why humans are so large compared to atoms. And of course the reason is that being so complicated, we contain many atoms. It is left as a simple exercise for the reader to explain why galaxies are large compared to humans.

This belief that qualities are corporeal appears again in the suggestion that there are other (unknowable?) worlds, in one of which atoms are the size of bumblebees. This would mean that a bumblebee would consist of one (or a few) atoms. An interesting animal indeed! Or perhaps it is meant to suggest that this bee living in this unknowable world is the size of an atom in ours. To give this sense an experiment would have to be performed comparing atoms in our world to objects in unknowable ones. A fascinating experiment if anyone can think of how to do it.

There are many, many other examples of qualities that are actually names for particular properties of objects, but are regarded as things independent of objects they help describe, some mysterious entities floating in the air, capable of being possessed even by objects not in our universe. This leads to many contentless, but quite beguiling, statements and questions.

IV.4.a.iii *Machines that think, and people who don't*

Can machines think (sec. III.4.c.ii, p. 135)? Can machines kniht? These sentences have exactly the same meaning; for the first it is just less obvious. This question assumes that thinking is some abstract thing, independent of human beings, floating in space and that we have to decide whether it can be attributed to computers (or perhaps have to decide whether computers are filled with it). But "think" is merely a word, a word that human beings have invented (as with all others) and whose definition is decided (too often fuzzily) by human beings. If we wish to apply the word "think" to the type of activities carried on by computers, we can (although it will likely confuse — people, not computers); if we do not so wish, there is nothing to force us to. We can also apply it to trees if we wish. There is no answer to the question of whether computers can think. It is merely a matter of how we wish to use the word. (The fascination that the word arouses in this context has lead to suggestions for tests about whether computers can think. Of course, this is nonsense, but since it is so popular, we suggest another test: If you can live with a computer for a year, and not be able to tell whether it is a computer or a human being, then it might be reasonable to say that the computer can think — or perhaps it might be more reasonable to say that you cannot.)

Words, really lines of letters, are often fascinating, a reason we so often misuse them.

Likewise the question whether computers can have emotions will lead to long philosophical discussions because "emotion" is undefined here (and undefinable). Emotions are results of certain activities of brains, not abstract properties that things may or may not possess. "Can computers have emotions?" is an excellent example of a sentence that seems to ask a question but is really a collection of words, elsewhere meaningful but that do not fit together, an empty collection, forming in a sense an empty set.

IV.4.b Does God have meaning?

Results giving, for example, the dimension of space are startling [Mirman (1995b), chap. 7, p. 122]. Why is it that everything works, but just barely? Many feel that God designed the universe in this way because he wanted intelligent beings, that is (fortunately) us (at least we like to think so — but too often do not act so), and so it made a universe in which we (happily) are possible. God wanted it that way.

But what is God? We tend to think of God as someone like ourselves (specifically white, European, male). God is believed to be some intelligent being who is able to construct a universe the way it wants. But intelligence is not some abstract quality floating around in air, but a property of our collection of neurons. Does God have neurons? Do they act like ours? Do they obey the same laws of physics? Do they also exist, must they exist, in a space of dimension 3+1?

IV.4.b.i *Does God have emotions?*

And "want" is an emotion. Does God have emotions? Apparently emotions are forms of activity in the limbic system [Shephard (1994), p. 608].

Does God have a limbic system? To take this view is to believe that God is also subject to laws of physics. And then we can ask why these laws hold, even for God? (Perhaps there is a superGod.) But if we do not take this view, then what is God? We are reduced to regarding it as a word which is comforting, but that refers to nothing (unless we believe that God really does have emotions, does has a limbic system).

To say that the universe is the way it is because that is what God wants, is to write a sentence seemingly intelligible but which cannot be, a collection of words that provides emotional comfort, but does not really say anything.

Some people believe that God is love. By that they mean that God is a particular pattern of excitations of neurons. Otherwise what else is love?

Likewise saying that the universe is as it is, that it is one in which humanity can exist, because it was designed that way also uses words appearing to explain, but really do not in this context. Designing is something that is an act of human beings (sec. IV.4.b.iii, p. 191). What does it mean to say that God designed the universe? That it drew blueprints; does God draw blueprints? That it wrote a computer program? Is God a computer programmer?

And we can ask "Why is there something, instead of nothing?" But if we say that it is because God wants it, then we can ask "Why is there God, instead of nothing?" Does God have emotions, as this answer implies? Are they produced in the same part of its brain as they are in ours?

IV.4.b.ii *Does life have meaning?*

As another example of people attributing emotions to God is the search for the meaning of life, the meaning of the universe. However it is not clear what the meaning of meaning is, what people want. They seem to feel that life is meaningless unless there is a God, sort of a superfather, who "loves" us. They look at the universe, with its immensity of galaxies, each with an immensity of stars, many perhaps with life, and feel "wow, this whole thing was created just for me" (of course it was not), thus displaying the humbleness called for by their religion. They seem to feel that unless the universe was created for them, so that they could exist, life has no value, they are unable to find value, meaning, in their own lives, their own joys, their relations with others, with the planet on which they live. Like adults unable to be satisfied by life, who have to look back to childhood for reassurance that their parents really did want them, really did love them, they feel that life is pointless unless there really is a superfather who really did want them, who really does love them. So they search for "meaning", for love not from, and for, their fellow human beings, and the world around them, but from something else, something vague and undefined, undefinable, something beyond humanity. Perhaps it is essential that what they want is undefined, and undefinable, beyond humanity.

And if they cannot find such a God, such a superfather, then life becomes worthless. This strange mixture of contempt for humanity — that it by itself is not enough to give life meaning — and themselves, and extraordinary hubris, in believing that the universe was created for the humanity of which they are part (but which they disdain), creates a yearning for something, which they cannot specify, to fill the emotional void left by their inability to value what they are, what they are part of.

This search for meaning is not a search for information about the universe, for understanding of it, but a search for emotional comfort. Looking at the universe and finding, or not finding, meaning, does not tell anything about the universe, but about human emotions.

IV.4.b.iii *Is the universe a product of design?*

Another question that fascinates people is whether the universe is designed? This is a deeply profound question, and it is because it is based on nothing but words, terms, phrases without substance. Thus it is possible to have many deeply profound philosophical discussions about it, many without end. In particular it is deep because that is where the vacuity of the subject is hidden by a pile of profoundly meaningless words.

Many people believe that the universe was created to serve a divine purpose, a purpose that is good. This not only supposes that God has emotions (purpose), but morals (good) — does God also have religion? But what do these words really refer to? And how did God design the universe, by (as we do) using coordinated actions of its neurons to create muscular movements leading to the planning and creation of the universe? And if not, what does "design" say here?

This belief that the universe is "designed" is based on a concept of existence of a super human being (subject to the same laws of physics as we are, living in a universe "designed" so these laws will hold?). But we have no way of knowing, of ever knowing, whether that is true — it makes absolutely no difference (except of course emotionally) whether it is true or not. Nothing that we observe, or can ever observe, depends on it. Thus the word "existence" has no denotation here, it has no existence. Nor therefore does "design".

There have been hundreds, thousands (perhaps far more) of learned treatises discussing whether the universe is "designed". These are (often huge) collections of impressive sentences, showing the vast erudition of their authors, but a vast erudition that is actually completely vacuous. Thus all these learned treatises are nothing but conglomerations of words, senseless, but often brilliant in the way they hide their emptiness.

Attempts have also been made to determine the probability that the universe is a product of design [Devlin (2000)]. But design is an empty word, one that cannot be filled with numbers. To ask "What is the probability that the universe a product of design?" is like asking "What is the probability that the universe a product of ngised?".

Here "design" is a word that has connotation, but no denotation, which, as we see, and not only from examples throughout, is very common, and very determinative of the ways that we think, the ways that we do not think, and quite often for our perplexity, and not only (perhaps not even mostly) in science.

IV.4.b.iv *The mystery of existence*

The universe is a mysterious place, mysterious because it is, and because it is simple and comprehensible. There are thus many questions — but unfortunately few answers are possible. If there is much that we can understand, there is much more that we cannot. We are part of nature, subject to its laws. It is human egomania, extreme hubris, to believe that we can understand everything, to believe that natural objects, humans, can even ask these questions. We can be impressed, even startled, by the universe, we can understand much about it, but we must accept that no matter how much we wish otherwise we cannot understand, we cannot even ask, why the universe is, and why it is what it is.

The mystery of existence is startling, the questions it raises are awesome. But mystery these must remain. Senseless words and phrases do not provide answers to the unanswerable.

IV.5 Empty words give emotional comfort not knowledge

To understand our universe requires statements that have denotation, questions that can be answered. And often we are deceived, we use words, sentences, questions, sensible elsewhere, but not in the utterances in which they appear. Thus we seem to be saying, or asking, something, but really are not. This distinction is essential to knowing whether we are actually knowing. It is this that so often tricks people — too often ourselves.

If we ask "What is onkly?" then we know that the question is senseless, for there is no such word "onkly". However suppose that we ask "What came before the beginning of the universe?". That question might seem to make more sense than the first, but it does not. It is a grammatically correct collection of words that elsewhere provide messages. However it does not refer to anything, indeed it is really an oxymoron — there is no way of answering it.

IV.5.a The need to explain creation

A question only has meaning if the sort of statements that could answer it can be specified (sec. IV.1.c.i, p. 147). If we ask "How do stars form?" it is possible to tell what statements will answer the question. For example we can say "a gas cloud collapses under its own gravitational attraction until it is so compact that nuclear reactions occur". This may be wrong (although in this case there is strong evidence of its correctness). But whether it is correct or not, meaningful statements can be given that provide answers; if incorrect then other such statements must be given. But the question has content, it asks something, because such statements can be exhibited.

But "What came before the beginning of the universe?". It has no meaning because there are no statements that answer it. Suppose we say "the universe was created from nothing by a strong life force"? What information does "created" or "nothing" supply in this context? What does "life force" mean, and not only in this context? There are no actual statements that can answer the question (but there are fake ones), thus it has no content, it verifiably asks nothing.

We may be uncomfortable not knowing "What came before the beginning of the universe?" or "How did the universe come to exist?" and such answers as God or life force may make use feel more comfortable. But they do not tell us anything, they are vacuous (the questions, not what came before the beginning of the universe), they do not provide information. They are useful, often quite useful, in providing emotional comfort, but are useless beyond that. Words like these are not statements, but rather incantations.

Statements about what came before the universe, or where the universe came from, or why it exists, and questions about these, are inherently nonsense. We are part of the universe, we cannot know what came before it since any information is also part of the universe — information consists of physical objects — thus information coming from before the universe was created was actually created as part of the universe, thus did not come before the creation of the universe. If there was information that we could receive, then there was a universe. Answers to such questions, and the questions themselves, are inherently circular, self-contradictory, oxymorons.

IV.5.b Equal emptiness in statements about quantum mechanics

There is a deep emotional need to explain the universe, to explain why it exists, why it is as it is, to find "meaning" in it. So many say things like the universe exists because God wants it, it has meaning because God loves us. These words provide emotional solace that so many so deeply desire. Yet these, despite their emotional value, clearly say nothing (perhaps they have emotional value because they say nothing).

Many physicists would recognize this, and realize the errors in regarding them as meaningful. It is surprising then that these same physicists have so much trouble in seeing the equivalent emptiness of so many statements about physics, about quantum mechanics and its interpretation, the emptiness of statements purporting to describe alternative theories to quantum mechanics, that they have so much trouble seeing that these statements exist, just as in other fields, not to provide information about nature, but only to provide emotional comfort. And it is surprising that they have such deep need for such comfort.

It is an interesting observation about human psychology that it is often so easy to see the lack of content in fields other than our own while being totally unable to see the equal lack of content in fields with which we should be the most familiar. This is a major reason for the befuddlement about, problems of interpretation of — unfortunately not only — quantum mechanics.

IV.5.c Limitations of science, of knowledge

Science is limited, being merely a way of carefully observing the universe, trying to understand it — as much as can be understood — of finding regularities, patterns. It cannot answer such questions as "why are we here?", "what is the meaning of life?", and so on. (Of course, "meaning" has no meaning here, nor does "why".) There are those who believe that there are other ways of knowing, other ways of answering such questions, other fields of inquiry capable of doing so. But do they really answer? What they do is provide statements and words that, while seeming to answer such questions, actually are empty, that say nothing. What these other fields of "inquiry" do is not supplement science, not answer questions it cannot, but rather help us to cover up the unpleasant reality that these questions are vacuous, unanswerable, that we do not know, cannot know, that humans, as part of nature, as physical objects, are quite limited, and must always be [Mirman (1995b), preface, p. vii].

These fields do not supplement science in providing knowledge, but only supplement it in providing emotional succor.

However, as another indication of the importance of language, the word "knowing" can be ambiguous. Other fields do provide knowledge of values, that is they provide sets of values that people "know" — that they accept. In that sense of the word, statements that these other fields provide different ways of "knowing" than science are meaningful, but because the set of letters "know" refers to something different, although spelled — the same set of letters in the same order — and pronounced the same way, it is actually a different word.

Such fields are not other ways of knowing — about nature, but ways of hiding our inability to know, ways of hiding the truth that knowledge we want is not merely unattainable, but nonexistent, has no sense, no content. Perhaps most of all they help hide our being part of nature, made of matter governed by its laws, not something remarkably different from the other parts of the universe in which we exist — despite our often intense desire to regard ourselves as special — but merely another piece of that universe, mere bits of matter, although with a remarkable ability to make sense, to some degree, of the other matter in the universe, and of the universe itself.

Chapter V

Errors, confusion and quantum mechanics

V.1 Why quantum mechanics often seems weird

Quantum mechanics is nonintuitive, sometimes highly so. Yet most difficulties people have with it are the result not of inexperience with the realm within which it is most relevant and thus with which we have little intuition, but rather such difficulties result from errors in understanding it, and often just carelessness, sloppiness, thoughtlessness. Much of this is due to improper use of, often meaningless, language. This is so important that we consider it separately (chap. IV, p. 142). But not all confusion and misunderstandings are caused by mistreating language. While we next look at other reasons for misinterpretations and misunderstandings that are all too prevalent, a large part of this also involves incorrect, misleading, or vacuous (often non-)language.

The number of errors in discussions of quantum mechanics is unfortunately quite vast. Here then only a few can be mentioned. It is hoped that by at least indicating a few, readers will be inspired (or provoked) to be more careful than others have been, leading (regrettably over a long period of time) to a decrease in the volume of confusion.

We shall find that once such errors and misunderstandings are removed, quantum mechanics nonintuitive as it may be, makes much sense, is not the least bit weird, nor is it difficult to comprehend. Difficulties, paradoxes, are due, not to quantum mechanics or the strangeness of nature, but purely to the confusion and carelessness with which (not only) quantum mechanics is too often unfortunately treated.

V.1.a Weirdness

It is disgraceful to call physics, or nature, weird. Weirdness comes only from unthinking and sloppy (what some call) thought. There is nothing weird about quantum mechanics. The only thing weird is that so many physicists do not understand it, and flaunt their incompetence by calling it weird. This encourages those who want to believe that the universe is weird, that science is just a social construct, and such other crackpots. It hurts science and physics.

What is weird is not nature, not physics, not the universe, but the obsession of so many physicists with demonstrating their inability to understand physics by regarding nature as spooky.

Physicists are often so obsessed with weirdness that they have come to regard it as a tool: A search engine that uses quantum weirdness has been experimentally proven, for example. This demonstrates nothing so much as the willingness of too many physicists to say something, no matter how absurd, just because others are saying similar things.

V.1.b Statements stand alone

Physics is (supposed to be) a science, not a religion. The correctness of its laws, and explanations of nature, are (supposed to be) determined by logic and experiment, not by authority, not by popular vote, not by the fame of some "experts". It is irrelevant what people think, no matter how great they are. Yet there is a tendency to excuse nonsense if put forward by a famous physicist, but to be strongly critical of it when someone who is not licensed to spout nonsense does so. Of course, this itself is nonsense. There is no difference between a licensed crackpot, and an unlicensed one. Nonsense must be fought no matter where it comes from. In fact, it is far worse coming from a physicist, especially a famous one, for then it is more likely to be accepted, to cause confusion, harm, more nonsense. Physicists must have higher standards than anyone else, at least when discussing physics, and must thus be held to higher standards — no matter who they are, or how famous they are. It is deeply regrettable that these requirements are so often rejected.

A well-known error in logic is *argumentum ad verecundiam*, appealing to authority in a field in which that person has no more standing than anyone or anything else. In physics, the only authorities with standing are nature and experiment. Appealing to human beings is an example of this logical fallacy.

And there is no excuse for poor writing. Having done some useful work (in physics) does not relieve a writer, a physicist or anyone else, of responsibility for insuring that what is written does what it is supposed to: communicate — clearly, and correctly.

If some comments here, at times but not often enough, seem to be suggested by views of "great authorities" of physics, and perhaps this is even explicitly stated, that is because these people must be held to the highest standard. Human beings, no matter how brilliant, will never be able to understand perfectly, and not just nature, we will always have difficulties. That is understood, and to some extent must be accepted. But irresponsibility should never be condoned.

V.1.c A fundamental problem is the rejection of Occam's razor

A fundamental reason for apparent problems with quantum mechanics is that too many physicists look for the most complicated, the most weird, the most unlikely, reasons for quantum mechanics, for physical laws, the most complicated, the most weird, the most unlikely, explanations of physical phenomena. The history of science shows that such approaches are unlikely to succeed [Mirman (1995b), sec. A.1.2, p. 178]. The simplest, most conservative approaches are the best. It may be surprising how simple fundamental laws are [Mirman (1995b,c)], and how much the most elementary reasoning can explain — those who look for simple answers are most like to find, not only simple ones, but correct ones.

But too often, such as with alternative theories to quantum mechanics, designed to fix its nonexistent flaws, Occam's razor is ignored — and frequently when it is most needed, and would do the greatest good.

V.1.d Be careful what you call things

Language and names are fundamentally important — and confusing, so are discussed in depth (sec. IV.1.a, p. 143). But it is necessary to mention this again to prevent misinterpretation of some of the terminology used.

The terms "quantum mechanics", "wavefunction" and "particle", for example, have greatly mislead far too many people. They emphasize how dangerous language, and names, are, how they can confuse and mislead. Unfortunately custom often must rule, so possibilities for improvement are limited. But there is at least a little room for change.

The name quantum mechanics is so deeply embedded that it is impossible to change (a better term might be functional mechanics, since objects are described by functions of space and time). But wavefunction we shall mostly avoid, using instead statefunction, the function giving the state of the system — a term that should carry no misleading connotations. And instead of particle we try to use a neutral term like "object", this carrying no implication of any properties, although for variety — only — we often use "particle", but with the understanding that this is merely a synonym for object.

V.1.e Quantum mechanics is not classical physics

One major reason for difficulties with the interpretation of quantum mechanics is that too many people (physicists!) are trying to find classical pictures for objects that are not classical. What too many are looking for are ways to comprehend quantum mechanics as if it were classical physics, they are looking for classical explanations, and these cannot possibly be correct. They are trying to understand quantum mechanics using classical concepts, that cannot hold, that are irrelevant or that themselves are meaningless, but whose emptiness is obscure in classical physics, but striking in quantum mechanics — although they refuse to see this meaninglessness.

This is recognized [Fuchs and Peres (2000); Zeh (1993)], but not explicitly, not often enough, not forcefully enough.

It is interesting to consider whether it would have been easier to understand quantum mechanics if it had been discovered before classical physics [Lamb (1994)]. Perhaps it would be better to teach quantum mechanics prior to teaching Newton's Laws.

There is, fundamentally, no such thing as classical physics [Mirman (1995b), chap. 1, p. 1] so that if a discussion uses classical physics it is phenomenological. Thus any problems or peculiarities can be assumed to arise from the phenomenological nature of the discussion. These should not be taken to indicate problems with quantum mechanics, for such discussions are not quantum mechanical, and involve inconsistent aspects, those that are classical.

V.1.e.i *Magnetic monopoles*

Thus the "still mysterious failure to detect a magnetic monopole" disturbs many people. This emphasizes, again, how strongly scientists are attached to classical physics, and are unable to deal with nature being quantum mechanical. There is much research showing that students have an Aristotelian view of the world, and how much difficulty they have in thinking in terms of Newtonian physics. Yet it is much worse for professional scientists who are unable to let go of classical physics, and accept quantum mechanics. The hole that the magnetic monopole is expected to fill does not exist — it appears purely because an incorrect, inconsistent, formalism, classical physics, is used. Anyone who does not believe this can try fitting the monopole into Dirac's equation [Mirman (1995c), sec. 7.3.3, p. 133].

V.1.e.ii *Why gauge noninvariance does not matter*

It is widely believed that the electromagnetic vector potential cannot be a physical object because it is not gauge invariant. This emphasizes in another way the compulsive attraction of classical physics and the confusion that it creates. Of course the vector potential A must be the physical field [Mirman (1995c), sec. 3.3.1, p. 37], while the non-gauge-invariant electromagnetic field cannot be, because, among other reasons, it is not measurable.

But if there is an object in an electromagnetic or gravitational field doesn't its potential energy change if the (arbitrary) origin is shifted? Classically it seems to, but classical physics is wrong. An object consists of spin-$\frac{1}{2}$ particles, electrons and nucleons. The fields do not act on classical objects but on particles making them up. The effect on an object is just the result of the sum of these. Thus the proper way of studying this question is to use Dirac's equation. If we change the vector potential, $A \Rightarrow A + \Lambda$, where Λ is an arbitrary function (except that it obeys the wave equation) then the phase of the statefunction of the electron, or nucleon, must also be changed, in the well-known way leading to minimal coupling. Doing this leaves the equation invariant. Thus the arbitrary change of A is irrelevant, it has no observable consequences.

What is wrong with the usual classical analysis? This changes the vector potential, but does not at the same time change the statefunctions of the particles making up the object. It is equivalent to moving or rotating an object in a field. We then find that the properties of the object (like its energy) change, so space is not homogeneous or isotropic, which of course is incorrect. If we move the object we must also move, say, the earth, and then we find space is actually homogeneous (ignoring other objects in the universe).

If we change only part of a system, the position or orientation of a body, or the origin from which we measure the potential, without the corresponding change of the rest of the system, the position of the earth, or phases of statefunctions of internal particles, then we find noninvariance. But this is a consequence, not of the unphysical nature of the potential, or the nonisotropy of space, but only of the incompleteness of the transformation.

Why is there such a strong conviction that the arbitrariness of the potential shows that it is not physical? Partly just carelessness and thoughtlessness. But also because of the inability to accept that nature is governed by quantum mechanics, and so the inability to stop thinking in terms of classical physics. This inability raises interesting psychological questions, whose answers might lead to ways of helping physicists understand quantum mechanics, physics, nature, better.

V.1.e.iii *So the Aharonov-Bohm effect does not imply nonlocality*

There is a strange belief that quantum mechanics is nonlocal. Part of this is due to misunderstanding the EPR experiment (sec. V.3, p. 208). There are of course many other misunderstandings that have lead to this confusion. One of these is the Aharonov-Bohm effect in which an object is affected by being in a region with no electromagnetic field [Anandan and Aharonov (1988)]. However there is, as we see, no physical object that is the electromagnetic field. It is the electromagnetic potential that is the physical object, and this is always nonzero in regions through which such particles traverse. Thus quantum mechanics is not nonlocal. It is only confusion that is nonlocal — and far too widely spread.

V.2 There are no waves, no particles, no wave-particle duality

Language can be dangerous, as quantum mechanics so often illustrates. One term that has caused much confusion is "particle". We have an intuitive idea of a particle as a little ball, a marble for example. But quantum mechanical objects are definitely not marbles, balls, or particles [Zeh (1993)]. Hence we use general terms like object and entity. And when for variety we use "particle" there is no implication that it is a little ball, or any implication of any other properties.

One problem with the term "particle" is that it likes to be modified to "point particle". But this is a meaningless term — even in classical physics [Mirman (1995b), sec. 3.1, p. 38]. We only use "point particle" when necessary to point out that it is nonsense.

In quantum mechanics we cannot say that an object is limited to (or even located at) a single point — a statement that is (almost) an oxymoron and which contains a physically undefinable term, which cannot be experimentally verified. We can limit a statefunction to be nonzero in only a finite region by putting an object in an infinitely deep potential well, which is itself impossible and has meaning only as an idealization. But while a statefunction can be so limited (in an ideal limit), it is thoughtless to say that an object has a finite extent, being nonzero in a finite region (what does this mean?) and zero outside it.

It is also wrong to say that an electron has, or has not, a nonzero radius. It has no radius, it is not a basketball. The concept just does not apply.

Objects cannot be considered as point particles, nor as waves. These are classical concepts (but not really both of them). There are no particles, no waves and so there is no wave-particle duality. Yet this idea seems to hold a peculiar fascination for too many people. It is useful to try to understand some of the reasons. Misunderstandings of experiments, and physical instruments, often give illusions that these are studying particles or waves, or that there are choices of which of the two to study. Thus we have to examine some examples.

The belief that objects are particles has caused many, many difficulties and mysteries with quantum mechanics [Silverman (1993)]. These could have all been avoided with a slight bit of thought, which would lead to the realization that the concepts of particles and waves are invalid.

V.2.a Dots are properties of screens, not objects

If a beam, say of photons or electrons, passes through slits and impinges on a screen, and the intensity is very low, dots appear, which over a period of time long enough for many of these objects to hit the screen become sufficiently numerous to form an interference pattern. Does not the appearance of dots, rather than darkening of a large part of the screen, show that these objects are particles, little balls? And does not the interference pattern show that these are also waves? Clearly the answer to both questions is no [Mirman (1995b), sec. 3.1.3, p. 41].

V.2.a.i *Dots are a property of the instrument, not the incoming object*

That a dot appears on the screen in, say, the double-slit experiment is a property, not of the object, but of the instrument, the screen: that its constituents are localizable (atoms are given by wavepackets with small extension), and that it has discrete energy states. Because it has discrete energy an object with small energy can excite only one atom, which then leads to a darkening in a small area (just around the single atom that was excited), a dot. Because of energy conservation it is impossible for the incoming object to give darkening except for only a small region.

The appearance of a single dot (rather than a darkening over a large area) does not indicate that the object is a point particle, nor anything else about it, besides its energy (nor does it imply that the screen consists of particles, points or not). No matter what the object is, only a single dot is possible. If the incoming object has small energy, it can excite merely one atom, so causing only one spot on the screen, even if the incoming object were a wave.

Also the incoming object might be described by a wavepacket, so localized. The positions of wavepackets of different incoming objects may be different, giving an interference pattern. But this implies nothing about the nature of the object.

V.2.a.ii *Why it is different classically*

Classically it is different — light carries more energy, especially since it is continuous, so it can excite many atoms quickly giving a pattern (the key word is "quickly"). Quantum mechanically, it is possible to get very low energy light which appears for a very short period of time (by definition this is not possible classically), so it (or equivalently one electron) is able to excite but a single atom. Thus there is a dot. That shows nothing about the impinging object, but rather is a property of atoms of the detector, the screen, that it consists of discrete objects (described by wavepackets, so localized), that it has discrete energy states, and that each excited atom affects only local objects, nearby atoms and molecules (it causes excitation or disassociation, thus a dark spot only around it, which, at least in part, is an implication of the fact that only a small amount of energy was delivered to it).

V.2.a.iii *Incoming objects with large energy*

If an incoming object has a lot of energy, enough to excite several atoms, or if it interacts with a molecule, with very closely spaced energy levels, that there is a dot does not tell anything about this object. A single atom might be excited to a higher level, so still causing only one spot, or the object might scatter and excite several, but nearby ones, giving, at most, a larger dot. This is similar to the production of a track in a cloud-chamber [Mirman (1995b), sec. 3.2.1, p. 43]. Essentially the state-function of the incoming object is converted to a (narrower) wavepacket. The atoms excited will be near the first one (the reason the track is, almost, a straight line). Thus they just contribute to the same dot as the first atom, or to contiguous ones — the darkening then will look like a single dot but perhaps a little larger one. Also the scattering further localizes the object, forcing it into a narrower wavepacket state, so it interacts with other atoms that are close together. It is difficult to think of a screen, or generally a detector, that can give, for a single incoming object that has to be taken quantum mechanical, a pattern spread over an entire screen.

V.2.a.iv *What if the atoms in the detector move?*

We might consider another way of getting a dot. Suppose the incoming object causes the struck atom to translate, which requires only a very small amount of energy, and this leads to a chemical reaction, giving a dot. It might seem that then the argument would not work — an incoming object could thus excite many atoms, giving many dots. However atoms of the screen are always in motion, thermal motion. And the film can be warmed, often quite a bit, before darkening. Thus to produce a chemical reaction, so a dot, much more energy than the thermal energy is needed, and even though the energy leading to a dot is continuous above that value, there is still a smallest energy, giving a similar argument. It is not the incoming particle, but the screen that results in a dot, rather than a continuous blackening (perhaps graying would be a better term).

V.2.a.v *Are there other ways of showing the object is not a point?*

Triggering of a particle detector is not experimental evidence of a particle. What other options are there? A realization that the concept of particle is wrong removes many of the difficulties with quantum mechanics.

It is important to emphasize that a dot on a screen does not show that the impinging object is a particle. It occurs because the detector, the atom, is localized — it is describable and described by a wavepacket that is almost zero everywhere except in a small region, and because it has discrete energy levels (although even without these it would be very difficult to get extended darkening from one object). Again a dot, the fact that only a single one appears, is not a property of the incoming object, but of the detector.

There is not the slightest reason to believe that physical objects are particles (whatever particle means), and much reason to believe that the term "point particle" is not only meaningless, but definitely misleading.

Thus there is no need for (nor possibility of) explaining particle motion as being guided by a "quantum potential". Nor is this needed to explain interference experiments.

V.2.b Quantum mechanical objects cannot be waves

An object is not a particle, and it is not a wave. We think of waves as say oscillations of water, a rope, or a line drawn on a paper — that is how we visualize them. But a statevector is a complex-valued function, and though it may be a function of position, it is thus not the same as a water wave. Both its amplitude (which is what we visualize) and its phase are functions of position. An electromagnetic wave is a better analogy. But this is quantum mechanical [Mirman (1995c), chap. 7, p. 123].

Also a statefunction of an object traversing a barrier is monotonically decreasing, thus is even less like a classical wave. There are no oscillations for such statefunctions. Moreover statefunctions in general — say those given by spherical harmonics or Bessel functions or other complicated functions describing objects in complicated potentials with complicated coordinate systems — while they may not be monotonic also do not vary in ways that as we visualize waves doing. And their amplitudes and phases can vary very differently.

Because objects are described by functions of space, they are extended, and are localized only in the sense that these statefunctions can be (and in realistic situations are) wavepackets. However quantum mechanics is local; equations governing statefunctions, including interaction terms, are all functions of the same position values, and of only a single value (sec. III.3.d, p. 127).

Interference patterns do not imply that objects are waves. Objects are described by statefunctions that are complex-valued functions of space and solutions of Schrödinger's equation. These, with the proper boundary conditions, say those given by slits, result in statefunctions that produce interference and diffraction patterns. That is all we can say about statefunctions and the objects which they are. Pictures, especially classical ones, can mislead, and have done so far too often.

It is thus incorrect to say that sometimes an object acts like a particle, sometimes like a wave. It acts like, and is, neither. These classical concepts do not apply. Attempts to explain the "wave-particle" duality fail to have content because objects are neither particles nor waves, and neither term is relevant to quantum mechanics. For it there are no particles, there are no waves, and there is no such thing as a "wave-particle" duality.

Particles and waves are classical concepts, and can be, and usually are, deceptive ones. A quantum mechanical object is neither. It must be understood in its own terms.

V.2.c Are leptons points?

People appear to believe that leptons behave like completely dimensionless points — without knowing what a completely dimensionless point is, and they cannot know since it isn't anything. Actually what this really means, even if those who make such statements do not know what they mean, is that a lepton is described by Dirac's equation with only the electromagnetic interaction (and weak and gravitational ones) included. If there are those who disagree with this, they should state explicitly what a completely dimensionless point is, how objects are determined as such, and how saying that a lepton is one differs from saying that its statefunction is a solution of Dirac's equation with only these interactions, and provide experiments that distinguish between these views of leptons.

V.2.d Are electrons points and protons extended?

There is a dogma that electrons, for example, as far as we can tell, are point particles (an undefined concept) while nucleons are extended, and have structure. There is no reason to believe this, much reason not to (sec. V.7.f.iv, p. 261); it is based on a complete misunderstanding [Mirman (1995b), sec. 6.4, p. 119].

Suppose that we have just discovered a new object, and one with spin-$\frac{1}{2}$. We expect it to obey Dirac's equation with the correct mass, but find that it does not satisfy the same equation as electrons since it also has other interactions — so, for example, its magnetic moment is different [Mirman (2001), chap. IV]. These other interactions do not imply that it has structure, or is extended. The most reasonable way of attempting to treat it is by using Dirac's equation, and including (necessarily) these other (strong) interactions (statefunctions of baryons are solutions of the equation with the electromagnetic and also strong interactions).

However these equations are impossible to solve in any reasonable approximation so we try to construct phenomenological models, with experimental parameters to represent the uncalculable terms. The obvious first model is an extended sphere, with a radius that is an experimental parameter (and only that). This does not mean that it is extended, or that it is a sphere, only that this is the simplest way of including a parameter. But this is too simple, so we next take a set of spheres, with different radii, giving more parameters. However this new object is coupled to other particles with various spins and isospins, and these must be included. Thus our phenomenological model becomes more and more complicated, regarding the new object as consisting of particles with these various attributes, giving more and more parameters to be determined experimentally, to represent the complicated interactions with the many different particles, with many different properties.

That we must resort to such models does not mean that the nucleon is other than an object that can be described (but unfortunately not computationally) by Dirac's equation with the inclusion of all interactions. There is no implication of anything beyond this, no implication of structure or extent. One reason for these beliefs is the view of objects as point particles, so more complicated ones are regarded as extended. But both these dogmas are wrong.

Another belief is that because Rutherford found from scattering experiments that atoms had most of their mass concentrated in a small region similarities in scattering from protons shows that it has most of its mass concentrated in a small region also, so it must be an extended object. There are two important differences however. First it was known that atoms had at least one constituent, the electron, that was observable as a free object, independent of its role in the atom. Also Rutherford had a formula for the cross section, since the scattering was due to the known Coulomb force, and this agreed quite well with experiment. It was not merely that some particles were scattered to large angles, but the angular function could be computed and found to agree with experiment. Thus his conclusion was strongly supported. But analogies that lack these two (as with scattering from nucleons) become quite weak, and conclusions drawn from them have to be looked at very carefully.

V.2.e Point particles (which do not exist) cannot cause difficulty in quantum electrodynamics

The word "point", as in point particle, has of course lead to much misunderstanding. For example there is a belief that there are problems with quantum field theory because of severe short-distance singularities. It would be interesting to find if those who believe this can specify what distance they are referring to, and what the singularities are. And they might be asked for evidence of such singularities, except that it is impossible to provide evidence for the existence of the undefined. There are ostensible problems in perturbation theory, although for quantum electrodynamics at least these appear to be purely a result of the approximation scheme, and this is likely true in general — there is no indication that the problems are in underlying theories themselves, and certainly not in undefined short distances, or that there are singularities that cause these (sec. II.5.f, p. 101).

V.2.e.i *Point-particle approximation*

Related to this is the view that quantum field theory is based on a point-particle approximation. It is difficult to base quantum field theory on point particles since there are none, nor does the concept have content, certainly at least in quantum mechanics and quantum field theory. And if anyone believes that there are short distances, or singularities caused by them, or point particles, they can undoubtedly state explicitly what distances they are discussing, how these cause singularities and what the term "point particle" means, and how it is related to quantum field theory. These people can undoubtedly give precise experimental meaning to such terms and concepts.

V.2.e.ii *The inverse square law*

The belief that the inverse square law of electromagnetism or gravity (which is purely a phenomenological expression [Mirman (1995c), sec. 7.2.2, p. 126]) says that if the distance between two objects becomes zero the force becomes infinite ignores the uncertainty principle. If positions of objects along a line perpendicular to that between them are known so precisely that we can say the distance becomes zero, then nothing is known about their momentum, so we cannot know where they are. Hence we cannot say the distance becomes zero. There is nothing wrong with the formula (except that it is phenomenological), but this statement shows that it is being applied improperly, especially as it is self-contradictory. And in quantum mechanics there are no point particles to apply it to.

V.2.e.iii *Problems in classical electromagnetism say nothing about quantum field theory*

It is widely assumed that problems in classical electromagnetic theory, such as radiation reaction, indicate there is something wrong because it uses point particles. Therefore there must also be something wrong (indicated by infinities in perturbation theory) with quantum electrodynamics. Beliefs that infinities in perturbation theory (which are probably due to the approximation scheme not to quantum electrodynamics) are the result of the electron being a point are based on meaninglessness, including experimental meaninglessness (sec. II.5.f, p. 101).

There are many problems with classical electrodynamics; don't they indicate that difficulties are caused by point particles, and these carry over to quantum electrodynamics? Such misconceptions are an additional cause for confusions about quantum electrodynamics: difficulties with classical electrodynamics and assumptions that these are due to "point particles" indicate the same (or worse) ones in quantum electrodynamics, and for the same reason.

However classical physics (all of it) is inherently inconsistent [Mirman (1995b), chap. 1, p. 1] — classical electromagnetism is inherently inconsistent, quantum electrodynamics is not — this has nothing to do with "point particles" (whatever these are), and it is not an indication that quantum mechanics, or any part of it, is inconsistent. Quantum mechanics is necessary (and fortunately is apparently consistent); classical physics is impossible. And quantum electrodynamics does not use point particles, a hollow concept. Problems with quantum electrodynamics almost certainly come from approximation methods. No conclusions can be drawn about quantum electrodynamics from classical electrodynamics.

Quantum mechanics is required and flaws in classical physics imply nothing about quantum mechanics. Nor is there reason to believe that they do. It is absurd to assume anything about a correct theory from a necessarily incorrect, and inconsistent, one.

These show once again how difficult it is for people (physicists!) to unlearn something that is wrong because they learned it before learning what is correct. It should not be surprising that students have difficulty with Newtonian physics.

V.3 The EPR "Paradox"

One of the enduring enigmas in quantum mechanics is induced by the various versions of experiments suggested by Einstein, Podolsky and Rosen [Einstein, Podolsky and Rosen (1935)]. This has lead to various strange ideas, EPR "Paradoxes" , like the belief that quantum mechanics is nonlocal and that a measurement on an object can affect another object in a different galaxy. Of course these are all nonsense. They arise only because such experiments and the theory believed to "explain" them are analyzed with much negligence.

In quantum mechanics, a statefunction gives probabilities. However if an object with zero angular momentum emits two spinning particles in opposite directions, and the direction of the spin of one is measured, then we immediately know that of the other. Some would then say that this shows that the spin of the second has a definite predetermined value (that is not included in the description of the system by the statefunction, and thus that the statefunction is not "complete"). This seems quite reasonable, provided, of course, "definite predetermined value" can be defined. If it cannot, then this statement allows quite interesting (infinitely long) philosophical discussions — and obviously ones also without content. While the statement can be made into a correct one, the implications that (too) many people believe follow from it actually do not, and such implications are fallacious. But unfortunately, because of the great confusion, they do require detailed discussion.

We first consider, as is always useful, classical analogs.

V.3.a The EPR "Paradox" and classical mechanics

The correlation in this EPR experiment (with opposite spins) results from conservation of angular momentum. Consider a spherical, non-spinning (spin-0) classical shell at rest. It explodes into two spinning shells; these of course have spins equal and opposite. Assume that in this classical case of two particles with opposite spin we determine whether the spin is up or down, but are unable to measure the component (so like the spin-$\frac{1}{2}$ case, we can only get one of two values). Then we find that the spins of the two particles are always opposite, no matter in which direction we pick the axis. It might be argued that a measurement along the line perpendicular to the actual spin gives zero. However since the angle giving that line is of measure zero, this is impossible.

The spin direction depends on details of the internal construction, which we have no way of knowing. Spins can point in any direction, but by conservation of angular momentum must be opposite. We now measure the direction of spin of one shell along an arbitrarily chosen x axis. We thus know, instantaneously, the direction of the other; the spin component of the other, now in a different galaxy, is then in some magic way immediately forced to be in the opposite direction from that of the first, along this arbitrary x axis. The other shell is instantaneously told what value its spin component must assume, along that arbitrarily chosen axis — and immediately changes it to that. This proves that there is an influence of one measurement on the other, thus that information travels faster than the speed of light showing the spooky action-at-a-distance of classical physics, and that it is nonlocal.

We simultaneously, in classical physics, measure the spin component along an arbitrary y axis. That of the other is then immediately forced to be opposite, along this arbitrary y axis, from the direction of the first. In addition we simultaneously measure spin along an arbitrary z axis (axes need not be orthogonal). The component of the other object is immediately forced to be in the opposite direction, along this arbitrary z axis, from the first. Thus measurements must be transmitted from the first particle to the second, instantaneously, faster than the speed of light. This is the argument that "shows" that quantum mechanics is nonlocal.

Undoubtedly many other examples from classical physics can be developed [Feldmann (1995)]. They should, in one way or another, help eliminate the confusion so rampant about these subjects.

V.3.a.i *Classical physics is even more "nonlocal"*

But as we see, for this EPR experiment classical results are even worse — for not only does measurement of spin along one axis immediately tell a far-away apparatus how it should respond, but also measurements along all three axes do: classical physics is three times worse. It is true that in quantum mechanics we can measure a spin along an arbitrary axis, and then know that of a measurement of the other particle along that axis. But in quantum mechanics we can only measure along a single axis, in classical physics we can measure along three, so know more about the other particle. Thus classical physics is even more "nonlocal" than quantum mechanics — it is even spookier than quantum mechanics.

How does this argument about classical physics differ from the corresponding one for quantum mechanics? Maybe physicists ought to report to the media the important discovery of spookiness and nonlocality in classical physics, since there is so much interest about this in quantum mechanics.

Of course classical physics is not spooky, but neither is quantum mechanics. It is true that classically both spins have been determined at the beginning, but that is exactly the same in quantum mechanics. It makes as much sense to say that the spins don't exist until we measure them in classical physics as in quantum mechanics.

V.3.a.ii *There are no acausal signals, ever*

There are, of course, no acausal signals — nor any other kinds of communication involved, even in classical physics. Directions of spins of the two shells are always found opposite. Thus when we measure that of one we immediately learn the direction of the other. No signal is required. This is a mistake that those considering the EPR "paradox" in quantum mechanics make. They think that we have to inform the second shell that we have measured the first. If the shells had rigidly attached fins which were thus rotating, we could find the directions of spin by just looking at them.

For a classical object torque and transfer of energy can be made vanishingly small, which is more difficult in quantum mechanics. However this is irrelevant since a measurement tells us the spin direction before that measurement, thus the spin direction of the other shell, very far away, instantaneously. And the two spinning classical shells are not in a spin 0 state originally for they did not exist. They are the two parts of the original sphere, so are created in the explosion.

V.3.a.iii *Does an object have spin before being measured?*

To say that a particle does not have spin until it is measured is like saying that it does not have mass until it is measured. Both statements are vapid. There is no experimental way of checking them, so they don't really say anything, and that is true in quantum mechanics and also classical physics. What can be said is that if we measure the spin of one particle along an arbitrary direction, then we know that if we measure the spin of the other along the same direction, no matter when, no matter where it is, the spin will be opposite. But this is exactly the same as in classical physics, where it is three times worse.

V.3.a.iv *Electromagnetic radiation gives another example*

Another classical example, but with a fixed axis, is an antenna that sends two electromagnetic beams, of opposite polarizations, in opposite directions. If we measure the polarization of one, we know immediately, as does the other beam, the polarization of the second, no matter how far away.

An electromagnetic wave can be either of two states, with clockwise or counterclockwise rotation [Mirman (1995c), sec. 2.4, p. 20]. A linearly, or elliptically, polarized wave is in a superposition of states. It is an entangled state, and provides a classical example of entanglement, for which the same argument holds.

V.3.a.v *Where is the money in the bank?*

It is not possible to assign a determinate spin to the second particle until after a measurement on the first. This is true in exactly the same way in both classical physics and quantum mechanics, except that for the latter we can say less. As an example of this consider having money in two banks, one in New York, the other in Switzerland, and knowing the total but not the amount in each bank. In that sense the amount in each bank is indeterminate. If the amount in one bank is determined, this instantaneously gives the amount in the other — so that the banking system is nonlocal.

V.3.a.vi *There is no actual physical state before measurement*

Another concept that has caused confusion is the "actual physical state ... " and the question of what is it before it is measured? This question asks nothing for the phrase says nothing — the state can't be measured before it is measured. What we can say is that if we measure the spin along any arbitrary axis then we know that of the other particle along that same axis (opposite). This is true both quantum mechanically and classically.

It makes no sense to say that the second particle had a spin direction before measurement but did not know what it was until we found the spin of the first.

V.3.b The basic reason for confusion about the EPR experiment

Much of the confusion about EPR type experiments comes from forgetting that expressions for statefunctions refer to ensembles (and this is not the only case in which such carelessness leads to confusion; it is quite common (sec. III.2.i, p. 116)). Thus consider a spin-0 object, at rest (an idealization that can itself lead to errors), decaying into two spin-$\frac{1}{2}$ objects, one with spin up, the other down, one moving right, the other left. We write the statefunction as (schematically)

$$|s\rangle = |l, u\rangle|r, d\rangle - |r, u\rangle|l, d\rangle, \qquad \text{(V.3.b-1)}$$

and say that when we measure the spin of one, we "force" the other into a state of opposite spin, no matter how far away it is, which seems to lead to all sorts of strange paradoxes, including beliefs that quantum mechanics is nonlocal. By conservation of angular momentum the statefunction must be that of a scalar, giving a coherent superposition, and it is on this that the arguments and "paradoxes" are based.

However expressions for statefunctions are for ensembles. A superposition has no meaning if used for a single case. Thus we do not know that this is the statefunction; we would have to measure it to show that it is. But then we would have to measure both spins, and also destroy the coherence — thus (unfortunately?) the paradoxes. For single events, statefunctions, including this, are vacuous.

We can say nothing about the system before measurement (except that the bodies have opposite spin and momenta, and that the total angular momentum is zero); we cannot give spin directions, we cannot give momentum directions, we cannot give statefunctions, a concept not relevant for a single event. We know nothing about the system before measurement, not the spin of the objects, not their momenta, not the statefunction of the system, nothing (sec. II.2.f, p. 73). Thus to say that a measurement "forces" the object into a state of definite spin direction, and simultaneously "forces" the other spin into the opposite direction is wrong. Since we do not have, cannot have, information about the initial state, there is no way of telling what the measurement does — such statements are senseless, just nonsense.

V.3.b.i *What does conservation of angular momentum require?*

Does not conservation of angular momentum require that the statefunction be a coherent superposition? Conservation of angular momentum is a consequence of the formalism and the commutivity of the Hamiltonian with the rotation operators of the Poincaré group (partly by definition, and also from space being invariant under these operators). But what this requires is that the statefunction, which gives an ensemble, be in a coherent superposition with two terms of opposite sign. In classical physics the two particles are in definite spin states, but opposite, and this is completely consistent with conservation of angular momentum. Why should it be different in quantum mechanics?

Conservation of angular momentum follows from the invariance of space (so the Hamiltonian) under rotations. Yet in classical physics that does not prevent a particle from having spin only along one line. And clearly if there is only a single pair, there can be no other possibility. But if there are a large number of shells with zero spin that explode, each into two particles (so with opposite spins), the number with spins along any axis is independent of the axis — by symmetry of space. In classical physics the isotropy of space requires that the distribution of spin directions of particles resulting from decays be circularly symmetric, but of course not the spins of a single pair of particles. And this is also the requirement in quantum mechanics, exactly the same.

Thus decay of a single object gives an example of broken symmetry. And it emphasizes that to show symmetry of space ensembles of particles are needed, both in classical physics and in quantum mechanics. It is impossible to look at a single decay and tell whether space is symmetric, so whether angular momentum must be conserved (although it is in each case studied, which does not show that it must be true for all).

If, in classical physics, the angular momentum vector, which is along z, is viewed in a different coordinate system, the component along z' is different. But if all that we can tell is whether the component is + or -, which is all that we can tell in quantum mechanics, this difference is meaningless — experimental results are the same in all coordinate systems for which the angle between z and z' is acute. In this sense classical physics and quantum mechanics are the same.

V.3.b.ii *Why coherent superposition cannot refer to a single object*

It is impossible to have a coherent superposition for a single system since "coherent" requires a definite phase, which can only be determined using ensembles. Also phase and particle number are conjugate variables [Mirman (1995b), sec. 8.4, p. 158] and if the spin component of a single particle is measured (in order to "force" the other single particle in another galaxy into a definite spin state), the particle number is determined — it is 1 — so the phase is completely unknown, which makes a coherent superposition impossible. Hence it is absurd to discuss forcing the other particle's spin to have a definite direction.

There is difficulty, as is well-known, in defining a phase operator [Barnett and Dalton (1993); Bialynicki-Birula, Freyberger and Schleich (1993); Carruthers and Nieto (1968); Leonhardt and Paul (1993); Vaccaro (1995)]. However this does not contradict the, here key, point that if we know either the phase, however it is defined, or particle number exactly, we cannot know the other.

Since phase and particle number are conjugate variables, using coherent superposition or entanglement for a single particle, or to describe an experiment involving only a single particle (or single pair), is inherently inconsistent — a direct violation of rules of quantum mechanics. It is no wonder that paradoxes and weirdness appear.

Those many experiments that purport to show that measuring the spin of an object forces the direction of spin of the other of the pair, in a different galaxy, into the opposite spin state actually measure correlations of probability distributions, as can be seen by carefully reading the descriptions of them, thus cannot apply to single pairs. That is they are measuring relations between ensembles. Although experiments are correct, and mathematical results drawn from them are correct, statements about them and conclusions drawn from them are in straightforward contradiction to the actual nature of such experiments.

While there are indications that some realize that correlations are necessary [Page (1982)], this does not seem to have been stated explicitly enough or often enough.

Use of a wavefunction for a single event is quite common, and is one of the causes of so many mistakes about quantum mechanics. Much of the discussions of "collapse of the wavefunction" really result from collapse of the understanding that a wavefunction applies to ensembles, and cannot apply to a single event. Thus statements [Dürr, Fusseder, Goldstein and Zanghi (1993)] like prior to the measurement of position (say in a double-slit experiment) the wavefunction is a coherent superposition of up and down parts and the measurement forces it to collapse into one of these pieces, are based solely on this misinterpretation of the meaning of a wavefunction. If there is a measurement of a single case it is impossible to discuss what the wavefunction was before that measurement. Such statements about collapse are self-contradictory.

Misinterpretations like these are in part a consequence of physicists repeating without thinking what others, no matter how far away, are saying. Thus they demonstrate spooky confusion-at-a-distance.

V.3.b.iii *What might the statefunction be?*

Why not say that the statefunction is either

$$|s) = |l,u)|r,d), \text{ or } |s)' = |r,u)|l,d), \qquad \text{(V.3.b.iii-1)}$$

and then say that when we measure the spin of one particle we discover which of the two statefunctions is correct, and that tells us the spin of the other, which is not forced into a state but was in it since the decay? Then there are no paradoxes. However since the z axis is arbitrary, up and down are also. Suppose that we rotate axes, and the correct state is

$$|s) = |l,u)|r,d) = \begin{pmatrix}1\\0\end{pmatrix}_l \begin{pmatrix}0\\1\end{pmatrix}_r . \qquad \text{(V.3.b.iii-2)}$$

What does it become? The rotation operator [Mirman (1995a), pb. X.5.a-1, p. 285; Varshalovich, Moskalev and Khersonskii (1988), p. 49] is

$$Q(\theta) = \begin{pmatrix} cos\frac{\theta}{2} & sin\frac{\theta}{2} \\ -sin\frac{\theta}{2} & cos\frac{\theta}{2} \end{pmatrix}, \qquad \text{(V.3.b.iii-3)}$$

giving the rotated state

$$|s_r) = \begin{pmatrix} cos\frac{\theta}{2} \\ -sin\frac{\theta}{2} \end{pmatrix}_l \begin{pmatrix} sin\frac{\theta}{2} \\ cos\frac{\theta}{2} \end{pmatrix}_r . \qquad \text{(V.3.b.iii-4)}$$

What are the probabilities of finding the left-moving object in the up state, and the right-moving one down, and the reverse? These are

$$P(u,d) = cos^2\frac{\theta}{2}, \quad P(d,u) = sin^2\frac{\theta}{2}. \qquad \text{(V.3.b.iii-5)}$$

Since we do not know which state $|s)$ or $|s)'$ is correct, to find these probabilities we have to add the ones for the latter, for which the sine and cosine are interchanged so that the probabilities are independent of angle. Probabilities are given by ensembles, refer to ensembles only, and ensembles contain both states.

For the rotated state of a single pair, what is the probability of finding the left-moving object in the up state, and the right-moving one also up? This appears to be

$$P(u,u) = (cos^2\frac{\theta}{2})(sin^2\frac{\theta}{2}). \qquad \text{(V.3.b.iii-6)}$$

In classical physics if two spins are opposite (giving spin 0), they will always be opposite no matter which axes are chosen. Yet here there is a probability of finding them in the same direction. However what is wrong with this is that we cannot talk about probability for a single case — an oxymoron. The state cannot be rotated, but only projected onto a new axis. A rotated state gives the probability as seen in the new coordinate system. But it has no meaning for a single event, so cannot be rotated because the state cannot be multiplied by a factor since in an experiment we can find only one of two values, say + or -, therefore the factor cannot change anything. Only if there are a large number of cases does it have meaning.

V.3.b.iv *What showing coherence requires*

That objects have spin along a definite line does not violate conservation of angular momentum, either in classical physics or quantum mechanics. There is however a difference in the (type of) expressions describing ensembles. But whether this difference is significant, or just a matter of formalism, depends on whether it is possible to measure the phase, say in an interference experiment, requiring ensembles, and thus show that the superposition is coherent, and with the proper sign.

What is the difference between these two states, the coherent superposition and the pair, which is not such a superposition? To distinguish them we have to show that there is coherence, and find the phase. But to do this we need an interference experiment, that is a large number of decaying objects with their decay products interfering. Coherent superposition refers, necessarily, to ensembles, only. If we use an ensemble then we cannot measure the spin of one object and tell the direction of spin of its partner decay product. To do that we would have to pick out, from the large number of particles, the one that is its partner, which would destroy coherence (as could a measurement of the spin).

However there is a more basic problem. For an interference experiment, phases of decay products of different decaying particles must be correlated. But this is impossible. If two different spin-0 particles decay into two pairs of spin-$\frac{1}{2}$, there is no relationship between the phases of the pairs. Objects resulting from the decay of different objects can have no meaningful relative phase [Mirman (1995b), chap. 4, p. 58].

Also it is generally believed that the statefunction of two fermions must be antisymmetric. But this does not, cannot, hold unless they interact [Mirman (1995b), p. 191].

V.3.b.v *Entanglement is not possible for a single object*

It is usually thought that a state resulting from a decay is an entangled state. But this is a superstition, not an experimentally verifiable, experimentally intelligible, or theoretically justifiable, conclusion. It cannot be theoretically justified because words, phrases and sentences used to do so, with content elsewhere, have none when used in this context, since "entangled state" and "a decay" contradict each other.

Related to this is the concept of "which-way" or "both-path" experiments [Kim, Yu, Kulik, Shih and Scully (2000)] giving a choice between "particle" and "wave" aspects of quantum mechanics. These are often written so as to imply that a decision about what to measure is made for a single particle suggesting strange properties of quantum mechanics. Actually they involve interference experiments so necessarily ensembles and correlations relating probability distributions (not single events). When considered what such experiments are actually about (ensembles) there is nothing strange or surprising in them. They have nothing to do with particles or waves, and using such irrelevant terms just leads to befuddlement.

We cannot discuss the state of the object before its spin is measured, we cannot say that it is in a coherent superposition, or that it is not. These are strings of words that have no denotation here. Thus it is logically impossible to predict results of a measurement from a state before the experiment, since we can have no knowledge of that state. All that we can say (ignoring uncertainties due to such things as not being able to determine the position and momentum of the decaying particle exactly) is that if we measure the spin of one (single) object along any axis, and then that of the other, they will be opposite — if we measure the spin of one the spin of the other is not exactly in the same direction. Also we can give probability distributions — for ensembles.

V.3.b.vi *In classical physics there is a meaningful state before measurement*

In classical physics it does make sense to consider a state before a measurement. It is possible that people (even ourselves) measured it before we (said we) did, and that possibility gives it meaning. That measurement could have been for all three components (not just one) along any set of axes, and with each component determined (in principle) with arbitrary precision, not merely its sign, and this could have been done with arbitrarily small effect on the object so on our measurement (in principle). It does not matter whether the experiment was actually carried out; it could have been.

In classical physics, from results of a complete measurement on an object we can retrodict results of a measurement should one have taken place before ours, so the state of the object before the measurement has meaning. Our determination of it can be verified experimentally by checking whether the measurement whose results we have retrodicted was made by anyone else (or even by ourselves).

None of this is true in quantum mechanics, say for a measurement of the spin of an electron. What can we say in quantum mechanics? If we measure the spin component then, by conservation of angular momentum, we take it to have had that component before the measurement. That is a state is what we measure it to be, but beyond that it has no independent significance. It is simply another way of giving results of an experiment. We know, and can know, nothing more about a state than this. Predictions based on nonexistent knowledge are, of course, likely to lead to strange views of nature.

For a single experiment, quantum mechanically we can say nothing about a system before measurement — there is no meaningful set of words that refers to it. Classically it can make sense to consider a single system before measurement. In quantum mechanics, for an ensemble of experiments, but only for an ensemble, the statefunction before measurement can be given because this gives the probability distribution of the results of these experiments, a concept that has no meaning for a single one.

V.3.b.vii *How do we know the position before measurement?*

However we can give the position of an object during some interval preceding the measurement, even though we could not have found it without preventing our measurement. Here we use theory, knowing the position and momentum (within the precision allowed by the uncertainty principle) at some time we can tell the (approximate) position at a different time (assuming, of course, that all interactions are known). It would not be possible to have a coherent view of nature without using theories like that giving position, though we could not check it experimentally in each detail (like the particular case of an object whose spin we measure). The difference between the spin component and position is that the latter does have an independent meaning. The value of the spin component immediately before measurement is the value obtained by the measurement, while for the position we can calculate a value, and we need a theory (like Newton's First Law, or Schrödinger's equation). And it is possible that values found are wrong, if, say, the potential used in finding them was wrong.

Thus results of experiments on position and momentum do have meaning, the position at any time before the measurement has meaning if it could have been measured (with some reasonable degree of accuracy), still has meaning if it could not have been, but less, while the spin component has no meaning beyond being a restatement of experimental results — there are no experiments or theories that might not be correct that we can use to give it an independent meaning.

V.3.b.viii *How the double-slit experiment differs*

The decay of a particle is different from a double-slit experiment for which a statefunction of a single particle passes through two slits and interferes with itself. The resulting interference pattern gives the probability of a dot being formed as a function of position. But since there is only a single dot, we cannot determine that there is an interference pattern. Only with an ensemble, a large number of particles, does the probability distribution (which has the form of an interference pattern) become manifest.

V.3.b.ix *The analysis is the same for photons*

Do photons differ? Their states are those of circular polarization [Mirman (1995c), sec. 2.5.3, p. 26], and a linearly polarized state is a sum of two circularly polarized ones with a definite relative phase. If an object emits two photons with opposite polarization, and the direction of linear polarization of one is measured, that of the other is known.

A state of circular polarization can be written as a sum of states of linear polarization [Jackson (1963), p. 205], and conversely. Writing a circularly polarized wave as a sum of two linearly polarized ones requires a phase difference between them. The value of this phase difference gives the direction of the polarization with respect to the x axis (so changes if that axis is changed). If a scalar object emits two opposite electromagnetic waves they have opposite circular polarization. They each have a relative phase difference between their x and y unit vectors which is opposite. Thus if the polarization of one wave is measured along some axis (the photon is sent through a polarizer aligned along that axis), the polarization of the other, along that same axis must be opposite.

V.3.b.x *Creation of linearly polarized beam*

While a photon is circularly polarized, a beam can be linearly polarized. How is such a beam created? For a polarizer to allow one direction of linear polarization to pass through, giving such a beam, incoming photons must interact with it (with its atoms). Thus an incoming photon is absorbed, and photons of opposite circular polarization emitted — with proper phase relationships so that the beam is linearly polarized and in the direction allowed by the polarizer.

If we model the polarizer as a set of oscillators that can oscillate only along one axis, an incoming circularly polarized photon can be thought of as having its electric field vector rotating and absorbed if it is in the direction of the oscillator, but not if in the perpendicular direction at the instant it passes the oscillator, and there will be a probability of absorption for other directions that depends on the angle. Of course since there are a large number of atoms, the photon will almost certainly be absorbed by the slab. After an oscillator absorbs a photon, it can then emit other photons, and because it can oscillate in only one direction, the ones of opposite polarization must have a relative phase giving linear polarization in the direction of oscillation.

The analysis is then the same as for two photons emitted by a scalar object in opposite directions.

V.3.c If quantum mechanics is time-symmetric, what is "before"?

One word that we have to be careful of is "before". Here "before" means during the time (and correspondingly space) interval between the decay and our measurement. It does not imply a direction of time. Is the procedure reversible? In classical physics, knowing exactly the complete state of the entire system, including that of every experimental apparatus, we can find the state at any other time, "before" or "after" our measurement. Here before and after means when the state is less and more complicated, if complicated is properly defined. The same is true for quantum mechanics (ignoring such things as decays of K mesons). Knowing the complete statefunction of the entire system at any time allows the determination of it at any other time. This is true whether we have obtained the instrument readings or not. If we have, then the statefunction of the complete system must include that of our bodies. Thus quantum mechanics is, like classical physics, reversible.

There is however a difference. In classical physics we can obtain (in principle) the state of a system with arbitrary precision, and of a single system. But determination of a statefunction requires repeating an experiment many, many times, with everything identical for each run — including our own bodies and brains, and all other external influences. In reality, but not in principle, time reversal is not possible, either in classical physics or quantum mechanics, though is more difficult to approximate in quantum mechanics.

V.3.d Is there a violation of causality?

One of the beliefs that confusion about experiments like these has lead to is that causality can be violated — signals can (or might) be able to travel faster than the speed of light. Thus if it were possible to decide on the direction of spin of one object of a pair and then immediately force the other object to have opposite spin, this would contradict causality. But, as we see, this is not possible. Spins of the two objects are correlated, but only statistically. Measuring the direction of one spin of a specific, single object tells us nothing about the other one of that particular pair, except that it is not in exactly (with measure zero) the same direction. If we measure the second spin we do not find it to be exactly in the same direction as the first. But we can tell nothing further about what the measurement might show.

Measuring probabilities for one tells us about probabilities for the other, but this requires many, large numbers of pairs. To use the statefunction for zero angular momentum, which requires a definite phase between the two terms, a minus sign, for a single event violates the uncertainty principle. Phase and particle number are conjugate variables.

Analyses that lead to beliefs that measurement of the spin direction of one particle of the pair produced by a decay forces the other into a definite spin state, even if it is in another galaxy, are inherently self-contradictory. They directly contradict the basic rules of quantum mechanics. It is not that quantum mechanics is weird, but conclusions that people draw from it are exactly the opposite of what it gives.

Thus there is no question of causality ever being endangered.

V.3.d.i *Bell's inequalities*

Bell's theorem has been called one of the most profound scientific discoveries of the century if not of all time [Aspect (1999); Whitaker (1998)] (the discovery being specifically that quantum mechanics is actually correct; our teachers really weren't lying to us after all). This leads to Bell's [Aspect (1998)], and related, inequalities [Yurke, Hillery, and Stoler (1999)]. These are undoubtedly important, and show that quantum mechanics holds true (of course), while other theories cannot properly describe nature. However it is essential to remember that they give correlations of probabilities. They do not, and cannot, say anything about a single event, but must refer to ensembles (being inequalities they necessarily require many experiments).

It has been suggested [Mermin (1985), p. 41] that people might be bothered by Bell's theorem and the EPR experiments, although some are not bothered but refuse to say why. There are those who are (or were until they were told they are not supposed to be) bothered by Newtonian physics, prefering the Aristotelian view of nature. Those believing that people who are not bothered by Bell's theorem should say why, first have to explain why they they are not bothered by Newtonian physics.

Bell's theorem and the EPR experiments involve space-like separated events, but only statistically. People forget this and get all confused. Even those who state this explicitly often appear not to believe it [Mermin (1981,1985)].

Many weird paradoxes arise from reasoning (?) that is inherently inconsistent, applying conditions for ensembles, or using coherent states, for single events. These lead to strange conclusions, like nonlocality of quantum mechanics, but these, and assumed implications of related experimental tests and similar experimental results, are wrong. Conclusions are based on using for a single event conditions on multiple ones, which are not relevant, nor can they be correct, since they necessarily refer to ensembles.

V.3.d.ii *Quantum mechanics is local*

Quantum theory does not imply correlations between spacelike measurements. This view has arisen only because of sloppy thinking. Quantum mechanics is not nonlocal, although statefunctions are defined over regions (as are electromagnetic and gravitational fields), but all terms in correct Hamiltonians are local. Quantum particles cannot instantaneously influence the behavior of distant particles. Nor is there the slightest experimental evidence that they can, or reason to believe so. Is quantum mechanics nonlocal? To say that it is means that we also have to say the classical electromagnetism is nonlocal. Their objects are both functions, the statefunction and the electromagnetic field (actually potential). There is no difference. But all interaction terms are local (all terms in products giving interactions are taken at the same point). It is in this sense, the only valid sense, that they are — both — local. Of course, many people believe that quantum mechanics is nonlocal (but not classical electromagnetism!) without saying what they mean by nonlocal. And if they mean that an observation at one point influences that at another spacelike separated point, obviously that is wrong, and in disagreement with experiment — and quantum mechanics.

There also seems to be an impression that quantum mechanics is a result of special relativity and nonlocality (which obviously it cannot be, since it is not nonlocal). But it really follows from geometry, and requirements of transformation properties, and the properties of the groups [Mirman (1995b)].

V.3.e Be careful with formalism

Formalism is a language, and like all languages is dangerous, so must be used with extreme care. Intended to convey information, it easily conveys much misinformation, leading to presumptions that are often quite strange, quite weird, highly paradoxical. But these do not reflect reality, but rather misuse of language.

Thus we cannot say that measurement of one spin forces the other to assume a definite value. There is no paradox, no nonlocality, no messages traveling faster than light, or any of the other problems people believe are shown by this type of experiment. It is important to be clear about differences between statefunctions and our expressions used for statefunctions in particular situations — and also whether these expressions are actually correct, or are merely based on formalism that is not relevant to the situations.

So much confusion, so many wrong beliefs like the one that quantum mechanics is nonlocal, have arisen because many discussions are inherently self-contradictory, for example ignoring the need for statefunctions on which they are based to refer to ensembles, while considering experiments that are for single measurements [Mermin (1981,1985)]. Both cannot be correct at once.

This type of experiment is one of the main reasons that so many people believe that quantum mechanics is "weird". But there is nothing weird about quantum mechanics. What is weird is that so many physicists are so careless, so thoughtless, and "explain" quantum mechanics on the basis of perceptions that are clearly wrong, clearly internally contradictory. Indeed these analyses show that the world is an extraordinarily peculiar place [Mermin (1981)], but not because of quantum mechanics but because so many physicists are so careless, so thoughtless. It is really extraordinarily peculiar that they are. There are even attempts to show how extraordinarily peculiar it is by using classical objects to prove that quantum mechanics is strange (and they seem carefully, and deliberately, designed to show how mysterious quantum mechanics is). Quite peculiar indeed.

While there has been much carelessness and confusion about these topics there have been more careful analyses, and with some views having a degree of overlap with ones presented here [Hartle (1968), p. 709]. But the number of careful analyses is far too small.

The obsession of (almost? all) physicists with the "weirdness of quantum mechanics" [Leggett (1999)], has resulted in much absurdity. An elementary physics student who calculated the time it takes for a ball to fall from a table to the floor and found it took a week would not dare hand the problem in. Something obviously was done wrong. And a professional physicist who received such a paper from a student would be very indignant. Yet that same professional physicist believes that measurement of the spin of one body immediately forces the spin of the other decay product into an opposite direction, even if that is in another galaxy, which is even more ridiculous than the week it takes for the ball to reach the floor. Something obviously was done wrong. It is unfortunate that physicists do not apply to themselves, and each other, at least the very minimal of standards that they require of their students.

Students are supposed to know if an answer is nonsense and to go back and check to find what was done incorrectly. Professional physicists are not supposed to know that.

V.4 Beliefs and statements weird and wild

Physicists, and journalists, say — and think — some of the most bizarre things about quantum mechanics and nature. If these weren't so harmful they would often be quite hilarious. For the amusement of those who enjoy absurdity, to try to restrain the damage that they do, and to discourage such nonsense, some (but only some) of these are collected here.

V.4.a Here, there and nowhere

One of the more preposterous statements (sec. IV.2.g, p. 160) about quantum mechanics [Alda (2000); Cirac and Zoller (1999); Percival and Strunz (1998), p. 1817; Rodgers (1998)] is that particles like electrons and photons are able to be in two or more places at the same time. (Are physicists, who are made of electrons, able to be in two places at the same time? If a physicist sailing on an empty sea does not know his position, is he everywhere at sea simultaneously? And a physicist who does not know in which bank his grant money is in presumably believes that it is in all banks at once.)

That a particle is in two places at the same time means that an experiment determining its position would get two different results simultaneously. Of course this has never been done, nor can it be. (However those who believe that objects can be in two places at once should be encouraged to perform, or at least suggest, such an experiment.) There is a probability distribution for finding the particle which is nonzero over a region of space, not merely at a point. But that does not mean that the particle is in many places at once — only that if we perform many experiments, an ensemble, the positions found differ.

This is not unlike classical probabilities, which give probabilities, until a measurement is made, and then the experimental value is determined. If a box contains a ball, which could be either white or black, would we say that it is both colors simultaneously? If the ball can be in either of two boxes, is it in both simultaneously? Undoubtedly those who believe that electrons and photons are able to be in two places at the same time would say that it is.

Some [Chang (2000)] believe that a quantum mechanical particle whose statefunction gives a nonzero probability of it being found anywhere in a range (as all do) differs from objects — buses, tacks, friends — that exist at one place at one time. Either they are here, or they are not. Those who believe this undoubtedly have never waited impatiently for a bus, which since they do not see it is everywhere on its route simultaneously and might appear (bizarrely?) in the next few seconds, or might not for the next hour. Quantum mechanical particles thus differ (how?) from a child swinging back and forth in the dark, who can be either on the right or left side of the arc. If someone shines a flashlight and the child is not on the right side then in quantum mechanics looking resets the swing to the left and alters the probabilities of where the child will be seen later. Of course if we know that the child is not on the right it must be on the left, so we have more information about where it is likely to be if we look again. It is however possible that those who take this view (of physics, not of the child) have children who are quantum mechanical objects, or ones who run off in the dark making it impossible to conclude that if a child is not on the right, it must be on the left.

Let us hope that the strangeness is only in their (bizarre) views of quantum mechanics, not in their children.

There is another concept related to the belief that a body can be in two places at the same time [Leggett 1999)]. This refers to a statefunction that is split into two well-separated parts, each a narrow wavepacket. In this case the value of the statefunction is almost zero except in these two regions.

Of course it can never be exactly zero as it must be analytic. Like an electromagnetic field it cannot suddenly disappear but must either remain slightly nonzero or terminate on a singularity, that is an object. However in quantum mechanics we cannot think of a body, as we can roughly in classical physics, as a point. Instead it also is an analytic function of space. Thus what it means for a statefunction to terminate at a physical object, if possible, remains to be investigated.

The problem with this view of the body being in two places at the same time is that it is not in two places, but (always) in a continuously infinite number of places — the statefunction is a continuous function of space. In this experiment it is approximately confined to two narrow, but not infinitely narrow, regions — these being finite give a continuously infinite number of places where the body might be.

Where then is the body, for example an electron? If we measure its position we get one single value (actually a range since measurements cannot be infinitely precise). We never get two values. But if the experiment is done many times, each gives a single value, with that value different for each repetition and it is almost always in one of the two regions, almost never outside them. The statefunction itself is not defined over a single region and, like an electromagnetic field, is therefore everywhere at once, so behaviors of other objects affected by this body are determined by the continuous function (but this must be shown by an ensemble of experiments).

Again these arguments apply a concept that holds only for ensembles to a single experiment (sec. V.3.b.v, p. 216), so are necessarily self-contradictory.

Statements about a body being in two places at the same time thus have an interpretation which is ridiculous, and wrong. Of course, they can also be a confusing form of shorthand for a correct statement. It is unlikely however that people who say such things really know what they want to say (if anything), or what they mean.

V.4.b Nature does not depend on human beings, but confusion does

Perhaps a paramount difficulty (appearing in many different ways) results from basing quantum mechanics on human beings — as if the laws of nature depended on our presence, on our consciousness. Suppose that life consisted (as it once did) only of archaea and bacteria. Would quantum mechanics be different? The implication from the centrality of humans in the measurement process is that it would. And when did it change to its present form? When worms arose? Fish? Amphibians? Mammals? Primates? The genus Homo? The species Homo Sapiens? Or perhaps when physicists first appeared (or only when those who study quantum mechanics did)? These are questions that should be answered by Homo Sapiens who base the properties of quantum mechanics (and more generally nature) on the nature of humanity, on human consciousness.

V.4.b.i *Our observations are not properties of systems*

There is another error, perhaps more subtle. If we do an experiment, say determine when atoms decay, we find fluctuations, atoms decay at different times, and the set of times differs for the various repetitions of the experiment. These are experimental results that we observe. But regarding our observations (such as fluctuations) as characteristics of the system is erroneous. These characteristics are given by statefunctions, and there are no fluctuations in statefunctions. Once again we regard properties as being determined by our observations, obviously absurd, rather than taking our observations to give information about properties and about nature.

V.4.b.ii *Confusion fluctuates, but spacetime does not*

Thus, for example, statements that quantum mechanics gives spacetime fluctuations (in, say, the metric) are not only wrong, but nonsensical (sec. V.7.c, p. 252). We can consider fluctuations of, say, heights or forces, as functions of space or time, or results as functions of experimental runs. But what does spacetime fluctuate with respect to? This view assumes that because experimenters get different values each time they make a measurement, actual physical values (which do not exist because they cannot be determined), like that of the metric, do actually fluctuate (which is meaningless since the "actual" fluctuations cannot be observed, but only variation of measured values).

Correctly quantum mechanics gives probability distributions for experimental values, and our observations do not make them vary.

V.4.b.iii *Can a system exist in all states at once?*

Another example of the belief that nature depends on human observation is the statement that until a system is observed it can exist in all quantum states at once. (How?) This is really an oxymoron since it says that before a system is observed we can observe it in all states at once. But if we wish to believe this then we have to say that it is true after the system is observed also — an object up along z, measured so, is in a superposition of up and down states along z' (sec. IV.2.b, p. 155).

If the statefunction of a system is a superposition of states (say eigenstates of an operator) that does not mean that the system is in all states simultaneously (with "in" and "simultaneously" having no meaning here), or has different properties (like up and down) given by two states, but only that the probability of finding the system in any of the states is nonzero. Of course, statefunctions are physical objects, so can give, say, interference terms if sums. But this does not say that a system has simultaneously different properties — for that to mean anything there would have to be a single experiment that gives contradictory results, which is quite unlikely (except that experiments such as listening to people talk about quantum mechanics do give contradictory results, as these statements illustrate).

And again underlying views such as these is the attitude that properties of a system, like the states it "exists" in, are determined by whether or not we look at it.

V.4.c Human egomania and discussions of quantum mechanics

What is wrong with the statement: "The Heisenberg uncertainty principle, stated simply" (too simply?), "holds that the act of observing an electron perturbs it so that its position and momentum cannot be measured simultaneously" [Jasny, Hanson and Bloom (1999)]? This essentially states that properties of an electron — whether it has an exact position and momentum simultaneously — are determined by actions of human beings observing it. The position and momentum of an electron are governed by laws which give the uncertainty principle because these are properties of electrons, in fact of all matter [Mirman (1995b)]. Whether human beings are involved or not is totally irrelevant. Electrons certainly do not care about us. Laws governing them have nothing to do with whether we exist or not.

V.4.c.i *What the uncertainty principle does, and does not, say*

Nor does the uncertainty principle say that a measurement changes the value measured. To say that we would have to know the value before measurement. Actually of course the uncertainty principle applies to ensembles, being a mathematical statement about them following from, for example, Fourier analysis, not a statement about the effect of an observation.

There is another implication of the statement that is wrong. It implies that the electron has simultaneously a definite position and a definite momentum, but we cannot find these. But as shown by our inability to find them, it does not, thus the statement that it has simultaneously a definite position and momentum is nonsense. Experiments that seem to show perturbation preventing simultaneous measurement do not actually show that the perturbation prevents simultaneous measurement, but rather illustrate how the properties of particles, and laws governing them, affect what experiments can be performed on them, how, and what their results are restricted to be.

Laws of nature govern what we can study and what we can find, our studies do not determine laws.

V.4.c.ii *Discussions, but not measurements, show human hubris*

Discussions of the measurement process often give examples of human hubris. All interactions are measurements, and all measurements interactions [Mirman (1995b), sec. 5.4.2, p. 94]. To believe that there is something special about interactions with human beings, that the wavefunction "collapses" when a person does something, is to believe that laws of nature are shaped by the presence of humans in the universe, and that they are designed to describe our interactions with nature. With people believing such nonsense, no wonder there is so many outlandish ideas about quantum mechanics.

The concept that quantum mechanics is based on human observers, an idea necessary for the confused belief in collapse of the wavefunction, is so strongly held that some even think that in certain conditions quantum mechanics must break down [Ghirardi, Pearle, Rimini (1990)]. Of course that would not be necessary if this strange idea that quantum mechanical laws are designed for the convenience of human beings were to break down, and be replaced by a little thought. There are even those who, to avoid this nonexistent problem with quantum mechanics, introduce the idea that the wavefunction at random times undergoes sudden collapse, that there really are quantum jumps that seem to occur without interactions. Thus human vanity leads to an abandonment of belief in physical laws, replacing them by some magical changes in physical systems. Perhaps it would be better to replace the magic by humility.

An observer is an object, an electron, a proton, a photon, for example, and each observation made by it is an interaction — and each interaction is an observation. But every particle in the universe is interacting with all others (actually with the various objects like photons and gravitational waves produced by others, that are at its position) continuously at all times. Thus it is continuously making measurements, a continuously infinite number of measurements in any time interval. And each particle is being observed continuously. If wavefunctions were to collapse upon observation, they would be collapsing continuously, giving a continuously infinite number of collapses in any time interval. If Schrödinger's equation did not hold when a wavefunction was collapsing, it would never hold. Quantum mechanics would have been fooling us by making us believe that it did hold.

V.4.c.iii *Was the universe designed for the pleasure of theoretical physicists?*

Another example is perturbation theory and the many attempts to use it as the criteria for laws of nature. This is based on the firm belief that the design of the universe was picked especially to make it possible for theoretical physicists to use their favorite approximation scheme for calculations; there are many illustrations of this (sec. V.4.j.iv, p. 242). Of course, it is well-known (but maybe not to those who are so attached to this approximation method) that there are many systems for which it cannot work [Anderson (2000)], so cannot be used in any fundamental way. Could it be that the universe was carefully designed to make the approximation scheme work, but only in a special set of cases?

Protagoras has been quoted as saying that man is the measure of all things. It is on this fundamental postulate that many physicists base their interpretation of quantum mechanics — and their entire world view.

V.4.d Entanglement of statefunctions and ideas

While entangled statefunctions are central in many physical phenomena, such as interference, this concept is often misapplied (sec. V.3.b.v, p. 216). It is impossible to discuss every example (unfortunately there are far too many), some are given above; here we list a few more.

It is believed that in quantum teleportation (sec. IV.2.t, p. 173), the measurement of spin of one photon instantaneously determines the spin of an entangled photon on the other side of the galaxy. But if the spin of a photon is quantized along z, then a measurement will give the spin of an entangled photon in another galaxy, along z. However classically, it will give the spin components along z, and also along x and y. Thus classical physics is far more weird (and entangled) than quantum mechanics (at least three times). Actually it really is, since it is necessarily inconsistent [Mirman (1995b), chap. 1, p. 1]. The problem of course is that the entanglement is the mixing of a concept, statefunction, that is relevant to ensembles with the discussion for a single pair of particles. It is the concepts that are entangled, not the particles, or their statefunctions.

V.4.d.i *Are all objects in the universe entangled?*

Another example is the conviction that every particle in the universe is entangled with every other one. Thus every detection event causes a simultaneous jump of every other particle in the universe (that is if a human being, or at least a physicist, finds the spin of an object then that action by a person affects all other objects in the universe). To say that particles are entangled means that the statefunction of the set is a sum of products of statefunctions. But to say this requires that we be able to find, experimentally, the statefunction of the set — so that if there are two particles, one in our laboratory, the other in a different galaxy, we must be able to determine the statefunction of the pair in order to say that they are entangled. An experiment that finds the statefunction of an object in another galaxy would be quite interesting. Anybody who believes the statement should be able to describe an experiment that reveals this entangled statefunction — and show that the statefunction of a pair of particles in a laboratory on earth and in, say, the sun is really a sum of products. But if there is no such experiment then the statement is nonsense. And if the statement seems weird, it of course is. Meaningless statements often seem weird.

One reason for this belief is the bias that statefunctions of identical particles must be symmetric or antisymmetric in their variables (and the unjustified extension of this to objects that are not identical). However this cannot be correct, and can have no content, unless the statefunction of the objects can be experimentally shown to have such symmetry. If objects interact, then the statefunction of the system resulting from the interaction must have this symmetry [Mirman (1995b), p. 191], provided of course that the experiment is repeated many times. But if they never interact, with themselves or with us, then we can say nothing, we can require nothing. Why should we be able to? And especially we cannot say that if we interact with (measure) one of a pair that measurement affects the other — which has not interacted with either the first nor with us.

Likewise to say that measurement on an object on earth causes a "jump" (whatever this means) in every other particle requires (for it to have content) that we be able to measure the statefunctions of all objects in the universe simultaneously. Again this would be a very interesting experiment. Perhaps more likely the sentence is a set of words that seem to say something, that is grammatically correct, but quickly jumps into senselessness — so is quite misleading.

V.4.d.ii *Entanglement of electromagnetic fields*

Another example of entanglement, classical of course, is that of two electromagnetic fields emitted in opposite directions by an object with zero angular momentum. These have circular polarizations, in opposite directions, so that if we measure the direction of polarization of one, we immediately know that of the other — even though it is a different cluster of galaxies. Also we can break the circularly polarized light into states of linear polarization, and measure the direction of linear polarization of one, giving immediately that of the other — in a different cluster of galaxies. Decisions of what to measure are not made until the waves are far apart. This is even more evident for elliptically polarized light. Here the electromagnetic potential is a superposition of states, the product of that with clockwise rotation moving right and counterclockwise one left, plus the other state with left and right interchanged. This extremely peculiar feature of classical electromagnetism shows the spooky action-at-a-distance, thus nonlocality, of classical physics. It shows the incompatibility of classical electromagnetism and local realism.

Those who say similar things about entanglement should have no difficulty in accepting these very strange properties of (bright) light.

While analogies between electromagnetism and quantum mechanics are often the most obvious, since we treat both as extended functions, much the same analysis holds for a shell that explodes into parts. Anyone who disagrees can undoubtedly show rigorous, relevant, differences between these examples, and ones using quantum mechanical terminology.

Or perhaps it is not the statefunctions that are entangled, just the thinking.

V.4.e Quantitization

Quantitization attempts to take an inconsistent, incorrect theory, classical physics, and turn it into a "quantum theory". This emphasizes, once again, the hold that history has on the present, and how people are programmed, rather than instructed, by it. The concept of quantitization is specious yet it is almost universally taken seriously, as seen in so many places [Echeverria-Enriquez, Munoz-Lecanda, Roman-Roy, Victoria-Monge (1998)]. Part of the reason of course is that it is not noticed, although it should be obvious, that classical physics is impossible [Mirman (1995b), chap. 1, p. 1].

When quantum mechanics was first developed it was not understood, so people had to guess how to use and apply it. What they did was take classical expressions and modify them, the guessing being called quantitization. This actually amounted to guessing correct operators knowing their eigenvalues (the classical expressions). Quantitization does not "promote classical to ... ", but rather finds operators whose eigenvalues the classical numbers are. However guessing is not the proper way, certainly if it is not necessary; it is not a good method guessing operators from their eigenvalues. There are reasons, usually geometrical, for operators, and it makes more sense to use them. Correct operators are mostly given by the relevant transformation groups for systems, plus their basic properties, like spins, masses and the set of objects they interact with. Once these operators are known the rest of the theory follows (in principle).

It is in particular common to "quantize" fields starting from the "classical" ones to make them "quantum mechanical". But this is absurd. There are no "classical" fields, and fields, being necessarily group basis vectors, are always already "quantized". Statefunctions, which fields are, must be group basis vectors because they have to transform under the group (rotations, Lorentz and Poincaré groups, perhaps others) — they must be in the domain of the group operators. Thus they must be basis vectors of irreducible (perhaps not decomposable) representations (sec. I.8.b, p. 48), or must be capable of being expanded in terms of these (if they cannot be so expanded they would be irreducible). Given the satisfaction of these requirements, which are difficult to avoid for a consistent theory, there is little, if any, freedom, or necessity, to "quantize".

We can introduce formalism to express this (and we assume no more than that it is formalism) with creation and annihilation operators. A function is a basis vector and we define creation operator a^*, and vacuum state $|0\rangle$, such that it can be written $a^*|0\rangle$, where a^* is the creation operator for that basis state [Mirman (2001), chap. IV]. If it is a product of basis vectors, we iterate. These operators are really defined by the group, given by its representations, and only provide a simple way of writing functions of its basis vectors. Use of this formalism is called quantitization. But it is pure formalism, there is no physics, no physical assumptions (beyond the requirements of group transformations) involved.

V.4.f Classical physics and quantum mechanics

The existence of classical physics (as a formalism) often seems to hinder the ability (of physicists!) to grasp quantum mechanics. Perhaps if we learned quantum mechanics first, and classical physics later as an approximate model useful in some cases, we would find quantum mechanics much easier to understand.

Thus many people believe that the world is divided into quantum and classical domains, requiring a fuzzy, impossible to define, border between the two. But this is (obviously) wrong. There can not be different (fundamental) laws of physics for objects of different sizes (fortunately).

Fortunately many realize the incorrectness of this [Englert (1999)], but surprisingly far too few.

V.4.f.i *There is no quantum-classical boundary*

There is no boundary between classical physics and quantum mechanics — there cannot be. Quantum mechanics holds always, everywhere, under all conditions, for stars, galaxies, for the entire universe. At one extreme classical physics is completely wrong, and quantum mechanics must be used for all computations and insights. At the other extreme classical physics provides a good approximation scheme, a useful approach, and quantum mechanics, while always completely correct, is useless as a calculational tool. In going from a small number of (microscopic) objects, in which quantum mechanics must be used, to a very large number, in which only classical physics is feasible, a statefunction becomes more and more localized (a wavepacket narrower and narrower in extent), for example, so taking it as if it were nonzero only at a point becomes a better and better approximation. But there is no region in which quantum mechanics suddenly becomes unusable and classical physics suddenly becomes correct — classical physics cannot become correct, either suddenly or sluggishly.

Quantum theory can be, must be, applied to classical objects. However because the deBroglie wavelength of a classical object is vastly smaller than its size, quantum effects are unnoticeable. Thus quantum mechanics approaches more and more the phenomenological classical physics. And this is merely a consequence of standard quantum mechanics.

V.4.f.ii *The transition from Newtonian physics to statistical mechanics is similar*

Deciding whether to use quantum mechanics or classical physics is like deciding whether to use Newton's laws or statistical mechanics. Classically, Newton's laws are always correct. However they become computationally impossible for more than a few particles (although with the advance of computers the word few is taking on new meaning). And we are not (usually) interested in positions and momenta of particles, but rather average values, giving quantities like pressure and temperature. So the choice of which to use is purely pragmatic, and this the same for quantum mechanics and classical physics. Classical physics may be more useful because it is easier to compute using it, and it may provide quantities that we can more easily deal with. And because we are familiar with it, have intuition about it, can form pictures of classical objects, it has great heuristic value. But it is always inconsistent, and quantum mechanics always correct, always applicable (in principle, although not necessarily in practice).

Going from quantum mechanics to classical physics is similar to use of Newtonian mechanics, and approximations appropriate for it, when there are only a small number of particles, but statistical mechanics for 10^{23} particles. Newtonian mechanics is always completely correct (classically), but is unworkable for a large number of particles. There is no boundary at which physics switches from it to statistical mechanics. Rather methods of one become less and less usable (Newtonian mechanics going to a larger number of particles, statistical mechanics going the reverse way), while that of the other becomes more and more so. There is a broad region in which neither works well. This is exactly the same for quantum mechanics and classical physics.

V.4.f.iii *Planck's constant and taking a limit*

It is often felt that classical physics holds, where it does, because the value of Planck's constant is so small. However this is a conversion factor because different systems of units are used in the same equation. It should always, except in calculations, be taken as 1 — units should be picked so that it is 1 (as is the speed of light c). This helps prevent such absurdities as taking the limit as $1 \Rightarrow 0$, or as $1 \Rightarrow \infty$ [Mirman (1995b), sec. A.1.1, p. 177]. Classical physics is a useful model, not because of the small value of 1, but because human beings, for whom it is useful and who regard it as useful, consist of many, many atoms. It is the number of atoms, or the ratio of the mass of a human being (or an instrument used by a human being) to that of an atom, that is the relevant factor, not the value of 1 (sec. IV.4.a.ii, p. 188).

Also people will be less tempted to suggest that the speed of light defines the relationship between space and time, between matter and energy (whatever these mean), and is intimately related (?) to questions of cause and effect [Collins (2000)]. How is the number 1 intimately related to questions of cause and effect? Of course it does define these relations, that is it relates these arbitrary units, just as a conversion factor would relate x and y if they were given different units.

Putting such constants to their correct values, 1, would reduce the number of questions such as what would the world look like if Planck's constant were 25 orders of magnitude larger [Rovelli (1996)]? It would certainly be a fascinating world in which $1 = 10^{25}$. Another question that would disappear if Planck's constant were taken equal to 1 is why is it universal for all types of particle? The answer of course is because it is a conversion factor [Bussey (1988)]. If the units in the x and y directions were defined differently there would be another "universal constant", the conversion factor. It would be quite surprising indeed if this were different for different objects.

These questions are equivalent to asking what would the world look like if the conversion factor between feet and meters were 25 orders of magnitude larger, or why is the conversion factor between feet and meters universal for all types of particles?

And those who wish to take some limit (as $h \Rightarrow 0$, that is $1 \Rightarrow 0$) and consider what the universe would appear to be like if we intuitively saw it governed by quantum mechanics, rather than classical physics, really want to consider a universe in which human beings consisted of two or three atoms. Limits like these do not change quantum mechanics, but rather the size of people (or at least physicists) compared to objects for which quantum mechanics has to be used, like electrons or atoms. If humans were made of two or three atoms, it is unlikely that they would have much intuition at all, or be able to care whether nature seemed classical or quantum mechanical.

V.4.g Constants

Many people regard it as a mystery that the constants of nature have the values that they do. But the constant that is the speed of light has the value 1 (in the proper units), and it would be very surprising if, as some apparently believe, 1 had a different value in other universes (whatever these are).

V.4.g.i *The fundamental constants have no physical significance*

Also there are not three constants. Both c and h should be taken to equal 1, being purely conversion factors, with no physical significance, but result from a purely arbitrary choice of units. We, and objects that we most easily deal with, have masses and lengths of the order of a kilogram and meter, so we choose units in which these are unity. But this is purely a matter of our own sizes, and we have no physical significance. To regard c and h as being physically important would mean that laws of physics are based on properties of human beings (which physicists, at least, should regard as ridiculous — although far too often physicists seem to feel that laws of physics actually are based on human beings, as least those human beings who are physicists, as has often been remarked here).

Thus numerical values of constants, like h and c, are properties, not of electrons, photons, or other such quantum mechanical objects — they have no fundamental meaning — but are merely consequences of the number of atoms of which human beings are composed. They tell something about people, not about laws of nature — and what human beings are like does not determine the laws of nature (no matter how much egotistical human beings wish them to, or believe that they do). The emphasis on values and meanings of these constants shows how deeply self-centered physicists are, how much they base their views of nature on their own existence and on themselves.

V.4.g.ii *The sizes of our bodies determine the values of fundamental physical constants*

To emphasize this let us consider such statements as the value of h is small, or the speed of light is very large. Of course, the question is small or large compared to what? We take $c = 1$, so units of distance and time are the same. A unit of distance can be taken to be the size of an atom.

Why are atoms, thus h, so small? Because they are small compared to us, and to construct something as complicated as a human being requires a very large number of atoms. Thus we are large compared to atoms — atoms are not small because the distance scale should be determined by them.

Why is the speed of light so large — that is why does it travel so far in what we regard as a reasonable time? The proper question is why is our time scale so long compared to the time it takes light to travel across an atom? The reason is that our sense of time is determined by the time scale of our neural impulses, and these need many atomic reactions, all happening slowly (the atoms move slowly) compared to the velocity of light and distance it travels when crossing an atom. Atoms in our bodies are essentially at rest (compared to light). One reason is that the biological processes give them little energy (compared to what they need to travel at close to the speed of light). Perhaps more basic is the strength of their interaction. If they did move at a speed about that of light they would be almost free — their kinetic energy would be vastly greater than their potential energy. But if they were free they could not constitute complex structures (which we are). Thus in the time interval in which a nerve impulse occurs light travels very far, compared to the size of an atom, and so of us.

A human being (certainly a physicist) would have a great deal of difficulty thinking about the speed of light if all its atoms were freely moving.

Questions about the size of atoms, of Planck's constant, the speed of light and so on, are not questions about physics, but about our own bodies. Because this distinction is often not made, some very strange ideas, statements and questions arise.

Many physicists are so enamored with the creed of Protagoras that man is the measure of all things that they base the values of their constants, and their views about the meaning of these, on it.

V.4.h Tunneling

An illustration of how people get themselves so befuddled about quantum mechanics is given by tunneling. One reason is that physicists have an intuitive idea that quantum mechanical objects are point particles.

Also many discussions of the phenomenon are based on an approximation scheme, which can be taken too seriously (something much too common).

V.4.h.i *What tunneling means*

Consider (since details are irrelevant here) a one-dimensional system with zero potential except for potential V between -b and -a, and between a and b — a potential well with barriers. At the initial time the wavefunction is extremely small at all points of space except within the well, and the expectation value of the energy is less than V. Then Schrödinger's equation (and experiment) give that the part of the wavefunction within the well decreases, those parts within the barriers and outside increase. Thus if we measure the position of the object, we find that at the initial time the probability of finding it outside the well is almost zero, but that this increases with time. Thus the object "tunnels" through the barrier, even though its energy is less than V (of course this is true for a classical electromagnetic wave also).

V.4.h.ii *Can the universe tunnel from nothing?*

This seems to lead to a belief [Barrow (1997), p. 172] that the universe is the result of a quantum mechanical tunneling process — that quantum mechanics allows it to tunnel from nothing. What this says is that the wavefunction of the universe (whatever this means) is exactly zero before a certain time (whatever this means) and then it becomes nonzero. That is the wavefunction is nonanalytic. However Schrödinger's equation (which is nonrelativistic) requires that the wavefunction be analytic (a reason that tunneling is necessary). Thus quantum mechanics, rather than allowing the universe to tunnel from nothing (whatever this means), forbids it.

Of course, Schrödinger's equation is rather doubtful as a way of giving the wavefunction of the universe — but then there is nothing that determines any such wavefunction. And if we accept that we cannot determine it, there is nothing that we can say about the wavefunction of the universe. But if we believe Schrödinger's equation governs this wavefunction we would have to explain why a nonrelativistic equation does so, what the wavefunction of the universe means (experimentally), and how it can be, in contradiction to the equation governing it, nonanalytic. (Of course, going to Dirac's equation would not remove the requirement of analyticity.) We would also have to give experimental tests to check whether all these ideas, which seem to come from nothing, are correct.

Of course phenomena that seem to be described by statements like these (but only in the sense that they lead to imaginative but misleading pictures since they do not describe anything, being empty), clearly are not analogous to tunneling through a barrier, except that the same words are unfortunately used — the words tunnel through a (regrettably not impassible) barrier from sense to nonsense.

V.4.h.iii *Does quantum weirdness produce tunneling?*

Tunneling has other interesting properties: "Quantum weirdness allows electrons to tunnel between the wells" [Beardsley (1999)]. Presumably what this is saying is that quantum weirdness is another kind of force (a fifth force?) which causes tunneling. Or could it be that quantum mechanics does not allow tunneling, but it is only allowed by quantum weirdness which has superseded quantum mechanics as the correct description of nature? It is indeed weird that presumably intelligent people would spout such nonsense. "Quantum interactions between the wells ... " reports an amazing discovery about nature — not only do objects interact but wells also do, and they have quantum interactions. Do wells have only quantum interactions but not classical ones?

Or perhaps none of the various statements are correct, perhaps they simply say nothing, perhaps those who come up with them do not know what they want to say, perhaps these exhibit nothing but how confused so many people are about quantum mechanics.

What would these people say about tunneling of a classical electromagnetic wave, something necessary for total internal reflection, so fiber optics — or is fiber optics also quite weird? Is fiber optics a consequence of classical weirdness? Are the huge amounts of money being made from fiber optics also a result of quantum weirdness (or perhaps classical weirdness)?

V.4.i Why do people believe in the cosmological constant?

One surprisingly enduring belief is that the value of the cosmological constant (believed to be almost 0) is a present day mystery [Weinberg (1989)]. It is not. What is surprising, and mysterious, about this belief is that it is so trivially wrong — and that so many people believe in it because so many other people do. But physics is not decided by majority vote, but by logic, mathematics and experiment — all of which require the cosmological constant to be zero.

V.4.i.i *Why the cosmological constant must be exactly 0*

It is very clear why cosmological constant Λ must be 0 — it transforms differently from the other term (G) in Einstein's equation, this coming from a massless representation, while Λ belongs to a momentum-zero one [Mirman (1995c), sec. 8.1.4, p. 139]. Putting it in is like equating a vector and a scalar.

Also it is a constant, while the other terms are functions of space. To put a cosmological constant in Einstein's equation is to set a function of space $G_{\mu\nu}$ equal to a constant Λ, which is obviously impossible. Thus to say that

$$G_{\mu\nu}(x) = \Lambda\delta_{\mu\nu}, \quad \text{(V.4.i.i-1)}$$

is like saying that

$$5x = 4 \quad \text{(V.4.i.i-2)}$$

for all values of x or

$$cos7x = 0.5, \quad \text{(V.4.i.i-3)}$$

for every value of x (a gravitational field, a gravitational wave say, is nonzero over all space, certainly over a finite region). And G is a sum of terms of the form $exp(ikx)$, which change as the (arbitrary) origin does.

Also $G_{\mu\nu}$, a function of waves, is complex (a sum of terms like $exp(ikx)$), while Λ is real. A complex number cannot equal a real one. This is like saying

$$5cos\theta + 2isin\theta = 4, \quad \text{(V.4.i.i-4)}$$

which cannot be true for any value of θ. It is absurd. No elementary physics student could ever get away with setting a complex number equal to a real one. Yet large numbers of professional (?) physicists are spending their careers doing just that. It is amazing.

Einstein's equation with Λ would be inconsistent. With it, an object would react to a gravitational wave an infinitely long time before the wave reaches it [Mirman (1995c), sec. 8.1.4, p. 139]. These are all consequences of inconsistency.

Such arguments against a cosmological constant are very similar to those showing why classical physics is not possible, and nature must be quantum mechanical [Mirman (1995b), sec. 1.4, p. 7].

Observations, no matter how compelling, cannot be explained by a theory that is mathematically inconsistent, however much too many people wish to do so.

V.4.i.ii *Obsession with subscripts*

Physicists seem to have an obsession with subscripts. But even for semisimple groups — and the Poincaré group is an inhomogeneous group — agreement in the number of subscripts is neither necessary nor sufficient for two terms to transform the same [Mirman (1995a), sec. XI.4.b.ii, p. 321; (1999), sec. IX.8, p. 505].

The history of physics has shown that the simplest, most obvious explanations are most likely correct. Yet here, as in far too many other places, physicists are searching for the wildest, most preposterous, most unlikely reasons, ignoring the obvious trivial one (sec. II.1, p. 54; sec. IV.2.c, p. 156).

V.4.j General relativity cannot be quantized, it is already

It is baffling that multitudes of physicists are trying to quantize general relativity. There is already a quantum theory of gravity, the only possible one: general relativity. It is required by geometry [Mirman (1995c), chap. 11, p. 183]. Perhaps people do not know what a quantum theory is. Those who are trying to do this do not seem to know what they are looking for, or why they dislike general relativity. And since physicists who are trying to quantize general relativity have no criteria for success, the effort is likely to go on forever.

The universe cannot go from quantum behavior to classical behavior, since classical mechanics is inherently inconsistent, and the universe cannot be inconsistent. Thus there can be no meaningful classical theory of gravity; such a theory is necessarily inconsistent. General relativity (which is very likely consistent) must (therefore) be the quantum theory of gravity. To call it a classical theory is to say that it is inconsistent, and there is no reason to believe that.

V.4.j.i *Gravitation, coordinates and momenta*

One argument [Wigner (1957), p. 255] is that in general relativity coordinates and momenta are only auxiliary quantities which can be given arbitrary values for every event. The measurement of position, say, has no significance since it is arbitrary. If this seems strange, it is, and in disagreement with reality. Any point in four-space can be taken as the origin, which is true for general relativity, as it is for nonrelativistic quantum mechanics and for classical physics.

It is certainly possible, for example, in one reference frame to measure the time it takes light to get from that point to any other. This defines the position of the second point along the light path. It is also possible to measure angles between light beams, so defining the positions of all points in space. Moreover from these it is possible to go to any other frame, by a well-defined transformation, so defining the coordinates in that frame by this transformation. That we can write these in other ways, using arbitrary functions of coordinates to get new ones, does not change the basic fact that we are always able to find where we are. We will never get lost in space because of arbitrariness of definitions.

Related to this is the statement [Wigner (1957), p. 260] that in general relativity coordinates are only labels to specify space-time points, and their values have no significance unless the coordinate system is anchored to events in space-time. But how else, in any theory, can we specify points in space-time besides events, and when we choose events, what more can coordinates be but their labels? (Of course there is a fundamental fact that events in space do specify a geometry, which is independent of how we label it or points, and properties of this geometry are essentially determined by there being physical points that do define it [Mirman (1995c), sec. 10.1.1, p. 166]. And the nature of this geometry controls the events within it [Mirman (1995b), chap. 5, p. 85].) These concerns about coordinates being only labels are based on experimentally meaningless statements.

There is another problem with this objection to general relativity as a quantum theory, the belief that quantum mechanics is based on position and momenta. Actually it is based on operators, the transformations of the group of space (sec. I.7.b.iv, p. 39), and states, the basis states of the representations of this group [Mirman (1995b), sec. 5.3, p. 100]. Position and momentum are merely operators, with (the exponentiation of) momentum being the translation operator, and the coordinate its dual. Thus to construct a quantum theory we need a realization of group operators (including momentum operators which includes the Hamiltonian), and (so) of states on which they act. This is exactly the same for gravitation as for the electromagnetic potential, and for electrons, nucleons, pions, and so on. The theory that this gives for gravitation is general relativity [Mirman (1995c), sec. 11.2, p. 186]. Why general relativity? Gravitation is a mass-zero spin-2 representation of the Poincaré group (and it must be a representation) and one coupled to matter — as is well-known. This then determines it.

V.4.j.ii *Can gravity modify spacetime?*

What does it mean to say that quantum gravitational effects modify spacetime structure? In what way can its structure be modified? How does that affect experimental results? Can it? Does the correct quantum theory of gravity, general relativity, show that? What does "spacetime fluctuation" denote? The word fluctuation implies that there is a (continual) change in space or time (or both). Can spacetime change with time? If not are these words anything more than strings of symbols (sec. V.7.c.i, p. 253)?

V.4.j.iii *What is the gravitational field?*

The gravitational field is not the metric tensor, but the connection — the fundamental, meaningful, physical quantity [Mirman (1995c), sec. 3.3.2, p. 39]. The metric tensor does not transform properly, as a mass-zero object; the connection does.

Why are gravitational statefunctions — connections Γ — and thus the curvature R tensor, the Ricci tensor, and $G_{\mu\nu}$, complex functions? Being Poincaré representation basis vectors, connection Γ must be. Also the connection is defined by

$$V_{\mu;\nu} = V_{\mu,\nu} + \Gamma^{\rho}_{\mu\nu}V_{\rho}. \qquad \text{(V.4.j.iii-1)}$$

But what is vector V which is used to define Γ? It is a statefunction, or a derivative of one, so is complex. What else could a statefunction be? When V is displaced in a field its phase (as well as direction) changes, thus Γ multiplies it by a phase, therefore Γ is complex. Now

$$[V_{\mu;\nu\rho}, V_{\mu;\rho\nu}] = R^{\zeta}_{\mu\nu\rho}V_{\zeta}, \qquad \text{(V.4.j.iii-2)}$$

and when V is taken around different paths in a field, its phase change is different, thus curvature R is complex. Also R is a function of Γ, and if it were real there would have to be cancellations that are impossible (except at best only under peculiar circumstances). This leads to the Ricci tensor and $G_{\mu\nu}$ necessarily being complex. Also by the Christoffel relation Γ is a function of derivatives of the metric, $g_{\mu\nu,\rho}$, which then must be complex. That is, since g transforms under a momentum-zero representation [Mirman (1995c), sec. 10.1.3, p. 169],

$$g_{\mu\rho;\nu} = g_{\mu\rho,\nu} - \Gamma^{\zeta}_{\mu\nu}g_{\rho\zeta} - \Gamma^{\zeta}_{\rho\nu}g_{\mu\zeta} = 0. \qquad \text{(V.4.j.iii-3)}$$

As g is real, Γ complex, $g_{\mu\rho,\nu}$ is complex. That is it depends on position through Γ, so (schematically)

$$\frac{dg}{dx} = \frac{dg}{d\Gamma}\frac{d\Gamma}{dx}, \qquad \text{(V.4.j.iii-4)}$$

thus is a complex function.

This also implies that the energy-momentum tensor, a function of derivatives of fields, is complex, as it must be for Einstein's equation to be consistent.

V.4.j.iv *General Relativity and quantum mechanics*

One of the most strongly, and strangely, held superstitions is that general relativity and quantum mechanics are logically irreconcilable at their very roots [Deutsch (1998)]. Yet it seems that no one has ever tried to justify this statement, nor given any reason why there is difficulty with general relativity beyond insulting it by calling it "classical" (however there is no reason to believe that general relativity is inconsistent, which is what is meant by saying that it is classical). It would be interesting to learn if anyone can actually show anything wrong with it, mathematically or physically. Of course, not only do people not regard it as a quantum theory, without saying why, but no one has ever stated what a quantum theory of gravitation, or even a quantum theory, is (except apparently in only one case [Mirman (1995b), sec. 1.4.4, p. 13], and it should be clear that general relativity is a, and in fact the, quantum theory of gravitation). Thus this is an open problem (only) for those who do not like general relativity [Smolin (1999)]. Why not?

Part of the belief that general relativity and quantum mechanics are mutually inconsistent comes from the nonsensical idea of space-time fluctuations. But people really have no clear conception of why they believe that general relativity and quantum mechanics are inconsistent, or what they are looking for when they search for a "quantum theory of gravity" (and since the goal is totally undefined, and undefinable, this topic provides the possibility of an infinite number of "scientific" (?) papers). There is no reason to believe that general relativity and quantum mechanics are inconsistent. Einstein was right: general relativity is the correct theory of gravity, and it is the quantum theory of gravity.

Another argument [Davies (1999)] is that general relativity is nonrenomalizable. This objection is based on the firm belief that the design of the universe was picked especially to make it possible for theoretical physicists to use their favorite approximation scheme for calculations (sec. V.4.c, p. 227). So it is felt that there is something basically wrong with general relativity, quantum mechanics or both. Perhaps more likely there is something basically wrong with physicists who think that the universe was designed for their purposes.

(Of course general relativity is Poincaré group representation theory for mass-zero spin-2 objects so Einstein's equation is a necessary condition, derived from Poincaré group theory, but not a sufficient condition [Mirman (1995c), sec. 6.1.3, p. 97] and it would be expected to have solutions that are physically irrelevant, perhaps even absurd. But these arise not because physics is weird, but only because the equations do not provide all the conditions. Thus any strange phenomena predicted from these only shows that not all conditions were used, not that they predict nonsense, or that the universe is nonsense.)

V.5 The measurement problem

A fundamental issue of quantum mechanics is the measurement problem. This complex question has many aspects, some real, some imaginary, due merely to fallacies and carelessness. Here we consider a few.

V.5.a Does measurement actually measure anything?

One statement of the "measurement problem" is [Griffiths and Omnes (1999)] to show that the results of the measurement are suitably correlated with the properties the measured system had before it was measured — that the measurement actually measured something. As emphasized words and phrases can often seem to be suitably correlated with meaning, but are not. What properties did the measured system have before a measurement? How can we tell? How do we know that the results are or are not suitably correlated? Can the results not be suitably correlated? And if we cannot tell, even in principle, can there be a measurement problem? Suppose they were not, what then?

Similarly we may ask whether when operator M measures "if the measurement has indeed really happened", or whether "we shall find that the measurement has happened if we check". Thus can we assign values to quantities independent of observations? It would be interesting to find a way to do so, to give values even if we cannot determine them — that is determine values that we cannot determine. Can we determine that the measurement has not actually happened?

In classical physics (sec. V.3.b.vi, p. 217) it is often possible to find these properties, thus we can ask whether our measurements agree with what we would have found had we experimented on a system before we actually did? In quantum mechanics this often cannot be done. So then how can we even give these questions content?

And the answer to the question whether the measurement actually measured something is that if we obtained an experimental result it clearly did (although perhaps not what we think it measured); but it would be interesting to find a measurement that did not measure anything. What other answers could we possible give, or want? We could, of course should, require that results be reproducible. Beyond that, what is the "something" that we wish to measure?

V.5.b Measurement and confused superposition

Another statement of a measurement problem is to show how a macroscopic quantum superposition appears. This phrase has meaning (only in theory), but it contains two words (macroscopic and superposition) that have opposite connotations, so is misleading. There is no real measurement problem given by this phrase. For a superposition (coherent is implied) relative phases of terms must be constant, and determinable. But for a statefunction of a system with a large number of subsystems this cannot be — phases vary wildly, and extremely quickly (sec. II.2.c.ii, p. 69). Thus superpositions for macroscopic objects are incoherent, that is classical, and states that we see are exactly the ones we should expect.

One (apparent?) measurement problem comes from the assumption that if a measuring apparatus has, say, two possible pointer positions then the superposition principle requires (says) that any superposition of the two must also be a possible state. But, of course, there are no such superpositions of macroscopic pointers, and no one has ever seen one. This is purely a result of applying the superposition principle wrongly. For such superpositions phases between states varies so wildly over such short periods of time that the term giving interference averages to zero. Quantum mechanics actually requires that a superposition of states of such classical objects is in reality impossible.

This illustrates the importance of knowing just what physical laws and definitions say, and how they are to be applied.

V.5.b.i *Does the double-slit experiment contradict the superposition principle?*

There is, for example, also a belief that for a single particle in a double-slit experiment the superposition principle says that the object must go through both slits, but the experimental result requires use of the word particle, and this contradicts the superposition principle, which requires waves. This measurement problem has nothing to do with quantum mechanics, but comes purely from using wrong words. The word particle is always wrong, it is incorrect for the description of the evolution of a quantum system, and its use causes misunderstanding as this indicates. Of course, patterns formed in such an experiment, and their development, do not in any way require, or indicate, that quantum mechanical objects are particles. Since there are no particles there cannot be a contradiction to the superposition principle, or any other part of quantum mechanics. Likewise it is incorrect to believe that a single object must go through both slits in order to give an interference pattern (sec. V.3.b.viii, p. 218). Statefunctions have nonzero values at both slits, and, when many objects are used, either simultaneously or sequentially, patterns form.

V.5.b.ii *Can a superposition of states decay into a classical state?*

Some people seem to have the idea that decay of a superposition into classical states has to be explained by general relativity. But no explanation is needed. Everything follows for Schrödinger's (correctly Dirac's) equation, and standard quantum mechanics, and that classical states describe macroscopic objects.

There is no such thing as decay of a superposition. Under certain conditions phases may be forced into wild oscillations, so average to zero, giving an effective incoherent superposition, but how this occurs is clear from Schrödinger's equation (provided the required calculations are possible).

V.6 Alternative theories

Many alternative theories have been postulated to solve the nonexistent problems of quantum mechanics. They provide some interesting topics of study, not necessarily about physics. Here we briefly mention some insights (not necessarily about physics) that they lead to.

Some of these theories may in fact be just quantum mechanics with extraneous concepts to help confuse what they really are. Or they may just be quantum mechanics with different terminology, which can also mislead. However many are incomplete. They do not take into account, for example, the extremely important fact that there is in the universe more than one observer (physical object) — we are not alone, and we cannot be for else the one that is alone would consist of but one object, one proton, one electron, one neutrino perhaps (but not one neutron for then there would not be only a single particle for all time). The observations of these different observers must be consistent, and this leads to group representation basis states, that is statefunctions [Mirman (1995b)]. Also there must be interactions. It is difficult to see how a theory could be complete, and consistent, without statefunctions — and physical objects, and interactions. And if it is not complete and consistent, can it be a fundamental physical theory?

V.6.a Deterministic theories that do not determine anything

An important source of empty words are those deterministic theories that do not determine anything, but are, at best, quantum mechanics disguised as something else. Many people find the inability to predict exact experimental results, only probability distributions, very uncomfortable.

In some such "deterministic" theories the position and momentum of an object at any time are completely determined, provided we know the initial conditions exactly. But unfortunately it is impossible to know the initial conditions exactly, thus they only give — just like quantum mechanics — probability distributions. These theories seem to be based on the belief that an electron knows exactly where it is and exactly where it is going, it just won't tell us. Thus "determine" here does not determine any meaning. But this is irrelevant, since the purpose of these theories is not physics, but providing emotional relief. Those people who find quantum mechanics so uncomfortable do feel better if they can rewrite it in slightly different notation, call it a different theory, and bless it with the chant "deterministic, deterministic". Thus these theories are useful, provided of course that the quantum mechanics is actually done properly.

V.6.a.i *Well-hidden variables*

A fundamental concept usually appearing in such theories is that of hidden variables. These variables are usually very well hidden. If a theory makes predictions different from quantum mechanics it is a legitimate alternative to it, and must be checked experimentally. Otherwise it is just quantum mechanics but with extra, hollow, terminology that has no content or relevance, except emotionally.

One type of such a statement [Goldstein (1998)] is that the wavefunction choreographs or governs the motion of more fundamental variables. Ignoring the question whether variables can move, how can there be any if all we can find is the wavefunction? Might these variables be forces exerted by angels on particles? Or might they be other variables, perhaps forces exerted by beings who are not angels (which is not meant to refer to physicists)? What is the difference? How can we tell?

Equivalent statements [Nogami, Toyama, van Dijk, (2000)] are that the position and velocity of a particle are both well-defined, that it moves along a Bohmian trajectory whether observed or not, and that the Schrödinger wavefunction accompanies and guides the particle. But how do we know whether the particle has a well-defined position or velocity if we cannot measure them, how do we know that it moves along a trajectory if we cannot check? Of course angels might know, but mere human beings do not. Thus perhaps these statements indicate a belief in angels. And how does a wavefunction accompany and guide a particle, how can we show that (or even define it)? Can a wavefunction interact with a particle; how? Does the wavefunction provide moral guidance or physical guidance — how can we distinguish them? Not only are the variables (well) hidden, but so are the meanings of these terms.

To predict when an atom decays we need to know the precise initial conditions — just like any other deterministic theory. But if a deterministic theory involves chaotic dynamics then knowledge of initial conditions must be very precise. Unfortunately for the ones given as alternatives to quantum mechanics, such knowledge is in principle unattainable. Thus in these theories it is impossible, in principle, to know initial conditions well enough to make predictions beyond those given by quantum mechanics. Hence, just as in quantum mechanics, all that they can give are probabilities — exactly the same ones as quantum mechanics gives. (These then are not alternative theories to quantum mechanics, but rather quantum mechanics with excuses.) They are deterministic only in the sense that the atom knows when it will decay, the atom knows the initial conditions exactly, and from that the atom has its decay time determined — but unfortunately, even though the atom knows, it will not tell us. Thus for human beings these theories are no more deterministic than quantum mechanics.

A religious person would say that God tells the atom when to decay, and then its decay time is completely determined — such a theory is also completely deterministic, although also giving only probabilities, as with quantum mechanics, because we are unable to know the mind of God. This attitude is perhaps not unknown. Werner Heisenberg is supposed to have written to Albert Einstein that even though causality (presumably the classical kind) does not hold in quantum mechanics, we can console ourselves that the dear Lord God would know the position of the particles, and thus He could let the causality principle continue to have validity [Holton (2000)].

Other people may prefer to believe that there are little green men standing on electrons in atoms, and when one decides to jump off, the electron drops to a lower orbit. Thus the theory is deterministic since the decay time is determined by what the little green man does. How do these various alternatives to quantum mechanics differ, these diverse hidden variable theories, with the hidden variables being in the mind of God, say instructions given by God to an electron, or the mind of a little green man, or perhaps unspecified? And aside from introducing meaningless concepts, what is the difference between them and quantum mechanics?

Words like "hidden" and "variable" have meaning, and even in some situations "hidden variable" does also, but they do not here. Yet because they have meaning elsewhere, their lack of content when used in quantum mechanics (or in theories that claim to substitute for it) is itself quite hidden. These other theories would be unchanged if "hidden variable" were replaced by, say, "neddih elbairav". This would be a great improvement for the distinction between these and quantum mechanics would then be clear.

One argument for such theories is that problems of quantum mechanics go away because these (well) hidden variables produce whatever is required to make them go away (that is they give believers in these theory the needed excuses to ignore what they do not understand so that they will not have to do the hard work of thinking about such unpleasant topics). But of course, nothing disappears because we ignore it.

V.6.a.ii *Psychological difficulties caused by quantum mechanics*

These alternatives emphasize the tremendous psychological difficulties (too) many physicists have in accepting the indeterminacy of nature, and their need to believe that it is really deterministic, even though their substitute theories are no more, nor less, deterministic than quantum mechanics. But just having an excuse to say "deterministic", which is all such theories are, makes them more comfortable.

This of course has nothing to do with physics, but detailed study of it by psychologists would undoubtedly contribute much to our knowledge of human psychology.

Scientists often disdain people like creationists who refuse to accept reality as it is but try to interpret it the way they wish it to be. But many physicists have even more trouble accepting the world as it is found to be, as shown by these various theories that are no more than quantum mechanics with the addition of empty concepts and phrases added for emotional reasons. Such people struggle to shape the world to their desires, acting as if they could decide what nature is like, instead of being able to accept what the universe actually is.

Perhaps it would help in the understanding of quantum mechanics if Occam's razor were vigorously applied to such alternatives [Mirman (1995b), sec. A.1.2, p. 178]. After all extraneous, irrelevant, empty concepts were cut away (and it is well-known that there are such [Englert, Scully, Sussman and Walther (1992)]), there might be nothing left but quantum mechanics — or possibly nothing at all. Or perhaps Occam's razor is not relevant here since the difference between these theories and quantum mechanics is not physical, but psychological.

V.6.b The many-worlds interpretation of quantum mechanics

The many-worlds interpretation [DeWitt and Graham (1973)] is one of the alternatives to the generally accepted interpretation of quantum mechanics (which is undefined since different people generally accept different interpretations, or more generally do not know what interpretation they accept).

Those who believe in this seem to feel that each measurement result occurs in a separate branch (component) of the superposition of "many worlds" giving the wavefunction of the system (the universe). They regard all branches of the "universal wavefunction" to have physical reality, but an individual's brain state in each branch is only conscious of the results of an observation in that branch, so is not aware of different results of that observation occurring in other branches. Another way of saying this is that "this" (unspecified) reality is composed of many worlds [DeWitt and Graham (1973), p. v]. Such a concept is difficult to argue with since there is no way to study all these other worlds to see if they are really there. Of course, we can also say that there are angels that decide where an electron goes, and with what momentum. There is no way of arguing with that either. So being in "many-worlds" the statevector decomposes to reflect a continual splitting of the universe into a multitude of mutually unobservable but equally real universes (sec. IV.2.b, p. 155).

Aside from the question of how universes split (and what mechanism causes that splitting) — into lots of little pieces which then move apart into different universes, or perhaps in other ways — this statement raises the fascinating question how these universes can be equally real (and indeed in what sense) if they are unobservable? Of course angels are equally real also. Discussions of those sentences that are mere vacuous collections of words seem quite relevant here (sec. IV.2, p. 154).

The universe seems to be about 10^{10} years old, the earth about 5×10^9 years, human beings about 10^5 yrs, and we have been studying microscopic objects, and quantum mechanics, for about 10^2 yrs. Thus if we accept the many-worlds interpretation, we must assume that the laws of physics changed about 10^2 yrs (or perhaps 10^5 yrs) ago, and now that humans are making observations the universe has decided to split into many branches each time a human being looks at something (or is only when a human being looks at some small thing, or only when a physicist does so?). Perhaps laws of physics, the nature of the universe, are governed by our presence and these laws were very different before we arrived. Certainly many egotistical people (and not only egotistical physicists) seem to want to think so. But it does appear rather implausible.

Actually measurements (interactions) have been happening since the birth of the universe [Mirman (1995b), sec. 5.1.3, p. 88], all over the universe, else there would be no universe.

A view that has lead to these beliefs is that a photon, say, is a particle, and that when it passes through two slits, the universe branches, depending on which it went through, and the wavefunction (of it, or of the universe?) collapses. But photons cannot be considered as particles, hence they do not go through two slits, and there need not be (and of course can not be) quantum branches, for there is no collapse.

This belief in the importance of humans in determining the properties of the universe is not limited to such "understanding" of quantum mechanics, but appears in many forms in different ways, unfortunately not only in interpreting quantum mechanics.

There is a non-anthropocentric version of this [DeWitt and Graham (1973), p. 161]. It holds that the universe is constantly splitting into a stupendous number — which (fortunately?) is undefined — of branches resulting from the measurementlike interactions between its myriads of components. How do interactions that are, and are not, measurementlike differ?

Of course each object in the universe is continually interacting with all others, continually making measurements (which is done by every object, not merely human beings, otherwise this again becomes an anthropocentric view in which people — or perhaps just physicists — are able to split the universe into many parts). An electron, say, has its path always being altered by the potential caused by other objects in the universe, thus is continually making measurements. This is true for each object in the universe (assuming that it interacts, otherwise it would not exist). Every particle in the universe in any finite (and really infinitesimal) time interval makes a number of quantum measurements at least of order $\aleph$ (although this seems never to have been calculated, but it might be interesting to do so). Thus in any time, no matter how small, the universe would split into an infinite number of branches (or maybe it is only a stupendous number, whose value seems to have no rationale). It is interesting to wonder where all these branches are (or go to).

Further every quantum transition of every object in the universe (every infinitesimal change in an object's path, or spin direction, say, in every star, in every galaxy) splits our local world (only?) into myriads of copies of itself. It is the usual view in physics that for one object to exert an influence on another it must interact with it. An electron in a distant galaxy must exert quite a force on earth if it can continually split it into (how?) many copies. But then the earth would be performing a quantum measurement on this electron, which, presumably, would cause all other objects in the universe (including itself?) to split even more. These splittings would then be measurements of the earth (and each object in it), causing even more splittings, and so on, thus giving an uncountable number of splittings in each infinitesimal interval of time.

Perhaps this all makes sense, or perhaps it is a sign of desperation by those who simply cannot accept the reality of quantum mechanics. These conceptions say nothing about physics, but a great deal about (pathological, and regrettably all too common) psychology.

V.7 The vacuum

The vacuum seems to fascinate people, even though there does not appear to be anything there to cause such glamour. Yet being empty it allows itself to be filled (and not only by physicists) with the strangest ideas. There appears to be something deep in human psychology that rejects nothingness, and fills what must be empty with concepts that range from implausible (and necessarily contradictory), to hollow, to absurd. Many of these concepts seem to be even more empty than the vacuum. This psychological yearning that there never be emptiness has lead to some of the wildest ideas in physics.

Many of these discussions of the vacuum illustrate the weirdness that, not quantum mechanics, but language without substance can result in (sec. IV.2, p. 154; sec. IV.2.x, p. 182).

A vacuum state is needed in the formalism [Mirman (2001), chap. IV], but this has nothing to do with all the strange ideas that the tantalizing word "vacuum" has fomented.

V.7.a The rejection of nothingness in history

Difficulties in dealing with nothingness go far back into human history. Of course belief by physicists in an ether is well-known (the word, though not the concept, is no longer used — although there are those who will not even give up the word [Finkelstein (1991), p. 251]). But long before physicists came into existence, Aristotle argued strongly against the existence of the vacuum [Grant (1981), p. 5], filling it with material surprisingly similar to the ether. Others at that time also could not accept it [Hiley (1991), p. 217].

In the Middle Ages, Thomas Bradwardine (writing around 1344) declared that it seems obvious that a void can exist without body, but in no manner can it exist without God [Grant (1981), p. 135, 141, 248]. Of course today no physicist would postulate God filling the vacuum, but replace God by "virtual particles" and the sentence becomes part of the canon almost universally accepted by modern physicists.

It is striking to consider how much discussion of the vacuum there has been throughout history [Grant (1981)], and how much difficulty there has been in accepting it. Now it is clear that all these discussions about the vacuum are themselves vacuous, much ado about nothing, mere strings of words without meaning, that cannot be given meaning, that have no relationship, near or distant, direct or extremely indirect, to reality. Yet what is even more striking is how many present-day discussions, of the vacuum, of quantum mechanics, and unfortunately of so much else, are equally much ado about nothing. While we can look back and appreciate the emptiness of strings of words that were so important, and senseless, in the past, it seems much more difficult to do the same for strings of words that are so important, and senseless, in the present. This, regrettably, we must leave to the future.

V.7.b There is no vacuum, only vacuity, in the discussions of the vacuum

The term "vacuum" has caused much befuddlement, much noise. One reason is that discussions of it, like those referring to particles popping in and out, consider situations in which objects (real ones) are present, so there is no vacuum. This of course is a fundamental error, resulting (as happens so often elsewhere) in arguments that are self-contradictory.

Particles do not pop in and out of the vacuum, even virtual ones (a meaningless physical term, being merely a name given to particular terms in an expansion, something that seems not generally realized, leading to further weirdness — but not in quantum mechanics). In an approximation expansion there are terms that can be visualized, as a way to help keep the bookkeeping straight, as virtual particles created from the vacuum. But this does not happen physically, it is merely a visualization of terms in an approximation scheme. They would not appear in different approximation schemes. They do not say anything about physics, but are merely a gimmick for a particular type of bookkeeping.

V.7.c Spacetime foam

Nothingness seems so scary that throughout history many concepts, really words (rows of letters), have been developed to repel it. Some have been quite picturesque, allowing us to fill our minds, if not space, so as to be able to avoid the dread idea of the vacuum.

The modern idea of spacetime foam provides quite pretty, attractive, pictures, some of the best in history, but no statements that are in any way attractive. It would be interesting to find if anyone could provide a precise mathematical definition of foam. We picture the foam as something bubbling in an underlying space — but those who bubble at the idea of such foam apparently do not believe in such an underlying space. Does it mean that positions of points change in time (?), or perhaps distances between points (?), or the metric, so distances change?

Perhaps, since this idea seems to have originated in a misunderstanding of quantum mechanics (and probability), those who believe in it are trying to say that if the universe were to be rerun again, starting from the big bang, with exactly the same initial conditions (?) the position of a point (?), at exactly the same time (?), would be different. Since probability is involved, the universe would have to be rerun many, many times. It would be an interesting experiment if believers in spacetime foam could rerun the universe.

V.7.c.i *Can violent fluctuations rip space apart?*

People seem to be obsessed with the idea that space is affected by quantum mechanics — without understanding what they are trying to say. For example violent fluctuations ... implied by the uncertainty principle (?) are at odds with the smooth geometry of space-time (whatever this means) that is the central feature of general relativity [Quigg (1999)]. Wow, that sounds exciting. How do violent fluctuations differ from nonviolent fluctuations, or peaceful fluctuations? What fluctuates? Indeed this is much more exciting: "we encounter the devastating quantum fluctuations that rend space-time". It sounds like a real battle. What is devastated? It seems unlikely that space-time will actually survive after it is so badly devastated. And how is it rended? It would be nice to find little pieces of space-time. How is space-time broken apart? What holds the little pieces; certainly space-time is too devastated to do so? Since space-time is broken into little pieces, these must be separated. What is the distance between them? And how can we tell what the distance is (or means) unless pieces of space-time are scattered over space (which then could not be broken apart)?

And how does this central feature of smooth geometry differ from bumpy geometry? Undoubtedly those who say things like this can produce, experimentally, pieces of space-time that we can examine to check that it has really been devastated.

What this emphasizes again is the power that classical pictures exert on people (physicists!), and how it prevents them from thinking (often not only about quantum mechanics). Of course such statements show a complete lack of understanding of quantum mechanics by those who say such things, and that the reasons might be more than misleading pictures.

V.7.c.ii *Can the vacuum be ripped apart?*

A similar statement [Voss (1999)] is that the very vacuum, which we think of as empty space, will be ripped apart, revealing its underlying fabric. This raises the fascinating question of what comes between pieces of the vacuum? Is it empty space? But if space is empty, and the vacuum is not, then what is the vacuum (traditionally defined as empty space — but, of course, using the understood definitions of words is much less exciting than saying something nonsensical). Those who hold that the uncertainty principle requires the vacuum to be teeming with whatever it is supposed to be teeming with obviously should believe that empty space must also be teeming, although perhaps with something other than what the vacuum teems with. Can empty space (which is thus not empty) also be ripped apart? But this reference to underlying fabric (presumably not cloth fabric) suggests that there is something else between pieces of the vacuum. Does this also teem? Can it be ripped apart? How then does the (cloth?) fabric between pieces of the vacuum differ from the vacuum itself? And does the uncertainty principle apply to the fabric that underlies the vacuum, as so many people believe that it does for the vacuum? If not, why should they differ? Is there a force on the vacuum that rips it apart? Obviously experiments that rip apart the vacuum would be quite interesting, especially to those who write things like this.

V.7.c.iii *On what scale does spacetime fluctuate?*

Also to say that space-time fluctuates on the Planck scale means that we can measure it (its metric or curvature, presumably) at such small scales, repeatedly, and find different values. But there are no (known?) ways of making such measurements, in principle. This is another reason why such statements are mindless. Although if people who say things like these want to make the measurements, they should certainly be encouraged to do so.

V.7.c.iv *What fluctuates in quantum fluctuations?*

What are quantum fluctuations? What fluctuates? In the approximation scheme of perturbation theory certain aspects can be viewed as fluctuations. But for an exact wavefunction of a gravitational field (the connection) there is nothing to fluctuate.

And those who disagree should be able to propose an experiment to show these fluctuations, and this foam — and tell what they mean, and how to distinguish violent from peaceful ones.

Related to these "concepts" is a superstition that there are physical phenomena such as quantum superposition of distinct spacetime geometries. It would be exciting to find experimental examples, or even explanations of what it means experimentally, and how we might study it. How are geometries superposed, and how do we know that they are distinct? Can we examine different superposed spacetime geometries? How?

V.7.c.v *The "miracle" of particles "popping out of the vacuum"*

A problem claimed solved by quantum field theory [von Baeyer (2000)] is the creation and annihilation of matter. When particles collide both may vanish while two new ones pop up out of nowhere (not even at the vanishing point?). No model can account for these miraculous events (some physicists seem to believe in religion more than physics) except quantum field theory. This miracle does not explain how quantum field theory "accounts" for creation and annihilation, although it does provide a formalism for their study. What accounts for them is not quantum field theory, but interactions, nonlinear terms in equations governing matter. As is well-known these lead to creation and annihilation, and they have to by the character of nonlinearities.

And why are there interactions? If particles did not interact we would not know of them, they would not exist. But the situation would be much worse if the particles in our bodies did not interact, for then we would not exist. If there were no interactions there would be no universe. Perhaps existence is a miracle, but it is not explained by quantum field theory, and in fact it is not explained, it cannot be explained.

V.7.c.vi *There is a difference between objects and the vacuum*

Here people's intuition and classical views leads them into absurdities. We can regard an object, a book, a piece of wood, a body of water, being broken into parts, pieces, bubbles, spray, for example. But these are pieces in space-time. How can space-time be broken up? Is it embedded in anything? Can we tell that pieces of space are actually not connected. How? What is between them?

Space-time is not like a sheet which can be ripped apart — rended — into little pieces in space. The sheet is embedded in space, but space-time is not embedded in anything. People have a picture (certainly not a thought) of space being broken into pieces, or a spray with bubbles, and are mislead by it into saying absurd things.

Fluctuations do seem to cause confusion (perhaps due to fluctuations in the care with which people use words). Thus we might consider a device that registers if the momentum of an object in it exceeds some value (it performs a measurement on the momentum to that extent). Since the device is of finite size, the momentum must fluctuate, and so exceed the threshold value to set off the instrument, and the device should register, at least eventually. But by conservation of momentum, momentum cannot vary (assuming that the object does not exchange momentum with the rest of the device). Thus it cannot fluctuate. Correctly, in constructing the instrument the momentum of an object must be less than the minimum or else the instrument responds no matter what the object it is studying does. But since the position of the object is known to be (very likely) within a finite region, it is impossible to construct a device with momentum less than given by the uncertainty principle. Specifically, in constructing many identical instruments, there will be some that register because the relevant momentum is greater than the minimum. There are no fluctuations in the momentum, rather the momentum of an object in an ensemble (of measuring instruments) cannot be determined more precisely than allowed by the uncertainty principle.

The problem here is again the word "fluctuation". It gives the impression of a value changing (suggesting a change in time), but actually what it means is that different members of an ensemble have different values, with the value for each constant.

V.7.c.vii *The Casimir effect*

One reason given for the bias that according to quantum mechanics the vacuum is not empty, but teeming with virtual particles that constantly wink in and out of existence, is that this sea of activity has a strange consequence, the Casimir effect: two flat metal surfaces attract one another if they get close enough. However the Casimir effect has been interpreted as a manifestation of the van der Waals force [Brevik, Marachevsky and Milton (1999)]. It is hardly likely that the van der Waals force, which is between systems like molecules, exists in the vacuum (although it is impossible to imagine what those who believe in a zero point energy of the vacuum, or that it is is not empty but teeming with virtual particles, think (?) in addition). Neither the Casimir effect nor the van der Waals force is strange nor do they imply anything about nothingness. Also the Casimir effect has been treated without the concept of the vacuum, and so without all those particles popping in and out of it [Schaden and Spruch (1998); Schaden, Spruch and Zhou (1998)].

The Casimir effect has also been calculated with the use of boundary conditions. But these boundaries must be physical objects. Hence the existence of the Casimir effect cannot be used to infer anything about the vacuum, since it occurs only when there is no vacuum.

V.7.d Does the vacuum have energy?

Related to these ideas is the concept that the vacuum has energy. How do we tell if a system has energy? The only way is to have an object interact with it and find that the energy of the object changes (the energy of a physical object can be found from its momentum and position in a field, these determined by conservation of energy and the form of the Hamiltonian). Thus to make sense of the statement that the vacuum has energy it is necessary to find physical systems, whose energies we can calculate, and whose energies change without a corresponding change of energies of other physical systems — that is we have to find that the law of conservation of energy fails. And if those who are pushing the concept that the vacuum has energy disagree with this, they should state what "the vacuum has energy" actually says — physically (experimentally).

A closely related belief many physicists have is that the cosmological constant (sec. V.4.i, p. 238) Λ (which cannot appear [Mirman (1995c), sec. 8.1.4, p. 139]) is a quantum mechanical effect: that the evanescent particles that flicker in and out of existence in "empty" space provide a well of energy ... [Glanz (1998b)]. Is it possible to draw energy from this well?

Apparently related to this (it is difficult to tell the meaning of senseless statements) is that [Weiss (2000)] the zero-temperature vacuum brings about the largest share by far of the mass of every nucleus. This states clearly that such an interesting "fact" has been experimentally demonstrated — the zero-temperature vacuum has been removed (?) and the mass of every nucleus found to decrease. And if it really has been proven experimentally (how?), it would be nice to have relevant references.

V.7.d.i *Is zero point energy of the vacuum pointless?*

One idea that these views about the vacuum has given rise to is that Heisenberg's uncertainty principle requires that the vacuum have zero point energy, so that its energy is not zero but finite or perhaps infinitely large. For an oscillator the position and momentum of the mass cannot simultaneously be known precisely, and since position determines potential energy, and momentum kinetic energy, the uncertainty principle requires that the oscillator have a minimum energy, the zero point energy (sec. IV.2.n, p. 169). Those who believe that the uncertainty principle gives a zero point energy of the vacuum (which presumably has neither a position nor momentum) undoubtedly also believe that if the oscillator did not exist the uncertainty principle would still require that it have a zero point energy. Thus since there are an infinite number of nonexistent oscillators the total zero point energy is infinite (which many people — whose energy is apparently so finite that there is not enough for careful thought — do seem to believe).

V.7.d.ii *Some interesting things to do with the vacuum*

People are so fascinated with the vacuum that they have thought of various things to do with it, and things that it can do. Certainly perhaps the most fascinating is to melt the vacuum [Quercigh and Rafelski (2000), p. 41]. It would be quite interesting to study a melted vacuum. How does it differ from a solid vacuum? Can it flow? Would someone immersed in a melted vacuum get wet? Is there also a gas phase for the vacuum?

It is believed that the cosmological constant provides energy that gives space its springiness, shoving it apart [Glanz (1998b)]. This clearly states that it has been shown that space is springy and can be shoved apart. Obviously we are all curious to get references to papers that report experiments that have shown this, to learn how it was shown, and to find out what comes between the pieces of space that have been shoved apart (?). Perhaps these pieces of space move back and forth (they are springy) and even collide, giving the gas phase (an ideal gas?) of the vacuum. What are the kinetic and potential energies of the pieces of space?

It is not surprising therefore that people (let us hope not physicists) have suggested that the zero-temperature vacuum exerts a pressure that keeps quarks and gluons cooped up [Weiss (2000)]. Is the zero-temperature vacuum a solid? What must be its temperature to melt it? How is the temperature of the vacuum measured?

Obviously the vacuum can lead to a large number of scientific (?) studies.

V.7.e Quantum mechanics, probability and space

There is also a belief that geometry resonates between configurations [Misner, Thorne and Wheeler (1973), p. 1193], that is each configuration has its own probability amplitude. This says that we can conduct many experiments on the same physical system (the same universe?) and find the geometry different for each. Those who say that geometry resonates should specify experiments to measure the probability amplitude.

If we measure the metric, for example, in three-space and measure it again at later times, then quantum mechanics says that we do not get the same value each time — it gives a probability distribution. One reason is that the metric is determined by masses, and their positions and momenta are not completely determinable. But this does not mean that the metric fluctuates (not even violently — the values might differ only slightly). It just means that measured values are not identical for the various repetitions.

If we measure the metric of space-time, then it is impossible to measure it again at the same space-time point (a fact so obvious that it should resonate even with those whose metric fluctuates). Hence it makes no sense to even discuss fluctuations or uncertainty for the metric of 3+1-space — this reduces quantum mechanics to silliness.

We can consider say a set of identical bodies (far from each other so they do not interact) and find the metric produced by each (experiments are done at different space or different time points, or both). Values for the metric will not be the same. Quantum mechanics gives probability distributions for these (as it must, because among other reasons distances to the particles and their momenta are not fully precise). This again does not mean that the metric fluctuates, but simply that quantum mechanics gives only probability.

If we measure the position and momentum of an object many times, we get different values. This does not mean that the object fluctuates. (Do physicists fluctuate when their students try to find where they are and where they are going?)

V.7.f How does the electromagnetic field affect an electron?

Another of these doctrines is that fluctuations in the electromagnetic field affect the motion of an electron in an atom [Misner, Thorne and Wheeler (1973), p. 1190] — and that these are vacuum fluctuations — which ignores the fact that if there is a field (and an atom) there is no vacuum. What people who say this do not specify is what fields fluctuate with respect to: space, time, or something else? Of course, fields do vary in space and time, but not in a vacuum, for by definition they do not exist there. An electron obeys Dirac's equation, with interactions, and its energy levels differ from those found from the equation without interactions, but with just a potential, not surprisingly [Mirman (1995b), sec. 6.4, p. 119]. A picturesque way of discussing these (electromagnetic) interactions is by talking about the electromagnetic field fluctuating. It is (too) picturesque, so clearly unfortunately quite misleading, and has mislead too many.

And it is said, and possibly even believed, that quantum fluctuations cause spontaneous decays, such as of atoms. But objects making up an atom obey nonlinear equations — with interactions — and these do not have solutions without decay (except for the ground state). Solutions of the nonlinear equations require decay, and the vacuum is completely irrelevant — if there are particles that decay there cannot be a vacuum.

V.7.f.i *Fluctuating vacuum field*

Another common allegation is that such effects as the Lamb shift are due to the fluctuating vacuum field (what is a vacuum field?). But for a Lamb shift there must be objects — for it is their energy that is changed. Thus there is no vacuum. (This is true for all cases in which "vacuum fluctuations" produce an effect — there is no vacuum in these cases.) To find the energy of a system, or of a particular part of it, the Hamiltonian expectation value must be calculated — that of the complete Hamiltonian, including all interactions. This value will of course be different from that found using only the Hamiltonian without interactions. This difference is the Lamb shift (in those cases for which this terminology is relevant). It has nothing to do with the vacuum, which is not present, and there can be no fluctuations in the vacuum — a totally silly concept.

It is disturbing, and baffling, that such an absurd notion that there are fluctuations in the vacuum (!) has been invented, and even claimed to have been experimentally verified [Kleppner and Jackiw (2000)] (!), to "explain" the triviality that two different equations, Dirac's equation without and with interactions, have different solution sets.

V.7.f.ii *Quantum electrodynamics rules out vacuum fluctuations*

The equations governing the behavior of an electron, with statefunction ψ, and an electromagnetic field, with potential A, are

$$i\gamma_\mu \frac{d\psi}{dx_\mu} + ie\gamma_\mu A_\mu \psi - m\psi = 0, \qquad \text{(V.7.f.ii-1)}$$

$$\frac{d^2 A_\mu}{dx^2} = ie\psi^* \gamma_\mu \psi. \qquad \text{(V.7.f.ii-2)}$$

It should be carefully noted that this has an exact solution

$$\psi = 0, \quad A_\mu = 0, \qquad \text{(V.7.f.ii-3)}$$

which is the vacuum. There are no fluctuations, there are no excitations, no virtual particles popping into and out of the vacuum, or any such things. And the expectation value of the Hamiltonian — the energy — for this solution is zero, exactly.

Nor is there a contradiction with the uncertainty principle, which is a consequence of these equations, not an independent superstition. In this case the equations give zero for all fields, and this is fully consistent. Requiring that these zero solutions give nonzero values is not only wrong, but absurd. It requests the equations to gives solutions that they do not have.

To say that particle number must fluctuate, due to the uncertainty principle, would mean that the uncertainty principle contradicts the equations governing the system, and thus that quantum electrodynamics is inconsistent — and in disagreement with quantum mechanics. The evidence is strong that it is consistent, and the evidence that there are fluctuations is (to put it mildly) nonexistent. It makes no sense to take a correct theory, and try to make it incorrect, which is what introduction of statements about what goes on in the vacuum (except nothing) does.

V.7.f.iii *Is the uncertainty principle a superstition?*

The uncertainty principle is too often taken as a superstition, not as it really is — a deduction from the mathematics of quantum mechanics, such as the transformations of space, including as a special case Fourier analysis. For position and momentum, Heisenberg's uncertainty principle comes from Fourier analysis, and relates the width of a wavepacket to the range of momentum values making a large contribution [Mirman (1995b), sec. 3.4.1 p. 52]. It cannot be applied to the number 0, which is what people who say that the uncertainty principle requires that the vacuum teems with virtual particles, or has energy, or things like that, are trying to do. Of course, physical principles are often treated as superstitions, rather than as consequences (sec. II.4.b, p. 81) of underlying theories [Mirman (2001), chap. I]. Obviously this causes much confusion.

It has been said that nature abhors a vacuum [Grant (1981), p. 67], which is certainly not true. But it seems that a lot of physicists do.

V.7.f.iv *Is the cloud surrounding particles that of photons, or of confusion?*

Another closely related notion [Weinberg (1997)] is that an electron appears as an elementary particle but is surrounded with a cloud of short-lived photons and electron-positron pairs which give it a finite extension. This then changes the magnetic moment from the value obtained from Dirac's equation with no interaction. Clearly those who believe this can give ways of measuring the extension and the lifetime of the short-lived photons and electron-positron pairs (otherwise how do they know that they are short-lived?), as well as show experimentally their existence, along with that of the cloud.

The magnetic moment of the electron calculated from Dirac's equation without interactions cannot possibly be correct of course — wavefunctions of electrons are not solutions of this equation (sec. V.2.d, p. 205). They are obviously solutions of the equation with interactions. These being different cannot give the same value of the magnetic moment (and of other such quantities). But this in no way implies a cloud (although it does lead to a cloud over people's judgment) or of photons or pairs of particles or other such pictorial fantasies.

It is important to remember (which many people apparently do not) that use of an approximation scheme, like perturbation theory, to analyze the correctness of a theory might very well show difficulties — but these can be in, perhaps are more likely to be in, the approximation scheme than the underlying theory. While difficulties might provide clues, and suggest ways of searching further, they cannot be used to definitively, even strongly, establish inconsistencies in the theory being investigated.

In particular discussions of apparitions like virtual particles in the vacuum, breaking up the vacuum, or fluctuations in it, do of course lead to inconsistencies, but not because of the vacuum or the theory, but because of the absurdities and vacuity of the discussions themselves.

Much labor in theoretical physics has been motivated by a desire to overcome the problems of perturbation theory. This is based on the firm conviction that the universe was carefully designed for the specific purpose of making theoretical physicists' favorite approximation scheme work.

V.7.g Does the uncertainty principle prevent conservation of energy and momentum?

Energy and momentum are subject to the uncertainty principle which has made many people uncertain about them.

One notion is that because there is an uncertainty in momentum or energy (and gravitation may place a minimum value on these uncertainties, independent of uncertainties in their conjugate variables [Mirman (1995c), sec. 11.3.6, p. 195]) the laws of conservation of energy and momentum do not hold precisely. However this is not correct.

What these uncertainties imply is that we cannot check laws precisely (at least simultaneously, and to verify them it is not necessary to do so). What is the difference? We can have confidence in a theory because of its interconnected diverse predictions and explanations, even if not all can be fully checked. Thus it is reasonable to say that energy and momentum are conserved, for this has not only have been found to be true in all the many situations in which it has been checked, but because these follow from general considerations, space and time translational invariance. If conservation did not hold many things would go wrong. They are part of a complex web, and it is for that reason that we regard them as true (sec. I.2.f, p. 8).

V.7.h Fluctuations, foam and rhetoric

The word "fluctuation" here has no meaning — an excellent example of a word that seems to have some but does not in the context in which it is used. Fluctuation means changes with respect to variables. But for space-time there are no variables. Each point is unique, it visited once and only once.

Violent fluctuations provide examples of people being carried away by their own rhetoric. This happens much too frequently, and not only in quantum mechanics. But it is a major cause of the bewilderment about quantum mechanics. It is quite common, and unfortunately in physics much too common, for people to repeat what others are saying (this shows that they are part of the in-group) without thinking. But to be noticed it is necessary for the statements to be louder and louder, sillier and sillier. Reading many of the papers in quantum mechanics emphasizes this.

Foam (in the vacuum) is so nonsensical that it does not imply anything about physics, but rather about psychology and sociology. And it emphasizes how people think the most absurd things, and say the most absurd things without thinking, because everyone else is saying the same absurd thing. It is quite common, and a major cause of so much nonsense, not only about quantum mechanics.

What this idea of foam also emphasizes is how obsessively physicists think in classical terms, and look for classical explanations and pictures, despite it being known for the last century that classical physics is wrong.

V.7.i The vacuum and vacuous statements

Many other obviously absurd, vacuous, pronouncements about the vacuum than those given above exist (in the vacuum left by lack of thought). Many are based on a logical error, that of taking the representation of a thing to be the thing itself, of regarding a thing as real because it can be pictured in a certain way, even though the pictures are merely bookkeeping devices. It is interesting to consider which of the following are based on that error, and whether there are any statements that are not.

One statement is that elementary particles ... can spontaneously pop out of nothingness (?) and disappear again ... [Krauss (1999), p. 55]. (This of course shows those who say things like this are in deep disagreement with the fundamental belief that all physicists are supposed to hold: the universe is governed by laws.) The unseen particles produce measurable effects, such as alterations in the energy levels of atoms .... Of course energy levels of atoms found in the approximation in which the electromagnetic field with which particles interact is ignored differ from those obtained when it is included. If calculations are done wrong, leaving out part of equations, then not surprisingly results are wrong, and disagree with ones found in a better approximation (and experiment). It is (unfortunately almost) beyond belief that people can think that because a better calculation gives better results particles ... can spontaneously pop out of nothingness.

"According to quantum theory the vacuum of space is far from empty. It seethes with particles and antiparticles constantly being created and destroyed. Energy from this vacuum can be tapped ... " [Cowan (1998)]. Might this allow a perpetual motion machine? Of course, quantum theory says no such thing, nor is there any reason to believe that it does. In solving the equations of quantum electrodynamics with interacting particles (so that there is no vacuum) using the approximation scheme called perturbation theory, there are terms that can be usefully — but for calculational purposes only — be pictured as particles and antiparticles being created and destroyed. But this is merely a way of keeping track of intermediate steps of an approximate calculation.

That the zero-point energy is now a scientific fact, "experimentally demonstrated" is shown by the statement that energy in the vacuum is very much real [Yam (1997)]. Have these experiments (?) been done in a world that is very much real, or perhaps like virtual particles, it and the experiments, pop out of nothing, and are thus also virtual? Maybe it is just the thoughts that pop out of nothing.

Of course there is a solution of the equations of quantum electrodynamics in which there are no particles or fields, and this has all values zero. In this case, the vacuum, there are no particles, no antiparticles, nothing being created or destroyed, and no energy (which might approximate the energy required for thinking by those with particles, but not thoughts, popping in and out of something). The energy calculated in this case, the vacuum, is zero. This befuddlement comes from taking pictures, useful for bookkeeping but no more, too seriously. Unfortunately (?) this solution with all values zero is quite real, and definitely not virtual.

However those who believe that this energy can be tapped should, if they do not want to explain their method, patent it, which would give them vast wealth. We are waiting to see the names of these people on the lists of the richest people in the world.

The inability to accept a vacuum is of course very old. Many of today's views of it are really a belief in a form of ether (or even God) — a word, but not a concept that disappeared. Indeed there are striking parallels between the many present attempts to interpret nature classically, despite the evidence that it is not possible, and that nature is based on, must be based on, quantum mechanics [Mirman (1995b), chap. 1, p. 1], and the many attempts to interpret the electromagnetic field mechanically. Many of today's physicists, like ones at that time, try to fit nature to their prejudices, rather than trying to understand the way it really is. And perhaps physicists really do include God in their theories; they just use a different name.

V.8 Checklist for determining why quantum mechanics appears weird

It is useful to summarize many of these arguments in a form that allows a quick check to determine reasons that quantum mechanics seems weird. Here then is a list of errors that lead to the appearance that nature is quite bizarre. Of course it is not (although many of the errors are), and care in avoiding errors like these makes all the nonsense disappear:

1. Is it weird because conclusions that people draw from it are exactly the opposite of what quantum mechanics gives, so directly contradict its fundamental principles?
2. Is it weird because arguments are mixtures of classical and quantum concepts?
3. Is it weird because attempts are made to interpret quantum mechanics in classical terms?
4. Is it weird because concepts have not been carefully defined (and perhaps are not definable)?
5. Is it weird because experimental procedures have not been (or cannot be) carefully specified?
6. Is it weird because symbols or words have no relationship to physical reality, to experimental schemes?
7. Is it weird because all steps have not been given?
8. Is it weird because a confused, sloppy newspaper story, written by a reporter who does not want to bother doing the work needed to understand what he is writing, makes it seems strange?
9. Is it weird because a journalist, editorializing in the news columns, to prevent readers from noticing his incompetence and inability to explain his story, uses meaningless, emotionally charged, melodramatic, gushy, flowery, sensationalist, exciting, so distracting, words: "astounding", "bizarre", "grotesque", "magic", "magical", "quirky", "outlandish", "sensational", "spooky", "strange", "weird", "unbelievable" (he doesn't understand it so he can't believe it), and so on?
10. Is it weird because it is mis-stated, mis-represented, by a physicist who wants to appear as if he is privy to the secrets of nature, able to understand things ordinary mortals cannot, be a high priest of the universe (or a high priest of confusion) so that he can be worshipped by all lesser human beings, get lots of government money

and plushy high-paying jobs with self-important titles at universities trying to prove how much better they are then everyone else?

11. Is it weird because a naive, gullible reporter falls for a trick pulled by a confused or fraudulent, manipulative scientist who wants to get publicity for himself by distorting reality to make it seem bizarre so it will be of interest to sensationalist media.

12. Is it weird because the local newspaper, out of sex and divorce stories to sensationalize, turns in desperation to quantum mechanics.

13. Is it weird because an author (or worse physicist) with a deep emotional, infantile, attachment to classical physics, and despite overwhelming experimental support for quantum mechanics, insists in thinking of the universe in classical terms, but uses the language of quantum mechanics, thus giving a discussion which sounds quite weird because the view of nature that is being expressed is all wrong, and because terms in the language are defined to mean one thing, and then used to mean another — language used improperly often sounds strange.

14. Is it weird because people (physicists!) think what they think not because it is true, not because it makes sense, but because other people think in the same way. People (physicists!) do not think about what they should believe, but take votes. And if others are spouting nonsense (not only) about quantum mechanics, there are too many who feel impelled to also spout nonsense. Many modern physicists are driven by mob psychology, but physics, ancient or modern, is not driven by mob psychology. Nature cares about neither mobs nor psychology.

15. Is it weird because physicists, seeing how much publicity and fascination crackpot subjects, astrology, alien abduction, flying saucers, and so on, get decide to also try to generate excitement and get publicity — by making quantum mechanics look as nonsensical as these. Others profit from exploiting human gullibility and fascination with the outlandish, the mystical, the incredible, the fantastic, the absurd, so why not physicists?

Indeed it is quite true [Johnson (2000a)] that knowing that there is a growing appetite for this stuff — the weirder the better — some physicists have taken their case directly to the public. And with theories getting wilder and wilder, real physics and fictional physics have become mashed into a new folklore. Public relations has become a consequential, and often harmful, part of science [Krauss (2000)]. Too often the most important, and the most competent, part of a scientific collaboration is the press office.

Nonsense sells, so why shouldn't physicists get part of the profit?

This leads to a fundamental rule for those who are writing about physics (and indeed, about many, many other things): If what you are writing seems like nonsense, it is.

The many absurd statements about quantum mechanics imply not merely misunderstanding or carelessness, but almost a compulsion to say the most ridiculous things, suggesting symptoms of a form of Tourette's syndrome peculiar to physicists.

No competent physicist, or any other competent person discussing it, finds quantum mechanics — or nature — weird or strange. Someone who says things such as quantum mechanics is weird or quantum mechanics makes strange predictions [Greenberger (1995)] or refers to, for example, quantum weirdness is simply using a euphemism for "I am incompetent" or "I am careless and totally confused".

V.9 Comprehensibility of a sensible universe

That we, mere conglomerations of matter, mere statefunctions, complicated though we may be as statefunctions, can understand the universe, how it works, why it works that way, so simply, in such depth, is awesome. Yet far too many, rather than exalting in this, in what they (should) have learned, in what they can achieve, go to great lengths not to understand, to see strangeness, confusion, where there is really sense and intelligibility — this is not merely a tragedy, but a desecration.

Rather than being appreciative, grateful, thankful, for what the universe is, for our ability to understand it, and for what we understand, for the simplicity, the sensibility, they try to make out of clarity darkness, out of grandeur tawdriness, out of sense the most utter nonsense. They aim, by flaunting their confusion, their misunderstandings, their vacuity, to destroy the elegance of the natural world of which we are so small a part. But fortunately, because they, we, are but mere pieces of matter, mere statefunctions, they cannot desecrate nature, but only demean themselves.

This then is the fundamental theme, moral, of not only what is considered here, but of the work that it follows. The universe is not mysterious, except perhaps in its lack of mystery, it is not weird, not strange, not beyond understanding. For those who want to see, for those who prefer to know rather than be baffled, its elegance, its simplicity, its comprehensibility, the sensibility of its laws and the reasons for them, is clear, and so striking. The universe then is not merely something we can know, but a world that we can deeply appreciate.

REFERENCES

Alda, Alan (2000), Curiosity Rhymed the Cat, *Scientific American* **282,** #5, May, p. 120.

Anandan, J., Y. Aharonov (1988), Geometric quantum phase and angles, *Phys. Rev.* **D38,** #6, 15 Sept., p. 1863-1870.

Anderson, Philip W. (2000), Brainwashed by Feynman?, *Physics Today* **53,** #2, Feb., p. 11-12.

Appleby, D. M. (1998), The Concept of Experimental Accuracy and Simultaneous Measurement of Position and Momentum, *Int. J. Theor. Phys.* **37,** #5, p. 1491-1510.

Appleby, D. M. (1999), Optimal Joint Measurements of Position and Momentum, *Int. J. Theor. Phys.* **38,** #3, p. 807-825.

APS News (1999), Physics in the 20th Century, **8,** #5, May, p. 11.

Aspect, Alain (1998), Experimental tests of Bell's inequalities with correlated photons, in Pratesi and Ronchi (1998), p. 345-375.

Aspect, Alain (1999), Bell's inequality test: more ideal than ever, *Nature* **398,** 18 Mar., p. 189-190.

Avron, J. E., E. Berg, D. Goldsmith, A. Gordon (1999), Is the number of photons a classical invariant? *Eur. J. Phys.* **20,** p. 153-159.

Ballentine, L. E. (1970), The Statistical Interpretation of Quantum Mechanics, *Rev. Mod. Phys.* **42,** #4, Oct., p. 358-381.

Ballentine, L. E., Jon P. Jarrett (1987), Bell's theorem: Does quantum mechanics contradict relativity?, *Am. J. Phys.* **55,** #8, Aug., p. 696-701.

Barnett, Stephen M., B. J. Dalton (1993), Conceptions of Quantum Optical Phase, *Physica Scripta* **T48,** p. 13-21.

Barrow, John D. (1997), Time in the Universe, in Burgen, McLaughlin, Mittelstrass (1997), p. 155-174.

Barrow, John D., Frank J. Tipler (1988), The Anthropic Cosmological Principle (Oxford: Oxford University Press).

Beardsley, Tim (1999), A laser in tune with itself, *Scientific American* **280,** #2, Feb., p. 41.

Bell, John (1990), Against 'Measurement', *Physics World* **3,** #8, Aug., p. 33-40.

Benioff, Paul (1999), Simple example of definitions of truth, validity, consistency, and completeness in quantum mechanics, *Phys. Rev.* **A59,** #6, June, p. 4223-4237.

Bialynicki-Birula, Iwo, Matthias Freyberger, Wolfgang Schleich (1993), Various Measures of Quantum Phase Uncertainty: A comparative Study, *Physica Scripta* **T48,** p. 113-118.

Bohr, N. (1935), Can Quantum-Mechanical Description of Physical Reality Be Considered Complete?, *Phys. Rev.* **48,** #8, Oct. 15, p. 696-702.

Brevik, Iver, Valery N. Marachevsky, Kimball A. Milton (1999), *Phys. Rev. Lett.* **82,** #20, 17 May, p. 3948-3951.

Bridgman, P. W. (1927), The Logic of Modern Physics (New York: The Macmillan Company).

Bridgman, P. W. (1960), The Logic of Modern Physics (New York: The Macmillan Company).

Brukner, Caslav, Anton Zeilinger (1999), Operationally Invariant Information in Quantum Measurements, *Phys. Rev. Lett.* **83,** #17, 25 Oct., p. 3354-3357.

Burgen, Arnold, Peter McLaughlin, Juergen Mittelstrass, (1997), The Idea of Progress (Berlin, Walter de Gruyter).

Bussey, P. J. (1988), The Foundations of Quantum Mechanics, *Foundations of Physics* **18,** #5, May, p. 491-528.

Cao, Tian Yu (1998), Conceptual Developments of 20th Century Field Theories (Cambridge: Cambridge University Press).

Carruthers, P., Michael Martin Nieto (1968), Phase and Angle Variables in Quantum Mechanics, *Rev. Mod. Phys.* **40,** #2, April, p. 411-440.

Caticha, Ariel (1998a), Consistency, amplitudes and probabilities in quantum theory, *Phys. Rev.* **A57,** #3, Mar. p. 1572-1582.

Caticha, Ariel (1998b), Consistency and linearity in quantum theory, *Phys. Lett.* **A244,** 13 July, p. 13-17.

Chang, Kenneth (2000), Here, There and Everywhere: A Quantum State of Mind, *New York Times* Science Times, July 11, p. F3.

Chevalley, Claude (1962), Theory of Lie Groups (Princeton, NJ: Princeton University Press).

Cirac, Ignacio, Peter Zoller (1999), Quantum engineering moves on, *Physics World* **12,** #1, Jan, p. 22-23.

Clauser, John F., Abner Shimony (1978), Bell's theorem: experimental tests and implications, *Rep. Prog. Phys.* **41,** p. 1881-1927.

Collins, Graham P. (2000), Einstein's Constant, *Scientific American* **283,** #1, July, p. 28.

Cooke, Roger, Michael Keane, William Moran (1985), An elementary proof of Gleason's theorem, *Math. Proc. Camb. Phil. Soc.* **98,** p. 117-128.

Cowan, Ron (1998), The greatest story ever told, *Science News* **154,** #25, December 19, p. 392-394.

Crease, Robert P. (2000), Case of the deadly strangelets, *Physics World* **13,** #7, July, p. 19-20.

d'Espagnat, Bernard (1990), Towards a Separable "Empirical Reality"?, *Foundations of Physics* **20,** #10, p. 1147-1172.

Davies, Paul (1999), Quantum gravity presents the ultimate challenge to theorists, *Physics World* **12,** #12, Dec., p. 21.

Deutsch, David (1998), Problems or prophecies?, *Physics World* **11,** #12, Dec., p. 41.

Devlin, Keith (2000), Snake Eyes in the Garden of Eden, *The Sciences* **40** #4, July/August, p. 14-17.

DeWitt, Bryce S., Neill Graham, eds. (1973), The Many-Worlds Interpretation of Quantum Mechanics (Princeton, NJ: Princeton University Press).

Dirac, P. A. M. (1956), The Principles of Quantum Mechanics (London: Oxford University Press, at the Clarendon Press).

Drummond, Peter (2000), Elegance: Keeping it Simple and Testable, *Physics Today* **53,** #8, Aug., p. 12-13.

Dürr, Detlef, Walter Fusseder, Sheldon Goldstein, Nino Zanghi (1993), Comment on "Surrealistic Bohm Trajectories", *Z. Naturforsch.* **48a,** p. 1261-1262.

Dürr, S., T. Nonn, G. Rempe (1998), Origin of quantum-mechanical complementarity probed by a 'which-way' experiment in an atom interferometer, *Nature* **395,** 3 Sept., p. 33-37.

Echeverria-Enriquez, Arturo, Miguel C. Munoz-Lecanda, Narciso Roman-Roy, Carles Victoria-Monge (1998), Mathematical Foundations of Geometric Quantization, *Extracta Mathematicae* **13,** #2, p. 135-238.

Einstein, A., B. Podolsky, N. Rosen (1935), Can Quantum-Mechanical Description of Physical Reality Be Considered Complete?, *Phys. Rev.* **47,** #10, May 15, p. 777-780.

Englert, Berthold-Georg (1999), Remark on Some Basic Issues in Quantum Mechanics, *Z. Naturforsch.* **54a,** p. 11-32.

Englert, Berthold-Georg, Marlan O. Scully, Herbert Walther (1999), Quantum erasure in double-slit interferometers with which-way detectors, *Am. J. Phys.* **67,** #4, April, p. 325-329.

Englert, Berthold-Georg, Marlan O. Scully, George Süssman, Herbert Walther (1992), Surrealistic Bohm Trajectories, *Z. Naturforsch.* **47a,** p. 1175-1186.

Everett, Hugh III (1973), The Theory of the Universal Wave Function, in DeWitt and Graham (1973), p. 3-140.

Farhi, Edward, Jeffrey Goldstone, Sam Gutmann (1989) How Probability Arises in Quantum Mechanics, *Annals of Physics* **192,** #2, June, p. 368-382.

Feldmann, Michel (1995), New loophole for Einstein-Podolsky-Rosen Paradox, *Foundation of Physics Letters* **8,** #1, p. 41-53.

Finkelstein, David (1991), Theory of Vacuum, in Saunders and Brown (1991), p. 251-274.

Fuchs, Christopher A., Asher Peres (2000), Quantum theory needs no 'interpretation', *Physics Today* **53,** #3, March, p. 70-71.

Ghirardi, Gian Carlo, Philip Pearle, Alberto Rimini (1990), Markov processes in Hilbert space and continuous spontaneous localization of systems of identical particles, *Phys. Rev.* **A42,** #1, July, p. 78-89.

Gilmore, Robert (1974), Lie Groups, Lie Algebras, and Some of Their Applications (New York: John Wiley and Sons, Inc.).

Gingerich, Owen, (2000), Copernicus and the Aesthetic Impulse, *APS News* **9,** #7, July, p. 8.

Glanz, J. (1998a), Does Science Know the Vital Statistics of the Cosmos?, *Science* **282,** 13 Nov., p. 1247-1248.

Glanz, James (1998b), Cosmic Motion Revealed, *Science* **282,** 18 Dec., p. 2156-2157.

Gleason, Andrew M. (1957), Measures on the Closed Subspaces of a Hilbert Space, *J. Math. Mech.* **6,** #6, p. 885-893.

Goldstein, Sheldon (1998), Quantum Theory Without Observers, *Physics Today* **51,** #4, April, p. 38-42.

Grant, Edward (1981), Much ado about nothing, Theories of space and vacuum from the Middle Ages to the Scientific Revolution (Cambridge: Cambridge University Press).

Greenberger, Daniel M. (1995), Preface, Fundamental Problems in Quantum Theory, Daniel M. Greenberger and Anton Zeilinger, Ed. *Annals of the New York Academy of Sciences* **755,** p. xiii-xiv.

Griffiths, Robert B., Roland Omnes (1999), Consistent Histories and Quantum Measurements, *Physics Today* **52,** #8, Aug. Part 1, p. 26-31.

Hartle, J. B. (1968), Quantum mechanics of individual systems, *Am. J. Phys.* **36** #8, Aug., p. 704-712.

Hiley, B. J. (1991), Vacuum or Holomovement, in Saunders and Brown (1991), p. 217-249.

Holton, Gerald (2000), Werner Heisenberg and Albert Einstein, *Physics Today* **53,** #7, July, p. 38-42.

Jackson, John David, (1963), Classical Electrodynamics (New York: John Wiley and Sons, Inc.).

Jancewicz, Bernard (1988), Multivectors and Clifford Algebra in Electrodynamics (Singapore: World Scientific Publishing Co.).

Jasny, Barbara, R. Brooks Hanson, Floyd E. Bloom (1999), A media uncertainty principle, *Science* **283,** 5 March, p. 1453.

Johnson, George (2000a), A Small Step for Man, A Giant Leap of Faith, *New York Times* Week in Review, February 20, p. 4.

Johnson, George (2000b), In Quantum Feat, Atom is Seen in 2 places at Once, *New York Times* Science Times, February 22, p. F1-F2.

Kayser, B., L. Stodolsky (1995), EPR experiments without "collapse of the wavefunction", *Phys. Lett.* **B359,** 12 Oct., p. 343-350.

Kempermann, Gerd, Fred H. Gage (1999), New Nerve Cells for the Adult Brain, *Scientific American* **280,** #5, May, p. 48-53.

Ketov, Sergei V. (1997), Conformal Field Theory (Singapore: World Scientific Publishing Co.).

Khrennikov, Andrew (1995), p-adic Probability Interpretation of Bell's Inequality, *Phys. Lett.* **A200,** p. 219-223.

Kim, Yoon-Ho, Rong Yu, Sergei P. Kulik, Yanhua Shih, Marlan O. Scully (2000), Delayed "Choice" Quantum Eraser, *Phys. Rev. Lett.* **84,** #1, 3 Jan., p. 1-5.

Kleppner, Daniel, Roman Jackiw (2000), One Hundred Years of Quantum Physics, *Science* **289,** 11 August, p. 893-898.

Koch, Christof, Gilles Laurent (1999), Complexity and the nervous system, *Science* **284,** 2 April, p. 96-98.

Krauss, Lawrence M. (1999), Cosmological Antigravity, *Scientific American* **280,** #1, Jan., p. 53-59.

Krauss, Lawrence M. (2000), Pushing the Limits of Science, and of Public Relations, *New York Times* Science Times, March 14, p. F5.

Lamb, Willis E. Jr. (1994), Suppose Newton had invented wave mechanics, *Am. J. Phys.* **62** #3, Mar. p. 201-206.

Leggett, A. J. (1984), Schrödinger's Cat and her Laboratory Cousins, *Contempt. Phys.* **25,** #6, p. 583-598.

Leggett, Tony (1999), Quantum theory: weird and wonderful, *Physics World* **12,** #12, Dec., p. 73-77.

Leonhardt, U., H. Paul (1993), Realistic Measurement of Phase, *Physica Scripta* **T48,** p. 45-48.

Mandel, L. (1999), Quantum effects in photon interference, *Rev. Mod. Phys.* **71,** #2, Mar., p. S274-S282.

Maxwell, Nicholas (2000), A new conception of science, *Physics World* **13,** #8, Aug., p. 17-18.

Mermin, N. D. (1981), Bringing home the atomic world: Quantum mysteries for anybody, *Am. J. Phys.* **49,** #10, Oct., p. 940-943.

Mermin, N. David (1985), Is the moon there when nobody looks? Reality and the quantum theory, *Physics Today* **38,** April, p. 38-47.

Mermin, N. David (1998), What is quantum mechanics trying to tell us? *Am. J. Phys.* **66** #9, Sept., p. 753-767.

Mermin, N. David (2000), The Contemplation of Quantum Computation, *Physics Today* **53,** #7, July, p. 11-12.

Miller, Julie Ann (1998), *Science News* **154,** #25 & 26, Dec. 19 & 26, p. 402.

Miller, Willard, Jr (1968), Lie Theory and Special Functions (New York: Academic Press).

Mirman, R. (1970), Analysis of the Experimental Meaning of Coherent Superposition and the Nonexistence of Superselection Rules, *Phys. Rev.* **D1,** #12, 15 June, p. 3349-3363.

Mirman, R. (1995a), Group Theory: An Intuitive Approach (Singapore: World Scientific Publishing Co.).

Mirman, R. (1995b), Group Theoretical Foundations of Quantum Mechanics (Commack, NY: Nova Science Publishers, Inc.).

Mirman, R. (1995c), Massless Representations of the Poincaré Group, electromagnetism, gravitation, quantum mechanics, geometry (Commack, NY: Nova Science Publishers, Inc.).

Mirman, R. (1999), Point Groups, Space Groups, Crystals, Molecules (Singapore: World Scientific Publishing Co.).

Mirman, R. (2001), Quantum Field Theory, Conformal Group Theory, Conformal Field Theory: Mathematical and conceptual foundations, physical and geometrical applications (Huntington, NY: Nova Science Publishers, Inc.).

Misner, Charles W., Kip S.Thorne, John Archibald Wheeler (1973), Gravitation (San Francisco: W. H. Freeman and Co.).

Mohrhoff, Ulrich (2000), What quantum mechanics is trying to tell us, *Am. J. Phys.* **68** #8, Aug., p. 728-745.

Nogami, Y., F. M. Toyama, W. van Dijk (2000), Bohmian description of decaying quantum system, *Phys. Lett.* **A270,** 12 June, p. 279-287.

Norton, John D. (1993), General covariance and the foundations of general relativity: eight decades of dispute, *Rep. Prog. Phys.* **56,** p. 791-858.

Omnes, Roland (1992), Consistent Interpretations of Quantum Mechanics, *Rev. Mod. Phys.* **64,** #2, April, p. 339-382.

Overbye, Dennis (1999), Did God have a choice?, *New York Times* Magazine, April 18, p. 180.

Page, Don N. (1982), The Einstein-Podolsky-Rosen Physical Reality is Completely Described by Quantum Mechanics, *Phys. Lett.* **A91,** #2, 23 Aug., p. 57-60.

Percival, Ian C., Walter T. Strunz (1998), Classical dynamics of quantum localization, *J. Phys. A: Math. Gen.* **31,** p. 1815-1830.

Peres, Asher (1979), Proposed Test for Complex versus Quaternion Quantum Theory, *Phys. Rev. Lett.* **42,** #11, 12 Mar., p. 683-686.

Peres, Asher (1984), What is a State Vector?, *Am. J. Phys.* **52** #7, July, p. 644-650.

Pitowsky, Itamar (1998), Infinite and finite Gleason's theorems and the logic of indeterminacy, *J. Math. Phys.* **39,** #1, Jan., p. 218-228.

Pratesi, R., L. Ronchi, eds. (1998), Waves, Information and Foundations of Physics, Conf. Proc. Vol. 60 (Bologna, Italian Physical Society).

Quercigh, Emanuele, Johann Rafelski (2000), A strange quark plasma, *Physics World* **13,** #10, Oct., p. 37-42.

Quigg, Chris (1999), Aesthetic Science, *Scientific American* **280,** #4, April, p. 125-127.

Rennie, John (2000), Quantum bits and reliable boats, *Scientific American* **282,** #4, April, p. 8.

Rodgers, Peter (1998), All features great and small, *Physics World* **11,** #5, May, p. 26-27.

Rovelli, Carlo (1996), Relational Quantum Mechanics, *Int. J. Theor. Phys.* **35,** #8, p. 1637-1678.

Saunders, Simon, Harvey R. Brown (1991), The Philosophy of Vacuum (Oxford: Clarendon Press).

Schaden, Martin, Larry Spruch (1998), Infinity-free semiclassical evaluation of Casimir effects, *Phys. Rev.* **A58,** #2, Aug. p. 935-953.

Schaden, Martin, Larry Spruch, Fei Zhou (1998), Unified treatment of some Casimir energies and Lamb shifts: A dielectric between two ideal conductors, *Phys. Rev.* **A57,** #2, Feb., p. 1108-1120.

Schiff, L. I. (1955), Quantum Mechanics (New York: McGraw-Hill Book Co.).

Schottenloher, Martin (1997), A Mathematical Introduction to Conformal Field Theory (Berlin: Springer-Verlag).

Schweber, Silvan S. (1962), An Introduction to Relativistic Quantum Field Theory (New York: Harper and Row).

Seife, Charles (2000), Cold Numbers Unmake the Quantum Mind, *Science* **287,** 4 Feb., p. 791.

Shephard, Gordon M. (1994), Neurobiology (New York: Oxford University Press).

Silverman, M. P. (1993), More than one mystery: Quantum interference with correlated charged particles and magnetic fields *Am. J. Phys.* **61,** #6, June, p. 514-523.

Smolin, Lee (1999), The new universe around the next corner, *Physics World* **12,** #12, Dec., p. 79-84.

Stephani, Hans (1994), General Relativity, an introduction to the theory of the gravitational field (Cambridge: Cambridge University Press).

Streater, R. F., A. S. Wightman (1964), PCT, SPIN AND STATISTICS, AND ALL THAT (New York: W. A. Benjamin, Inc.).

Tegmark, Max (2000), Importance of quantum decoherence in brain processes, *Phys. Rev.* **E61,** #4, April, p. 4194-4206.

Teller, Paul (1997), An Interpretive Introduction to Quantum Field Theory (Princeton, NJ: Princeton University Press).

Trimble, Virginia (2000), Powers of Ten: Astronomy's Greatest Hits, *APS News* **9,** #9, Oct. p. 3.

Vaccaro, John (1995), Number-phase Wigner function on Fock Space, *Phys. Rev.* **A52,** #5, Nov., p. 3474-3488.

van Kampen, N. G. (1990), Quantum criticism, *Physics World* **3,** #10, Oct., p. 20.

van Kampen, N. G. (1991), Mystery of Quantum Measurement, *Physics World* **4,** #12, Dec., p. 16-17.

van Kampen, N. G. (2000), Quantum Histories, Mysteries, and Measurements, *Physics Today* **53,** #5, May, p. 76-80.

Varadarajan, V. S. (1993), Quantum Theory and Geometry: Sixty Years After von Neumann, *Int. J. Theor. Phys.* **32,** #10, 1815-1834.

Varshalovich, D. A., A. N. Moskalev, V. K. Khersonskii (1988), Quantum Theory of Angular Momentum: Irreducible Tensors, Spherical Harmonics, Vector Coupling Coefficients, 3nj Symbols (Singapore: World Scientific Publishing Co.).

von Baeyer, Hans Christian (2000), Quantum Mechanics at the End of the 20th Century, *Science* **287,** 17 March, p. 1935.

Voss, David (1999), Making the stuff of the big bang, *Science* **285,** 20 Aug., p. 1194-1197.

Wallace, Philip R. (1996), Paradox Lost. Images of the Quantum (New York: Springer-Verlag).

Weinberg, Steven (1989), The cosmological constant problem, *Rev. Mod. Phys.* **61,** #1, Jan., p. 1-23.

Weinberg, Steven (1997), The first elementary particle, *Nature* **386,** 20 March, p. 213-215.

Weiss, Peter (2000), Seeking the Mother of all Matter, *Science News* **158,** #9, August 26, p. 136-138.

Weng, Gezhi, Upinder S. Bhalla, Ravi Iyengar (1999), Complexity in biological signaling systems, *Science* **284,** 2 April, p. 92-96.

Whitaker, Andrew (1998), John Bell and the most profound discovery of science, *Physics World* **11,** #12, Dec., p. 29-34.

Wigner, Eugene P. (1957), Relativistic Invariance and Quantum Phenomena, *Rev. Mod. Phys.* **29,** #3, July, p. 255-268.

Wilczek, Frank (1999a), The Persistence of Ether, *Physics Today* **52,** #1, Jan., p. 11-13.

Wilczek, Frank (1999b), Quantum field theory, *Rev. Mod. Phys.* **71,** #2, Mar., p. S85-S95.

Wilford, John Noble (1999), New Measurement Suggests Universe is Younger Than Believed, *New York Times* June 2, p. A21.

Yam, Philip (1997), Exploiting zero-point energy, *Scientific American* **277,** #6, Dec., p. 82-85.

Yurke, B., D. Stoler (1997), Observing Local Realism Violations with a Combination of Sensitive and Insensitive Detectors, *Phys. Rev. Lett.* **79,** #25, 22 Dec., p. 4941-4945.

Yurke, Bernard, David Stoler (1995), Bell's-inequality experiment employing four harmonic oscillators, *Phys. Rev.* **A51,** #5, May, p. 3437-3444.

Yurke, Bernard, Mark Hillery, David Stoler (1999), Position-momentum local-realism violation of the Hardy type, *Phys. Rev.* **A60,** Nov., #5, p. 3444-3447.

Zeh, H. D. (1993), There are no quantum jumps, nor are there particles!, *Phys. Lett.* **A172,** p. 189-192.

Zeilinger, Anton (1999), A Foundational Principle for Quantum Mechanics, *Foundations of Physics* **29,** #4, p. 631-643.

Zeilinger, Anton (2000), Quantum Teleportation, *Scientific American* **282,** #4, April, p. 50-59.

Zurek, Wojciech H. (1997), Probing quantum origins of the classical, *Physics World* **10,** #1, Jan., p. 24-25.

Index

Definitions are listed in bold

ℵ, 250
abbreviation, 7
abbreviations, 127, 180
abbreviations, symbols as, 78
abduction, alien, 266
Abelian group, 139
Abelian groups, representations of, 56
abnormal psychology, 106
absolute square of statefunctions, 60
absolute time, Newton's, 164
absolute value, 63
abstract conception of nature, 122
abstract mathematics, 80
abstractly, 115
absurd, vi, 266
absurd connotation, 173
absurdity, 223
absurdity, underlying, 20
acausal signals, 210
accelerator, 6
accelerators, 19
accept, 20
accident, 110
action principle, 137
action, quantum of, 171
action-at-a-distance of classical physics, spooky, 210
action-at-a-distance, spooky, 231
activity, neural, 176
actual physical processes, 80
actual physical situation, 147
actual physical state, **184**, 212
actual physical values, 226
actually happens, 184
adjectives, 187
adults, 190
adventurous, xii
aesthetics, xi
affine group, 38
after, **220**
Aharonov, Y., 201
Aharonov-Bohm effect, 200
Ahluwalia, Dharam V., xiii
air, qualities floating in the, 187
air, vibrations in, 12
Alda, Alan, 224
algebra, Euclidean, 39
algorithm, brilliant, 134
alien abduction, v, 266
alien creatures, 27
alive, 70
all steps, 265
allure, 148
allure of equations, 154
alone, we are not, 246
alternative theories, 194, 245
alternative theory to quantum mechanics, 158
ambiguity of language, 119
ambiguous, 153
American Physical Society, 162
amphibians, 226
amplitude of a single basis vector, **90**
amplitude, square of the, 61
amusement, 223
analog, classical, 79
analogies, overreliance on, 169
analogy, 146
analogy, false, 135
analogy, method of false, 168, 170
analytic, 225, 237
analytic functions, statefunctions are, 158
Anandan, J., 201
ancient, 146

ancient rocks, 185
Anderson, Philip W., 229
anesthesiologist, 156
angels, 247, 249
angle, 139
angles, 28
angles of a triangle, sum of the, 35
angular momentum, **136**
angular momentum, conservation of, 213
angular momentum, selection rules for, 94
animal, 27
annihilation, 255
anomalous, 109
answers, emotionally comfortable, 150
ant, 133
antenna, 211
anthropic cosmological principle, 174
anthropocentric, 250
anthropocentrism, 119
anthropology, v
anti-matter, x
anti-rationality, vii
antiparticles, 163
antonym, 179
apparition, ghostly, 168
apparitions, 262
appetite, 266
Appleby, D. M., xiii, 112
approach, safest, x
approaches, conservative, 198
approximation expansion, 252
approximation method, 208
approximation scheme, 233, 252
approximation scheme, favorite, 229, 243
APS News, 162
arbitrariness of definitions, 240
arbitrary choice of units, 235
arc, 136
archaea, 226
arguments, philosophical, 147, 158
argumentum ad verecundiam, 197
Aristotelian, 106, 199, 221
Aristotle, 251
artificial canals, 143
artificial distinction, v
artificial intelligence, **133**
Aspect, Alain, 221
assumption, 17
assumptions of the theory, **16**
assumptions, no, 22
assumptions, underlying, 20
astounding, 265
astrology, v, 266
astronomical observations, 86
astronomy, ix
astronomy, workable, 4
atom, **132**
atom, Bohr's theory of the, 78
atom, classical model of the, x
atom, nonrelativistic hydrogen, 50
atoms are small, 188
atoms, excited, 62
attraction of classical physics, compulsive, 199
author, 42
authority, 197
auxiliary quantities, 240
Avron, J. E., 88
awareness, 118
awesome, 192
axes, different, **44**
axioms of the geometry, **29**

babbling, 166
bacteria, 226
Baleanu, Dan, xiii
ball, 224
ball hitting a wall, 65, 79
ball, little, 201
Ballentine, L. E., 105, 108
balls, 172
bands in crystals, 143
bank, 211, 212, 224
Barnett, Stephen M., 214
barrier, 94, 204, 237
barrier, potential, 144
barriers, 143
Barrow, John D., xiii, 111, 174, 237
baryons, statefunctions of, 206
baseball bats, 20, 184
basic objects, 84
basic rules of quantum mechanics, 221
basis function, representation, **15**

basis functions of a single representation, 46
basis state, **57**
basis state of a group representation, **46**
basis states, **44**
basis vector, **83**
basis vectors, integral powers of, 91
basketball, 201
battle, 253
Beardsley, Tim, 238
beautiful, xi
beauty, xi
bee, 188
before, **220**
beginning of the universe, 192
beguiling statements, 188
behavior of the electron, **126**
beings, cognizant, thinking, 122
beings, intelligent, 189
belief, misguided, vi
believe in, **19**
Bell, John, 178
Bell's inequalities, 221
Bell's theorem, 221
beloved, 155
Benioff, Paul, 152
Berg, E., 88
Bessel functions, 204
better, **18**
better framework, **18**
better than physics, 153
Bhalla, Upinder S., 133
Bialynicki-Birula, Iwo, 214
big bang view, 16
billiard balls, 172
bimodal, 91
biological processes, 236
biology, 15, 26, 27, 176
birth, 134
birth of the universe, 250
bits, 144
bizarre, v, 155, 223, 265
bizarre delusions, 155
black, 224
blackbody radiation, x
blackening, 203
blades, 73
bless, 246
Bloom, Floyd E., 227
blue, 148
blueness, 176
blueprints, 190
Bohmian trajectory, 247
Bohr, x
Bohr, N., 170, 180
Bohr's theory of the atom, 78
book, 255
bookkeeping, 252, 264
bookkeeping device, 8
border, 232
bored, 163
Bose-Einstein statistics, 53
boson, 53
both-path, 217
bothered, 221
bottle, 160
boundary conditions, 76, 256
box, 224
box, closed, 73
Bradwardine, Thomas, 252
brain, 67, 71, 133
brain, data set for a, 134
brain, quantum states of the, 156
brain state, 249
brains, 131
brains, no, 71
branch, 249
Brans, Carl H., xiii
Brevik, Iver, 256
Bridgman, P. W., 146, 150
bright line, 61
brilliant, 191
brilliant algorithm, 134
brilliant people, xi
broken into parts, 255
broken symmetry, 213
Brukner, Caslav, 54
bubble, 71
bubbles, 255
bubbles, spray with, 255
bubbling in an underlying space, 253
building, top of a, 4
buildings, 19
bumblebees, 188
bumpy geometry, 253
buses, 224
Bussey, P. J., 234

Caenorhabditis elegans, 133
calculational tool, 233
camouflage, 165
canals, artificial, 143
cannot know, 194
canon, 252
Cao, Tian Yu, 34, 98
capacitance, 143
captivated with equations, 153
car, 110
career, 19
careful, 196
carefully defined, 265
carelessness, vii, 64, 196
Carruthers, P., 214
Cartesian coordinates, 35
Casimir effect, 256
Cat, Schrödinger's, 74
cathode ray tubes, 7
Caticha, Ariel, xiii, 106
causality, **31**, 220, 247
cause, common, 171
causes, 159
cellular, 145
century, 263
certainty, 117
chain, 122
chain of reasoning, 120
chair, 101, 130
Chang, Kenneth, 224
change of our knowledge, sudden, 118
chaos, v
Chapline, George, xiii
charge conservation, 96
charge, effective, **34**
chartreuse, 146
checklist of errors, 265
chemical reactions, 68
Chevalley, Claude, 57
child, 224
childhood, 190
chimpanzee, 26
chordates, 35
choreographs, 246
Christoffel relation, 242
chronological, **163**
Cirac, Ignacio, 224
circular, 184, 193
circular definition, 145
circular reasoning, 136
circularly polarized, 94, 219
circularly symmetric, 213
circumstances, similar, 178
class of statefunctions, 126
classical, **242**
classical analog, 79
classical and quantum concepts, mixtures of, 265
classical electrodynamics, inconsistencies of, 102
classical electrodynamics is inconsistent, 157
classical electromagnetic theory, 208
classical electromagnetic wave, 237
classical electromagnetism, 61
classical electromagnetism and local realism, incompatibility of, 231
classical electromagnetism is nonlocal, 222
classical example of entanglement, 211
classical explanation, 105
classical explanations, 263
classical expressions, **231**
classical groups, 38
classical intuition, 122
classical light wave, 156
classical model of the atom, x
classical object, **70, 74**
classical objects, 73
classical objects, statefunctions of, 122
classical physics, 17, 19
classical physics as a framework, 15
classical physics, compulsive attraction of, 199
classical physics, infantile attachment to, 266
classical physics, nonlocality of, 231
classical physics, quantitization in, 144
classical physics, spooky action-at-a-distance of, 210
classical pictures, 199, 253

classical probability, 107
classical shell, 209
classical statefunction, 73
classical theory of gravity, 240
classical variables, **98**
classical weirdness, 238
Clauser, John F., 179
Cleve, Richard, xiii
clicks, 178
climbers, mountain, 157
cling, 157
clock, 36, 59, 150
clocks, 11
closed box, 73
cloth fabric, 254
cloud-chamber, 60, 71, 203
cloud of electrons and positrons, **34**
cloud of short-lived photons, 261
cluster of galaxies, 231
coefficients that are time dependent, 77
cognizant, thinking beings, 122
coherence, 3, 178
coherent, meaningful picture, 131
coherent picture, **5**, 18
coherent superposition, 75, 212
coherent superposition of a macroscopic object, 70
coherent superposition of states, 70
coherent superpositions, 73
coherent view of nature, 7
collapse, 155
collapse of the wavefunction, 64, 160, 228
collapse of velocity, 65
collapses, 228
collection of neurons, 189
Collins, Graham P., 234
collision, 65
collisions by point particles, 182
column of mercury, 9
comfort, emotional, 5, 190, 191, 193
comfort, finding emotional, 125
comfort, psychological, 147
comfortable, 193, 248
comfortable answers, emotionally, 150
comforting, 154
comforting, emotionally, 158
common cause, 171
common sense, v, 64
communicate, 125
communication, 210
commutation relations, 139, 171
compact semisimple group, single irreducible representation of a, 49
comparing a galaxy with a pebble, 134
competent, 267
competent physicist, vi
competing theories, 16
complementarity, 161
complete description, **120**
complete description of reality, 181
complete set for statefunctions, 45
completely dimensionless point, 205
completely predictable, 117
complex function, 241
complex number field, 55
complex web, 262
composite, **48, 50**
Compton radius of the electron, 30
Compton wavelengths, 6
compulsion, vi, 267
compulsive attraction of classical physics, 199
computational mechanism, 118
computer, 130
computer languages, 187
computer, live with a, 189
computer program, 187
computer programmers, conceited, 135
computers, 188
conceited computer programmers, 135
concept, correctness of a, **8**
concept, meaning of a, **8**
conception of nature, abstract, 122
concepts, experimental meanings of, viii
concepts, extraneous, 245
concepts, primitive, 104
concepts, strange, xi

concepts, web of, 8, 10
conclusions, method of jumping to, 138
conditions, boundary, 76
conditions, regularity, 76
conductance, 18
conduits, 4
configuration space, 127
conformal field theory, 13
confused, 238
confusion, vi, vii
confusion-at-a-distance, spooky, 215
conglomerations of words, 191
Congress, vi
conjectural, x
conjugate, 107
conjugate variables, 108, 214
connection, gravitational, 95
connection, **241**
connotation, 142, 154
connotation, absurd, 173
connotation, has, 192
connotation overwhelms denotation, 174
connotations, 180
connote, 173
consciousness perceptions, unfathomable, 176
consciousness, 118, 147, 156, 175, 225
consciousness, human, 116
consensus, 3, 178
consensus, physical, 4
consensus, social, 4
conservation, energy, 59
conservation, momentum, 59
conservation of angular momentum, 213
conservation of energy, 257
conservation of probability, 58
conservative, ix
conservative approaches, 198
conserved, energy and momentum are, 262
conserved, energy not, 58
conserved locally, 96
conserved, why probability must be, 58
consistency, 20, 178
consistency under transformations, 43
consistent, 42
consistent physical universe, 27
consistent picture, 133
consistent, theory is, **102**
conspire, 40
constancy, 3
constant, Planck's, 234
constants, numerical values of, 235
construct, social, 197
constructs, **5**
contempt for humanity, 191
content, physical, 151
contentless string of words, 112
continuous function of time, probability is a, 77
continuous set of coordinates, **6**
continuum, 6
continuum, infinite, invisible, 165
contradict the fundamental principles, 265
contradictory results, 227
control, 5
conviction, 200
conviction, firm, 262
Cooke, Roger, 63
coordinate systems, 35
coordinate transformations, general, 95
coordinates, **6**
coordinates, Cartesian, 35
coordinates, continuous set of, **6**
coordinates, polar, 35
Copernician view, 15
corporeal, 188
correct, dramatically, 230
correct, grammatically, 146
correct operators, 231
correct question, 148
correct scientific paper, 183
correctness of a concept, **8**
correlated, 166, 220
correlated with, 124
correlations between spacelike-separated measurements, 171
correlations of probability distributions, 214

corresponds to reality, **18**
corresponds to reality, theory, 18
cosmological constant, 238
cosmological principle, anthropic, 174
Coulomb potential, 83
coupled, nonlinear, equations, 40
coupling constant, **33**
coupling constant, phenomenological, 33
coupling constant, variable, 34
courses in quantum weirdness, viii
covariance, general, 41
covariant derivative, 96
Cowan, Ron, 264
CPT theorem, 55, 143
crackpot, licensed, 197
crackpot subjects, v, 266
crackpots, vi, 197
Crease, Robert P., vi
created from nothing, 193
creation, 255
creation, immaculate, 168
creationism, v
creationists, 248
creatures, alien, 27
creatures, intelligent, 143
credibility, v, vi
criteria, xi
criteria for laws of nature, 229
criteria for reality, **3**
criteria for success, 240
criteria, mathematical, 153
cross terms, 70, 74
crystal, 119
crystal, electron in a, 46
crystal in an external field, 46
crystals, electrons in, 18
culture, xi
curvature, 28, 29, 38, 241, 254
curved space, 29
cytoskeleton, 133

Δ resonance, 6
d'Espagnat, Bernard, 2
Dalton, B. J., 214
dangerous, definitions are, 121
dangerous instrument, 7
data set for a brain, 134
database, 134
Davies, Paul, 243
daydreams, x
dead, 70
deal with, 5
deBroglie wavelength, 233
decay, 59, 226
decay of a superposition, 245
decays, 39
decays are uncorrelated, 106
decays, spontaneous, 259
deceiving society, vi
decision, 217
decoherence, 70, 72, 73
decrease, exponentially, 144
decreasing, monotonically, 204
deep, 181, **191**
deep philosophical problems, 186
deeply profound question, 191
deeply rooted, 41
deeply-held positions, 158
define reality, 178
defined, carefully, 265
defined, words are, 146
defining representation, 39
definite phase, 214
definite physical state, 182
definite position, 158
definite predetermined value, 209
definition, **132**
definition, circular, 145
definition of probability, 110
definition of reality, 4
definition of space, 141
definitions, arbitrariness of, 240
definitions are dangerous, 121
definitions, physical, 36
degenerate, **46**
delusions, bizarre, 155
denotation, 142, 154
denotation, no, 192
denotations, 180
denote, 173
density, energy, 61
derivative, covariant, 96
describe, **1**
description, complete, **120**
description of nature, intelligible, 186
descriptions of nature, local fantastic, 179

description of nature, local realistic, 179
description of reality, complete, 181
description, useful, 74
design of the universe, 229, 243
designed, 190
designed, universe is, 191
desire, 149
desires, 120, 248
desperate, 75
desperation, 251
detected, 124
detector, 178
deterministic theories, 246
deuteron, 53
Deutsch, David, 242
devastating quantum fluctuations, 253
Devlin, Keith, 192
DeWitt, Bryce S., 249, 250
diamond, 119
dictionaries, 126
dictionary, viii
difference between science and religion, 19
different axes, **44**
different frequency, photons of, **90**
different observers, 39
different properties, simultaneously, 227
different questions, mixing, 113
differential equation, 123
differential equations, 10
difficulties, psychological, 248
diffraction, 115
diffraction grating, 87
dilemmas, 162
dimension, 189
dimension 3+1, 27
dimensionless point, completely, 205
Dirac's equation, 138
Dirac, x
Dirac, P. A. M., 93, 98
direction of polarization, 219
direction, intrinsic, 42
disappointing, 121
discomfort, emotional, 159
discontinuity, 64
discoveries, xi
discrete, 91
discrete energy states, 202
discrete objects, 203
discreteness, 76, 143
discussions of quantum mechanics, errors in, 196
discussions, philosophical, 191
discussions of the measurement process, 228
disgraceful, 197
disguise, 246
displacement operator, 171
distance function, 30
distance, minimum uncertainty, 11
distance, small, 10
distances, 28
distances, small, **11**
distant fantasy, 179
distant realism, 179
distinction, artificial, v
distinction between matter and fields, 101
distinction between particles and forces, 98
distinctions between space and time, 31
distracting, 265
distress, v
distribution of dots, 61
distributions, correlations of probability, 214
distributions, probability, 217
divine purpose, 191
Dodelson, Scott, xiii
does happen, 4
dogmas, 206
dolphin, 26
domain, 15
domain of operators, **46**
domain of the group operators, 45, 232
dot, 60
dots, distribution of, 61
double-slit, 93
double-slit experiment, 202, 218
dramatically correct, 230
dread idea of the vacuum, 253
dreams, 3
driver, 110

Drummond, Peter, xi
dual space, 138, 140
duality, 139
duality of electric and magnetic fields, 97
duality, wave-particle, 97, 201
due to, 169
dust particles, 100
Dürr, Detlef, 214
Dürr, S., 182

ears, 129
earth, 200
earthquakes, 162
eat, 19
Echeverria-Enriquez, Arturo, 231
economic, 144
editorializing in the news columns, 265
educational, 154
effect on our senses, 125
effective charge, **34**
egomania, human, 185, 192, 227
egotistical, 235, 250
eigenstate, exact momentum, 78
Einstein, x
Einstein, A., 180, 208, 243
Einstein, Albert, 247
Einstein's equation, 41, 81
Einstein's equation is a necessary condition, 243
Einstein, Podolsky and Rosen, 208
elbairav, neddih, 248
electric and magnetic fields, duality of, 97
electrodynamics is inconsistent, classical, 157
electrodynamics, inconsistencies of classical, 102
electromagnetic beams, 211
electromagnetic field, 100, 200
electromagnetic field, external, 83
electromagnetic fields, 230
electromagnetic interaction, 145
electromagnetic interactions, 81
electromagnetic theory, 162
electromagnetic theory, classical, 208
electromagnetic vector potential, 199
electromagnetic wave, 204
electromagnetic wave, classical, 237
electromagnetism, 39
electromagnetism, classical, 61
electromagnetism is nonlocal, classical, 222
electron, **7**, **124**, 126, 132, 150, 166, 172
electron, Compton radius of the, 30
electron field, 165
electron in a crystal, 46
electron in a star, statefunction of an, 119
electron knows exactly where it is, 246
electron, mass of an, 168
electron-positron pairs, 261
electrons and positrons, cloud of, **34**
electrons in crystals, 18
electrons in stars, 18
elegant, xi
element of reality, 179, 187
elementary geometry, 55
elementary particle, 157
elementary physics student, 223, 239
elliptically polarized, 211
emergent properties, 84
emotion, 189
emotional, 176
emotional comfort, 5, 190, 191, 193
emotional comfort, finding, 125
emotional discomfort, 159
emotional feeling, 147
emotional reasons, 248
emotional relief, 148, 246
emotional rewards, 153
emotional satisfaction, 105
emotional solace, 194
emotional succor, 194
emotional value, 194
emotional void, 191
emotionally, 15, 105, 110, 184, 191, 246
emotionally charged, 265

emotionally comfortable answers, 150
emotionally comforting, 158
emotionally satisfying, 147
emotionally unsatisfying, 148, 176
emotions, human, 191
emptiness, 194
empty concepts and phrases, 248
empty sea, 224
empty set, 189
empty space, 254
empty space, fields in, 181
empty string of words, 147
empty word, 192
energy, **59, 62**, 86, 91, 99, 136
energy and momentum are conserved, 262
energy conservation, 59, 135, 202
energy, conservation of, 257
energy density, 61
energy, infinite, 169
energy, infinite gravitational, 169
energy is the Hamiltonian eigenvalue, 136
energy not conserved, 58
energy, potential, 200
energy, pure, 99
energy state, **45**
energy states, discrete, 202
energy, total, 58
energy, vacuum has, 257
energy, well of, 257
energy, zero point, 257
energy-momentum tensor, 242
Englert, Berthold-Georg, 70, 118, 142, 232, 249
English professors, ix
enjoy, 158
ensemble, 71, 108, 112, 117, 256
ensembles of universes, 112
entangled state, **211, 216**
entangled statefunctions, 229
entanglement for a single particle, 214
entanglement, quantum, 183
entity, **157**, 201
entity, fundamental, 175
entropy, 116
environment, 3, 67, 70, 74
enzymes, 19
EPR experiment, 160, 179
EPR experiments, 151
EPR "Paradox", vii, 208
equating a vector and a scalar, 238
equations of motion, 138
equations, allure of, 154
equations, captivated by, 153
equations, coupled nonlinear, 40
equations, differential, 10
equations, Maxwell's, 15
equations, nonlinear, 259
equations, relativistic wave, 138
equations, substitute numbers in, 154
equivalence, principle of, 41
error, logical, 82
errors, checklist of, 265
errors in discussions of quantum mechanics, 196
errors, linguistic, 154
erudition, 191
escape from mathematics, 124
esoteric explanations, 55
esoteric words, 153
ether, 144, 181, 251, 264
Euclidean, 30
Euclidean 3-space, transformation group of, 47
Euclidean algebra, 39
Euclidean space, 29, 38
eukaryotes, 35
euphemism, 267
European, male, white, 189
evanescent particles, 257
evasive, 148
events, **6**, 240
Everett, Hugh III, 159
every particle in the universe, 229
everything works, 189
evolution, theory of, 15
exact momentum eigenstate, 78
exchange, particle, 145
excitations of neurons, pattern of, 190
excited, xi
excited atoms, 62
excited state, 42
excitement, 148, 266
exciting, 253, 265
excuse nonsense, 197

excuses, 248
excuses, quantum mechanics with, 247
exist, 155, 175
exist physically, **160**
exist, universe to, 125
existence, 159, 191
existence is the greatest mystery, 28
existence, mystery of, 192
exists, 159
expandable, **57**
expansion, approximation, 252
expectation value of an operator, 62
expectation value of the Hamiltonian, 62
experiment, **178**
experiment, EPR, 160
experiment, realistic physical, 80
experimental consequences, 126
experimental fact, 106
experimental meanings of concepts, viii
experimental predictions, 5
experimental procedures, 265
experts, 197
explaining, **18**
explanation, classical, 105
explanation, wild, 156
explanations, esoteric, 55
exploit, 266
exponentially decrease, 144
expressions, classical, **231**
expressions for statefunctions, 109
extended, nucleons are, 205
extension, finite, 261
extent, finite, 201
external electromagnetic field, 83
external field, crystal in an, 46
external nature, 125
external reality, 131, 179
external reality, model of an, 131
external world, xii, 18, 118
external world exists, **125**
extraneous concepts, 245
extraordinarily peculiar, 223
extraordinary hubris, 191
extreme implausibility, 156
extremely peculiar, 183
eye, 134
eyes, 129, 176

fabric, cloth, 254
fabric, underlying, 254
face, 167
fact, experimental, 106
factual reality, 179
fad, 65
faith, science and, 125
fallacy, logical, 197
false, 8, 17
false analogy, 135
false analogy, method of, 168, 170
falsifiable, 17
falsified, 15
falsify, 17
fame, 197
famous physicist, 197
fantasies, pictorial, 261
fantastic, 266
fantastic descriptions of nature, local, 179
fantasy, distant, 179
fantasy, local, 179
fantasy of our intuition, 3
faraway realism, 179
Farhi, Edward, 109
fascination, 154, 266
fascination, peculiar, 201
faster than the speed of light, 220
father, 190
favorite approximation scheme, 229, 243, 262
feature, intrinsic objective, 110
features of the physical world, objective, 187
feel, 120
feeling, emotional, 147
feeling, intuitive, 180
feeling, nice, warm, 154
feeling, subconscious, 179
feeling, subjective, 187
feeling, vague, 187
feelings, subjective, 177
Feldmann, Michel, 210
fellow human beings, 190
fermions, 53

Feynman path integral method, 136
fiat, 76
fiber optics, 238
fiction, science, 173
fictional physics, 266
field, **32, 83**
field, complex number, 55
field, electron, 165
field, real number, 55
field theory, **32**
fields, **84, 232**
fields and matter, distinction between, 101
fields, electromagnetic, 230
fields in empty space, 181
fields of inquiry, 194
fifth force, 238
film, 71
finding emotional comfort, 125
finding knowledge, 125
fine-structure constant, 33
finger, 133
fingers, 26, 129
finite extension, 261
finite extent, 201
Finkelstein, David, 251
firm conviction, 262
fish, 133, 226
fit nature to their prejudices, 264
flame, 132
flashlight, 224
flat, 30
flat space, 29
flaunt, vi, 197
flexibility, 18
flicker in and out of existence, 257
floating in the air, qualities, 187
floor, 223
flow, **164**
flow of time, 164
flowery, 265
fluctuating vacuum field, 260
fluctuation, **256**
fluctuation, spacetime, 241
fluctuations, 226
fluctuations, nonviolent, 253
fluctuations, peaceful, 253
fluctuations, spacetime, 226
fluctuations, violent, 253
flying saucers, v, 266
foam, 254
foam, spacetime, 253
Fock representation, 85
folklore, 266
food, 122, 178
fool, 167
fooling us, 229
force, 145
force becomes infinite, 207
formalism is a language, 222
formalism is an idealization, 78
formula, **123**
formulas, 154
formulations of quantum mechanics, rigorous, 11
foundation of our discussions, 40
foundation of physical theories, 151
foundations of general relativity, 41
Fourier analysis, 88, 112, 114, 261
Fourier series, 85
fragmentation, 65
framework, **14**
framework, better, **18**
framework, classical physics as a, 15
framework, phenomenological, 20
framework, quantum mechanics as a, 15
fraudulent, manipulative scientist, 266
free will, 187
freedom, no, 22
frequency, kinematical meaning of, 86
Freyberger, Matthias, 214
friction, 20
friends, 224
fruitful, 15, 16
frustrations, 154
Fuchs, Christopher A., 118, 178, 199
function of time, probability is a continuous, 77
function, complex, 241
function, distance, 30
functional mechanics, 143, 198

fundamental aspect of geometry and physics, most, **37**
fundamental entity, 175
fundamental limitation on meaningful knowledge, 107
fundamental methods of modern science, 138
fundamental object, 48
fundamental object, statefunction of a, 48
fundamental objects, **90**
fundamental postulate, 44
fundamental postulates, 81
fundamental principles, contradict the, 265
fundamental problem, 173
fundamental question, 147
fundamental requirement on physical theories, 177
fundamental technique, 138
funding, 112
funds, 19
Fusseder, Walter, 214
future, 31, 252
fuzzy, 6, 153, 232

Gage, Fred H., 134
galaxies are large, 188
galaxies, cluster of, 231
galaxies, local group of, 79
galaxy, 229
galaxy with a pebble, comparing a, 134
Galilean group, 98
game, mathematical, 111
gas, 72, 258
gas cloud, 193
gauge covariance, 41
gauge invariance, 95
gauge noninvariance, 199
gauge transformations, 95
gaze, 185
general coordinate transformations, 95
general covariance, 41
general relativity, **41**, 81
general relativity, foundations of, 41
general relativity, quantize, 240
generalize from a single case, 138
generator, translation, 56
genes, 162
genius, **xi**
geometers, 35
geometrical objects, **36**
geometry, **28**, 139
geometry and physics, most fundamental aspect of, **37**
geometry, axioms of the, **29**
geometry, bumpy, 253
geometry, elementary, 55
geometry, groups of, 38
geometry is a fact of physics, 35
geometry of space-time, smooth, 253
geometry, quantum field theory as a property of, 1
geometry resonates between configurations, 258
geometry, transformation group of, 13
geometry, transformation group of the, 1
Ghirardi, Gian Carlo, 228
ghostly apparition, 168
ghosts, 160
Gilmore, Robert, 57
gimmick, 252
Gingerich, Owen, xi
give up, 174
glamour, 251
Glanz, James, 164, 257, 258
Gleason, Andrew M., 63
Gleason's theorem, 63
glia, 133
glue, 157
gluons, 258
gobbledygook, 148
God, 174, 189, 194, 247, 252, 264
God loves us, 194
God, playing, vi
God's thinking, vi
Goldsmith, D., 88
Goldstein, S., xiii
Goldstein, Sheldon, 214, 246
Goldstone, Jeffrey, 109
goofy, vi
Gordon, A., 88
governing objects, **123**
government money, 265

governments, viii
Graham, Neill, 249, 250
graininess, 132
grammatically correct, 146, 192
grandiose name, 98
grant money, 224
Grant, Edward, 251, 252, 261
grants, xi
grasshopper, 178
gravitation, 39, **41**
gravitation is necessarily nonlinear, 81
gravitational connection, 95
gravitational effects, quantum, 241
gravitational energy, infinite, 169
gravitational field, 239, **241**
gravitational statefunction, **241**
gravitational wave, 239
gravity, classical theory of, 240
gravity, quantize, viii
gravity, quantum theory of, 240
graying, 203
greatest mystery, existence is the, 28
Greek letters, 154
green men, little, 20, 184, 248
Greenberger, Daniel M., 267
Griffiths, Robert B., 243
grotesque, 265
ground state of a harmonic oscillator, 169
group, Abelian, 139
group, affine, 38
group, Galilean, 98
group, inhomogeneous, **39, 93**, 139
group, inhomogeneous pseudo-orthogonal, 39
group, noncompact simple, 13
group of Euclidean 3-space, transformation, 47
group of the geometry, transformation, 1
group operators, domain of the, 45, 232
group, Poincaré, 39
group representation, basis state of a, **46**
group representation, meaning of, **46**
group representations, linear, 81
group, rotation, 47
group, transformation, **40**
group, transformation of geometry, 13
groups of geometry, 38
groups, representations of Abelian, 56
guess, 231
guessing, 98
guesswork, 19
gullibility, 266
gushy, 265
Gutmann, Sam, 109

habits of thought, unexamined, 4
Hadley, Mark J., xiii
hadrons, 33
half-integer spin, 82
half-life, 109
Hamiltonian, 45, 51, **59, 135**, 138
Hamiltonian eigenvalue, energy is, 136
Hamiltonian, expectation value of the, 62
Hanson, R. Brooks, 227
happen, does, 4
happen, should, 4
happens, actually, 184
happiness, 149
hard question, 176
hard work, 248
hardness, 132
harmonic oscillator, ground state of, 169
harmonic oscillators, 85, 169
Hartle, J. B., 223
has connotation, 192
he or she, 176
heat, 132
Heisenberg uncertainty principle, 227
Heisenberg's formulation, 136
Heisenberg, Werner, 247
helicity-2, 95
helium atom, 128
helpful, most, 55
hermitian, 58
heuristic, 15
heuristic value, 20, 233

hidden variables, 246
hide problems, 180
high-energy scattering, 53
high priest, 265
highly rigorous mathematically, 37
hilarious, 223
Hilbert space, 63
Hilborn, Robert C., xiii
Hiley, B. J., 251
Hillery, Mark, 221
historical development, 55
historical reasons, 81
historically, 82
history has on the present, hold that, 231
history of physics, x, 239
history of science, xi
history of time, 163
history, v, 102, 110, **163**, 251, 253
Hofer, Werner, xiii
hold that history has on the present, 231
Holton, Gerald, 247
Homo, 226
homogeneous, 200
homomorphic, 82
hope, 19
hubris, 192, 228
hubris, extraordinary, 191
hubris, human, 118
human being, 35
human beings, 118, 187, 197, 227, 234
human beings, fellow, 190
human beings, mere, 247
human beings, properties of, 235
human consciousness, 116
human egomania, 185, 192, 227
human emotions, 191
human hubris, 118
human languages, 187
human psychology, 194, 251
humanity, 190
humanity, contempt for, 191
humbleness, 190
humility, 228
hydrogen atom, x
hydrogen atom, nonrelativistic, 50
I am careless and totally confused, 267
I am incompetent, 267
idea, intuitive, 236
idealization, ix, 78
idealization, formalism is an, 78
idealizations, improper, 80
ideas, strange, 79
identical bodies, 259
identical particles, 230
ignorance, 106
ignorance, necessity for, 107
ignorant, not, 107
ignored, 36
illusion, 168
images, mental, **133**
images, subjective, 130
immaculate creation, 168
immense storage capacity, 134
implausibility, extreme, 156
implausible, xi
implications, subjective, 187
impossible, logically, 217
imprecise, 153
impressive, 191
improper idealizations, 80
improvements, vii
in-group, 263
inability, 105, 200
inaccessible information, 161
incantations, 193
incoherent sum of states, 75
incoherent superposition, 73
incompatibility of classical electromagnetism and local realism, 231
incompatible questions, 117
incompetence, vi, 197
incompetent, I am, 267
incomplete, 245
inconsistencies of classical electrodynamics, 102
inconsistent, 43, 47
inconsistent, classical electrodynamics is, 157
inconsistent, inherently, 214
inconsistent, postulates can be, 24
incredible, 266
independent superstition, 260
indeterminacy of nature, 248

individual systems, 109
inequalities, Bell's, 221
inertia, 22
inertia, necessity for, 23
infantile attachment to classical physics, 266
infinite energy, 169
infinite, force becomes, 207
infinite gravitational energy, 169
infinite, invisible continuum, 165
infinite precision, 156
infinite regression, 72, 144
infinite-speed, 161
infinitely deep potential well, 201
infinities, 101, 208
inflammatory language, 183
information consists of physical objects, 193
information, 55
information, inaccessible, 161
information, transmission of, 173
infuriate, v, 12
inherently inconsistent, 214
inhomogeneous group, **39, 93**, 139
inhomogeneous operator, **92**
inhomogeneous pseudo-orthogonal group, 39
inhomogeneous rotation group, 38
inhomogeneous subgroup, 13
initial conditions, 253
ink, spots of, 12
inquiry, fields of, 194
insect, 178
inside a star, 10
inspire, 196
instantaneous state vector reduction, 78
instinct, 3
instructions, mathematical, 131
instrument, dangerous, 7
insult, 183
insulting, 242
insulting classical physics, 183
insulting quantum mechanics, 183
integral powers of basis vectors, 91
intelligence, 133
intelligence, artificial, **133, 133**
intelligent beings, 174, 189
intelligent creatures, 143
intelligible description of nature, 186
intensity, 61
interaction, electromagnetic, 145
interaction, **33, 145, 228**
interactions, 246
interactions, electromagnetic, 81
interconnected diverse predictions, 262
interconnected observations, web of, 131
interference, 61, 73
interference pattern, 70, 82, 202
interference terms, 227
interior of stars, 184
interior, stellar, 131
intermediate steps, 264
internal structure, 53
interprets, **132**
intrinsic direction, 42
intrinsic objective feature, 110
intuition, 20, 132, 255
intuition, classical, 122
intuition, fantasy of our, 3
intuitive, 80
intuitive feeling, 180
intuitive idea, 236
intuitive meaning, 36
intuitively pleasing, 118, 121
invariance, **40**
invariant under transformations, 43
inverse square law, 207
invisible continuum, infinite, 165
IO(k,l), **39**
irrational, 106
irrationalism, v
irrationality, vi
irreconcilable, logically, 242
irreducible representation of a compact semisimple group, single, 49
irrelevant, physically, 243
irresponsible, 173
irreversibility, 31, **71**, 116
irritation, 2
isotropic, 200
isotropy of space, 213
Italian, 143
itself, photon interferes with, 93

Iyengar, Ravi, 133

jack-in-the-box, 124
Jackiw, Roman, 165, 260
Jackson, J. D., 219
Jancewicz Bernard, xiii, 138
Jarrett, Jon P., 105
Jasny, Barbara, 227
jobs, plushy high-paying, 266
Johnson, George, 157, 160, 266
Josephson junction, 74
journalist, 265
journalists, 156, 223
judgment, 19
judgment, people's, 261
jumble of sensations, 18
jump off a building, 26
jump, quantum, 76
jumping to conclusions, method of, 138
jumps, quantum, 118, 228
justification, 54

K mesons, 220
Kastner, Ruth E., xiii
Kayser, B., 64
Keane, Michael, 63
Kempermann, Gerd, 134
Ketov, Sergei V., 13
key requirement, 178
Khersonskii, V. K., 215
Khrennikov, Andrei, 56
killed, 110
kilogram, 235
Kim, Yoon-Ho, 162, 217
kinematical meaning of frequency, 86
kinetic energy, 236
Kisil, Vladimir, xiii
Kleppner, Daniel, 165, 260
kniht, 188
know, **195**
know, cannot, 194
knowability of our world, 23
knowable, 119
knowing, other ways of, 194
knowledge, 118, 194
knowledge, finding, 125
knowledge, fundamental limitation on meaningful, 107
knowledge, nonexistent, 218
knowledge, objective, 19
knowledge of the world, 129
knowledge, sudden change of our, 118
Koch, Christof, 134
Krauss, Lawrence M., 263, 266
Kuckert, Bernd, xiii
Kulik, Sergei P., 162, 217

laboratory, 178
Lagrangian, 137
Lamb shift, 260
Lamb, Willis E. Jr., 199
language, viii, 125, **179**
language, ambiguity of, 119
language can mislead, 77
language, formalism is a, 222
language, inflammatory, 183
language of nature, 172
languages, computer, 187
languages, human, 187
large, galaxies are, 188
lattice, orthocomplemented, orthomodular, separable, 152
Laurent, Gilles, 134
law of nature, **31**
law, inverse square, 207
laws, 23
laws, mysterious set of, 54
laws, Newton's, 15, 65, 118
laws of nature, 35, 119
laws of nature, criteria for, 229
laws of physics, **3**, **125**, 189
laws of physics, realistic possible, 21
laws, universe is governed by, 263
laziness, 166
leap, quantum, 76
leaping without thinking, 77
learned treatises, 191
Leggett, A. J., 74
Leggett, Tony, 64, 223, 225
Leonhardt, U., 214
leptons, 205
Leslie, John, xiii
letters, Greek, 154
letters, linear set of, 181
letters, sequence of, 143

letters, sequences of, 104
letters, strings of, 179
liberal art, ix
licensed crackpot, 197
licensed to spout nonsense, 197
lie, vi
life, 19, 177
life force, 193
life, meaning of, 190
lifetime, 261
light, **101**
light carries more energy, 202
light, speed of, 234
light wave, classical, 156
limbic system, 190
limitation on meaningful knowledge, fundamental, 107
line, bright, 61
linear group representations, 81
linear set of letters, 181
linearly polarized, 94, 219
linguistic, 14
linguistic errors, 154
linguistic inaccuracies, 142
linguists, ix
lists of the richest people, 264
literature, v
little ball, 201
little green men, 20, 184, 248
little pieces of space-time, 253
live with a computer, 189
lobby, vi
local, **128, 204**
local fantastic descriptions of nature, 179
local fantasy, 179
local group of galaxies, 79
local, physics is, **97**
local, quantum mechanics is, 204
local realism, 179
local realism, incompatibility of classical electromagnetism and, 231
local realistic description of nature, 179
locality, **97**
localizable, **202**
localized, **204**
logic, 153, 238
logical error, 82
logical fallacy, 197
logical structure, 151
logical symbols, 151
logically impossible, 217
logically irreconcilable, 242
logorrhea, **163**
look, 3
Lord God, 247
Lorentz invariance, 158
Lorentz transformations, x
lost in space, 240
louder and louder, 263
Lounesto, Pertti, xiii
love, 190
loves us, God, 194
Lubkin, Elihu, xiii
Lubkin, Gloria, xiii
lying, viii, 173, 221

MacCallum, Malcolm, xiii
macroscopic, 61
macroscopic object, **70, 73,** 74
macroscopic object, coherent superposition of, 70
macroscopic objects, 64
macroscopic physical object, 122
macroscopic quantum superposition, 244
macroscopic state, 68
magic, 265
magic way, 209
magical, 228
magnetic and electric fields, duality of, 97
magnetic field, 40
magnetic moment, 261
magnetic monopole, 199
major tool of modern science, 170
majority vote, 118, 238
makes sense, **5**
male, white, European, 189
maliciously, 77
mammal, 26
mammals, 226
man is the measure of all things, 229, 236
man, little green, 248
Mandel, L., 118
manifold, 30

manipulation of symbols, rigorous, 115
Mannheim, Philip, xiii
many-body aspects, 55
many types of nothingness, 182
many universes, 112
many-worlds interpretation, 156, 249
Marachevsky, Valery N., 256
marble, 201
marry, 19
Mars, 143
mass of an electron, 168
mass of every nucleus, 257
mass-zero spin-2 representation, 41
massless objects, 41
massless representations, 39
match, 178
mathematical criteria, 153
mathematical game, 111
mathematical instructions, 131
mathematical proofs, rigorous, 63
mathematical, reality is, 129
mathematical symbols, 154
mathematical, universe is purely, 130
mathematically, highly rigorous, 37
mathematics, **2**
mathematics, abstract, 80
mathematics, escape from, 124
matter and fields, distinction between, 101
matter, 99
Maxwell's equations, 15
Maxwell, Nicholas, 17
meaning, **5**
meaning, intuitive, 36
meaning of a concept, **8**
meaning of group representations, **46**
meaning of life, 190
meaning of probability, 109
meaning of the statefunction, 158
meaning, question has, **147**
meaning, ultimate, **7**
meaningful, **9**
meaningful question, 111
meaningful questions, **149**
meaningful, physical theories, 125
meaningless properties, 186
meaningless question, 111
meaningless questions, **146, 149**
meaningless terms, 187
meaningless verbiage, 167, 175, 186
meaningless, probability becomes, 110
meaninglessness, **5**
meanings of concepts, experimental, viii
meanings, well-hidden, 111
measure, 184
measure zero, 182, 220
measured system, 243
measurement, 24, **66**, 183
measurement problem, 67, 243
measurement process, discussions of the, 228
measurements actually measure something, 243
measurements, correlations between spacelike-separated, 171
measurements, spacelike, 222
measuring rods, 11
mechanical oscillations, 181
mechanics, functional, 143, 198
mechanics, transformational, 143
mechanism, computational, 118
media, 210
media, sensationalist, 266
medicine, vii
medium, mysterious, 181
melodramatic, 265
melt the vacuum, 258
memory, 130
men, little green, 20
mental construct, 100
mental images, **133**
mental pictures, 4, 14
mercury, column of, 9
mere human beings, 247
Mermin, N. D., 221, 223
Mermin, N. David, xiii, 110, 148, 177, 181, 185, 221
mesoscopic, 74
metaphorical, 173
metaphysical, 14

meter, 235
meter stick, 9
method of false analogy, 135, 168, 170
method of jumping to conclusions, 138
methods of modern science, fundamental, 138
metric, 28, 29, 226, 254
metric tensor, 241
microscope, 108, 113
microscopes, 19
Middle Ages, 252
Miller, Julie Ann, 174
Miller, Willard, Jr., 140
Milton, Kimball A., 256
mind, 118
mind of God, 247
mind-teasing, 148
minimal coupling, 95, 200
minimal of standards, 223
minimum uncertainty distance, 11
miracle, 28, 255
misconceptions, vii
misguided belief, vi
mislead, language can, 77
misleading terminology, 77
Misner, Charles W., 258, 259
mixing different questions, 113
mixtures of classical and quantum concepts, 265
mnemonic, 8
mob psychology, xi, 266
model of an external reality, 131
models, simplified, 9
modern physics, 183
modern science, fundamental methods of, 138
modifier, 146
Mohrhoff, Ulrich, 3, 185
molecule, **119, 132**
mollify, v
momentum, **136**
momentum conservation, 59
momentum eigenstate, exact, 78
momentum operator, 171
momentum operators, 41
money, viii, 211, 238
money, government, 265
money, grant, 224
monopole, magnetic, 199
monotonically decreasing, 204
Moon, 185
morals, 191
Moran, William, 63
mortals, 153
mortals, ordinary, 265
Moskalev, A. N., 215
most fundamental aspect of geometry and physics, **37**
most helpful, 55
mother tongue of nature, 172
motion, equations of, 138
motion, random, 60
motion, thermal, 203
motion, unlawful, 22
motions of pointers, 131
mountain climbers, 157
movements, muscular, 191
much ado about nothing, 252
muddle, 177
multi-particle state, **88**
multiparticle states, 127
multitude of universes, 5, 174, 182
Munoz-Lecanda, Miguel C., 231
muscular movements, 191
musical instruments, 144
myriads of components, 250
mysteries of the universe, 153
mysterious, vi, 105
mysterious medium, 181
mysterious place, 192
mysterious quantity, 135
mysterious set of laws, 54
mystery, 23, 238
mystery, existence is the greatest, 28
mystery of existence, 192
mystical, 266

n photons, **86**
naive, gullible reporter, 266
name, 143
name, grandiose, 98
name, unfortunate, 76
names, viii, 198
names, qualities that are, 188
natural objects, 192
nature, **3**, 195
nature abhors a vacuum, 261

nature, abstract conception of, 122
nature, coherent view of, 7
nature, external, 125
nature, intelligible description of, 186
nature, language of, 172
nature, law of, **31**
nature, laws of, 35, 119
nature, local fantastic descriptions of, 179
nature, local realistic description of, 179
nature, mother tongue of, 172
nature of time, 162
nature, privy to the secrets of, 265
nature, strangeness of, 196
nature, understand, 17
necessity for ignorance, 107
necessity for inertia, 23
neddih elbairav, 248
negative statements, 178
net, 17
neural activity, 131, 133, 147, 176
neural impulses, time scale of, 236
neural processes, 125
neural system, 132
neuron, 131
neurons, 133
neurons, collection of, 189
neurons, pattern of excitations of, 190
neurons, sensory, 11
neurophysical, 150
neurotransmitters, 134
neutron, 246
never see a table, 132
New York, 211
newborn universe, 164
news columns, editorializing in the, 265
newspaper, 266
newspaper story, 265
Newton's absolute time, 164
Newton's First Law, 218
Newton's laws, 15, 65, 118
Newtonian physics, 199, 208, 221
nice, warm feeling, 154
Nieto, Michael Martin, 214
no assumptions, 22
no brains, 71
no denotation, 192
no freedom, 22
Nogami, Y., 247
noise, 12
nomenclature, 41
non-anthropocentric, 250
nonanalytic, 237
noncompact simple group, 13
nonexistent knowledge, 218
nonexistent problem, viii
nonexistent problems, 158
nonexistent problems of quantum mechanics, 245
nonintuitive, 196
noninvariance, 40
nonlinear equations, 259
nonlinear equations, coupled, 40
nonlinear representation, 84
nonlinear representations, 140
nonlinear terms, 33, **83**
nonlinear, gravitation is necessarily, 81
nonlocal, 171, 200
nonlocal, classical electromagnetism is, 222
nonlocality of classical physics, 231
Nonn, T., 182
nonobjective reality, 177
nonrelativistic approximation, 137
nonrelativistic hydrogen atom, 50
nonrelativistic limit, 25
nonrenomalizable, 243
nonsense, vi, 7, 80, 146
nonsense, excuse, 197
nonsense, licensed to spout, 197
nonsense, obliterated, 73
nonsense, predicts, 243
nonsense, seems like, 267
nonsense, spout, 266
nonsense, worse than, 153
nonsensical, 254
nontrivial universe, 22
nonviolent fluctuations, 253
nonzero radius, 201
Norton, John D., xiii, 22, 41
not ignorant, 107
nothing, created from, 193
nothing, tunnel from, 237
nothingness, 251

nothingness, many types of, 182
nothingness, spontaneously pop out of, 263
now, 31
nuclear reactions, 193
nucleon, 49
nucleons, 20
nucleons are extended, 205
nucleons by nucleons, scattering of, 20
number field, complex, 55
number field, real, 55
number of photons, **86**
number operator, 85
number systems, 56
number, particle, 214
number, stupendous, 250
numbers in equations, substitute, 154
numerical values of constants, 235

object, **157, 198**
object, classical, **74, 70**
object, fundamental, 48
object, macroscopic, **122, 70, 73**
object, physical, **126**
object, statefunction of a fundamental, 48
objective, 159
objective feature, intrinsic, 110
objective features of the physical world, 187
objective knowledge, 19
objective occurrence, 178
objective reality, **177**
objects, **124**
objects, basic, 84
objects, classical, 73
objects, discrete, 203
objects, fundamental, **90**
objects, governing, **123**
objects, natural, 192
objects, primary, 84
objects, statefunctions of classical, 122
obligation, 115
obliterated nonsense, 73
observation, **42, 228**
observational science, ix
observations, **40**
observations, organize our, 5
observations, web of interconnected, 131
observe, **46**
observer, **35, 67, 246**
observer, possible, 45
observers, 176
observers, different, 39
obsessed, 43, 197
obsession, 223, 239
obsession of so many physicists, 197
Occam's razor, 9, 198, 249
occult, 55
occurrence, objective, 178
ocean, 165
octonion, 56
octonions, 55
Omnes, Roland, 142, 151, 243
onkly, 192
ontology, primary, 84
opaque screen, 93
operational, ix
operator, expectation value of an, 62
operator, inhomogeneous, **92**
operator, number, 85
operator, phase, 214
operators, correct, 231
operators, domain of, **46**
operators, domain of the group, 232
operators, position, 139
opposed, strongly, v
opposite of what quantum mechanics, 265
ordinary mortals, 265
ordinary reality, 181
organize, 15
organize our observations, 5
orthocomplemented, orthomodular, separable lattice, 152
orthodox view, 170
orthomodular, 152
oscillation, 143, 204
oscillations, mechanical, 181
oscillator, 257
oscillators, 219
other ways of knowing, 194

out there, 4, 159
outlandish, vi, 80, 265, 266
Overbye, Dennis, 188
overdefinitions, 121
overreliance on analogies, 169
oversight, 79
oxygen, 188
oxymoron, 108, **146**, 161, 182, 193, 201, 216, 226

Page, Don N., 214
pain, 186
pain of quantum mechanics, 110
paleontology, 138
pander, vi
paper, correct scientific, 183
papers, pile of, 73
papers, production of scientific, 111
papers, scientific, 243
paradoxes, vii, 186, 196
paradoxical results, 80
parameters, transformation, 107
parents, 190
Park, Robert L., xiii
particle, 198
particle, elementary, 157
particle exchange, 145
particle number, 214
particle number must fluctuate, 260
particle, point, 157, 201
particle, single, **86**
particle, virtual, **7**
particles and forces, distinction between, 98
particles, evanescent, 257
particles, identical, 230
particles, point, 236
particles, virtual, 252
past, 31
patent, 264
path integral method, Feynman, 136
pathological, 251
pattern of excitations of neurons, 190
Paul, H., 214
PCT theorem, 55, 143
peaceful fluctuations, 253
peak, 117
Pearle, Philip, 228
pebble, comparing a galaxy with a, 134
peculiar, extraordinarily, 223
peculiar, extremely, 183
peculiar fascination, 201
peculiarities of quantum mechanics, 67
people, 101, 235
people, brilliant, xi
people, religious, 175
people, roomful of, 184
people's judgment, 261
perceived, 100
perception, 3, 130
perception, similar, 178
perceptions, 8
perceptions, unfathomable consciousness, 176
Percival, Ian C., 224
Peres, Asher, 57, 118, 178, 199
perpetual motion, 264
perturbation theory, 101, 208, 229, 254, 262, 264
phase, 60
phase changes, 74
phase, definite, 214
phase operator, 214
phases vary wildly, 244
phenomena, strange, 243
phenomenological, 17
phenomenological coupling constant, 33
phenomenological framework, 20
phenomenological representation, **83**
phenomenological theories, 33
phenomenological theory, **20**
philosophical arguments, 147, 158
philosophical discussions, 187, 191
philosophical problems, deep, 186
philosophical view, 179
philosophy of science, ix
philosophy, underlying, 34
photon, **86, 166**
photon interferes with itself, 93
photon, single, **86**
photons of different frequency, **90**

photons, n, **86**
photons, number of, **86**
physical, 180
physical consensus, 4
physical content, 151
physical definitions, 36
physical experiment, realistic, 80
physical object, **126**
physical object, macroscopic, 122
physical objects, information consists of, 193
physical processes, actual, 80
physical reality, 118, 177, 187
physical sense, 153
physical senselessness, 37
physical situation, actual, 147
Physical Society, American, 162
physical state, actual, **184**, 212
physical state, definite, 182
physical theories meaningful, 125
physical theories, foundation of, 151
physical theories, fundamental requirement on, 177
physical theory, **11**
physical universe, consistent, 27
physical values, actual, 226
physical world, objective features of the, 187
physically irrelevant, 243
physically real, **8**
physically, exist, **160**
physicist, competent, vi
physicist. famous, 197
physicist, professional, 223
physicists, 156, 266
physicists, obsession of so many, 197
physicists, professional, 239
physicists, self-centered, 235
physicists, theoretical, 229, 243, 262
physics, better than, 153
physics, fictional, 266
physics, history of, x, 239
physics is local, **97**
physics, laws of, **3, 125**, 189
physics, modern, 183
physics, realistic possible laws of, 21
physics, rigor in, 11
physics student, elementary, 223
physics teachers, 154
pictorial fantasies, 261
picture, 172
picture, coherent, **5**
picture, coherent, meaningful, 131
picture, consistent, 133
pictures of reality, **133**
pictures, 129
pictures, classical, 199, 253
pictures, coherent, 18
pictures, mental, 4, 14
pile of papers, 73
Pitowsky, Itamar, 63
place, mysterious, 192
places, two or more, 224
Planck, x
Planck scale, 254
Planck's constant, 234
planet, 190
playing God, vi
plushy high-paying jobs, 266
Podolsky, B., 180, 208
poetic, 173
Poincaré group, 39
Poincaré group representation, mass-zero spin-2, 41
Poincaré representation, single, 89
point, **28**
point, completely dimensionless, 205
point particle, 102, 157, 201
point particles, 236
point particles, collisions by, 182
pointer positions, 244
pointer, 69
pointer, statefunction of, 69
pointers, motions of, 131
pointers, statefunction of, 72
Poisson bracket, 136
polar coordinates, 35
polarization, 230
polarization, direction of, 219
polarized, 156
polarized, circularly, 94, 219
polarized, elliptically, 211
polarized, linearly, 94, 219
polarizer, 219
polemical, v

poor writing, 197
pop out of nothingness, spontaneously, 263
pop up, 255
popular vote, 197
position, 10, 139
position, definite, 158
positions, deeply-held, 158
positions, pointer, 244
positions, special, 46
possible observer, 45
position operators, 139
post hoc ergo propter hoc, 82
postmodern, v
postulate, fundamental, 44
postulated, 35
postulates can be inconsistent, 24
postulates, fundamental, 81
postulates of quantum mechanics, 25
potential, 83
potential barrier, 144
potential, Coulomb, 83
potential, electromagnetic vector, 199
potential energy, 200, 236
potential, quantum, 204
potential, vector, 97
potential well, 237
potential well, infinitely deep, 201
power structure, v
pragmatic, 7, 19, 20, 144, 233
precision, infinite, 156
predator, 178
predetermined value, definite, 209
predictable, completely, 117
predictions, experimental, 5
predictions, interconnected diverse, 262
predicts nonsense, 243
prejudice, 4, 81, 102
prejudices, fit nature to their, 264
preparation, 112
preparing systems, procedure for, 118
preposterous, 224
present, hold that history has on the, 231
press office, 266
pressure, 233
priest, high, 265
priests, 153
primary objects, 84
primary ontology, 84
primates, 226
primitive concepts, 104
principle of equivalence, 41
principle, action, 137
principle, anthropic cosmological, 174
principle, superposition, 81, 244
principle, uncertainty, 161
privy to the secrets of nature, 265
probabilistic character, 107
probability, **109**
probability becomes meaningless, 110
probability, classical, 107
probability, conservation of, 58
probability, definition of, 110
probability distributions, 115, 217
probability distributions, correlations of, 214
probability for the universe, 111
probability is a continuous function of time, 77
probability, meaning of, 109
probability must be conserved, why, 58
probability, quantum, 107
problem, fundamental, 173
problem, nonexistent, viii
problems, hide, 180
problems of quantum mechanics, nonexistent, 245
problems, nonexistent, 158
procedure for preparing systems, 118
procedures, experimental, 265
process of science, 146
production of scientific papers, 111
professional physicist, 223
professional physicists, 239
professor, 109
profit, 267
profound question, deeply, 191
programmed, 231
programmer, conceited computer, 135

prokaryotes, 35
proofs, rigorous mathematical, 63
proper universe, 21
properties, emergent, 84
properties, meaningless, 186
properties of human beings, 235
properties, set of, 126
properties, simultaneously different, 227
property of geometry, quantum field theory is a, 1
Protagoras, 229, 236
proton, **124**
provide, 181
provoke, 196
pseudo-Euclidean space, **38**
pseudo-orthogonal group, inhomogeneous, 39
psychiatrists, vi
psychiatry, 110
psychological, 187
psychological comfort, 147
psychological difficulties, 248
psychological question, 142
psychological questions, 200
psychological reasons, 137
psychological succor, 149
psychologically, 9
psychologists, vi
psychology, 248, 251, 263
psychology, abnormal, 106
psychology, human, 194, 251
psychology, mob, 266
Ptolemic view, 15
public, 266
public relations, 266
publicity, 266
publishable, 178
pure energy, 99
pure state, 72
purpose, divine, 191
puzzlement, 177

qualities floating in the air, 187
qualities that are names, 188
quanta, virtual, 83
quantities, auxiliary, 240
quantitization, **98**, 143, 231
quantitization in classical physics, 144
quantitization, second, 98
quantity, mysterious, 135
quantize general relativity, 240
quantize gravity, viii
quantum electrodynamics, 81, 102, 208
quantum electrodynamics is fully consistent, 103
quantum entanglement, 183
quantum field theory, 1, **32**, 83
quantum field theory is a property of geometry, 1
quantum fluctuations, devastating, 253
quantum gravitational effects, 241
quantum jump, 76
quantum jumps, 118, 228
quantum leap, 76
quantum mechanical statefunction, 1
quantum mechanics, **32**, 198
quantum mechanics, alternative theory to, 158
quantum mechanics as a framework, 15
quantum mechanics, basic rules of, 221
quantum mechanics, errors in discussions of, 196
quantum mechanics is local, 204
quantum mechanics is weird, 221
quantum mechanics, nonexistent problems of, 245
quantum mechanics, opposite of what gives, 265
quantum mechanics, pain of, 110
quantum mechanics, peculiarities of, 67
quantum mechanics, postulates of, 25
quantum mechanics, rigorous formulations of, 11
quantum mechanics to classical mechanics, transition from, 74
quantum mechanics with excuses, 247
quantum mechanics, weirdness of, 223

quantum of action, 171
quantum potential, 204
quantum probability, 107
quantum states of the brain, 156
quantum teleportation, 173, 229
quantum theory of gravity, 240
quantum uncertainty, waves of, 164
quantum weirdness, viii, 197
quantum weirdness allows electrons to tunnel, 238
quantum weirdness, courses in, viii
quarks, 258
quaternion, 56
quaternions, 55
qubits, **144**
Quercigh, Emanuele, 258
question, **147**
question, correct, 148
question, deeply profound, 191
question, fundamental, 147
question, hard, 176
question has meaning, **176**
question, meaningful, 111
question, meaningless, 111
questions, incompatible, 117
questions, meaningful, **149**
questions, meaningless, **146, 149**
questions, mixing different, 113
questions, psychological, 200
quicksand, xi
Quigg, Chris, 253
quirky, 265

radiation, 42
radiation reaction, 208
radio, 4
radioactive, 73
radius, nonzero, 201
Rafelski, Johann, 258
random motion, 60
rapidity, 74
razor, Occam's, 198
reactions, chemical, 68
reader, 42, 180
real, **3**, 118
real number field, 55
real, physically, **8**
real world, 2, 177
realism, distant, 179
realism, faraway, 179
realism, local, 179
realistic description of nature, local, 179
realistic physical experiment, 80
realistic possible laws of physics, 21
reality, 3, **3**, 180, 187
reality, complete description of, 181
reality, corresponds to, **18**
reality, criteria for, **3**
reality, define, 178
reality, definition of, 4
reality, external, 131, 179
reality, factual, 179
reality independent, 178
reality is mathematical, 129
reality, model of an external, 131
reality, nonobjective, 177
reality, objective, **177**
reality, ordinary, 181
reality, physical, 118, 177, 187
reality, pictures of, **133**
reality, theory corresponds to, 18
reality, unpleasant, 194
realization, 29
really, 184
really are, 2
reasoning, chain of, 120
reasoning, circular, 136
reasons, historical, 81
reasons, psychological, 137
reduction of the wavefunction, 175
reduction, instantaneous state vector, 78
reduction, state vector, 76
reference systems, 35
referents, 187
reflection, total internal, 238
regularity conditions, 76, 171
reject, 20
related, 169
relativistic extension, 55
relativistic wave equations, 138
relief, emotional, 148
religion, 190, 191, 197, 255
religion, difference between science and, 19
religious people, 175

religious person, 247
Rempe, G., 182
rend space-time, 253
Rennie, John, 173
reporter, 265
reporter, naive, gullible, 266
representation, **44**
representation basis function, **15**
representation, basis functions of, single, 46
representation, basis state of a group, **46**
representation, defining, 39
representation, Fock, 85
representation labels, 139
representation, mass-zero spin-2, 41
representation, meaning of group, **46**
representation, phenomenological, **83**
representations of Abelian groups, 56
representations, linear group, 81
representations, massless, 39
representations, nonlinear, 84, 140
reproducibility, 150
reproducible, 244
reputation, v
requirement, key, 178
requirement on physical theories, fundamental, 177
resolving power, 115
resonance, 6
responsibility, 116
results, contradictory, 227
results, paradoxical, 80
retina, 66
retrodict, 31, 217
rewards, emotional, 153
rhetoric, 263
Ricci tensor, 241
richest people, 264
ridiculous, 223, 267
right place at the right time, xi
rigid rod, 36
rigor, 153
rigor in physics, 11
rigorous formulations of quantum mechanics, 11
rigorous manipulation of symbols, 115
rigorous mathematical proofs, 63
rigorous mathematically, highly, 37
rigorous systems, 104
Rimini, Alberto, 228
ripped apart, 254
ripples, 165
risks, x
robot, 129, 133
rocks, 106
rocks, ancient, 185
rod, rigid, 36
Rodgers, Peter, 224
Roman-Roy, Narciso, 231
roomful of people, 184
rooted, deeply, 41
rope, 204
ropes, 165
Rorschach test, 181
Rosen, N., 180, 208
rotation group, 47
rotation group, inhomogeneous, 38
rotons, 16
Rovelli, Carlo, 234
rules of quantum mechanics, basic, 221
rules, set of, 126
Rutherford, 206

Σ hyperon, 169
safest approach, x
sailing, 224
Santiago, David I., xiii
satisfaction, emotional, 105
satisfying, emotionally, 147
saucers, flying, 266
scalar equaled a vector, 43
scatter, 172
scattering, high-energy, 53
scattering of nucleons by nucleons, 20
Schaden, Martin, 256
scheme, approximation, 233, 252
Schewe, Phillip F., xiii
Schiaparelli, 143
Schiff, L. I., 98, 108
Schleich, Wolfgang, 214

Schottenloher, Martin, 13
Schrödinger's cat, vi, vii, 70, 74
Schrödinger's equation, time-dependent, **136**, 137
Schrödinger's equation, time-independent, **137**
Schweber, Silvan S., 33, 98
science and faith, 125
science and religion, difference between, 19
science fiction, 173
science, fundamental methods of modern, 138
science, history of, xi
science is a tool, 19
science, major tool of modern, 170
science, observational, ix
science, philosophy of, ix
science, process of, 146
science, standards of, v
science, supplement, 194
science writers, 142
scientific paper, correct, 183
scientific papers, 243
scientific papers, production of, 111
scientific theory, 14
scientist, fraudulent, manipulative, 266
screen, 202
screen, opaque, 93
Scully, Marlan O., 118, 142, 162, 217, 249
SE(2), 39
sea, empty, 224
sea of activity, 256
search engine, 197
second quantitization, 98
secrets of nature, privy to the, 265
seductive, xi, 20, 153, 174
see, 8, 120
see a table, never, 132
seems like nonsense, 267
seethes with particles and antiparticles, 264
Seife, Charles, 155
selection rules for angular momentum, 94
self-centered physicists, 235
self-contradictory, 115, 146, 193, 221, 222, 252
self-energy, 101, 168
self-important titles, 266
sells, 267
semi-direct sum, 39
semi-meaningful, 150
semisimple group, single irreducible representation of a compact, 49
sensation, 148
sensation of blueness, 176
sensational, 265
sensationalist, 265
sensationalist media, 266
sensationalize, 266
sensations, 122, 129
sensations, jumble of, 18
sense, 7
sense, makes, **5**
sense, physical, 153
senselessness, physical, 37
senses, 122, 125
senses, effect on our, 125
sensors, 129
sensory inputs, 125
sensory neurons, 11
separable, 152
sequence of letters, 143
sequences of letters, 104
set of coordinates, continuous, **6**
set, empty, 189
set of laws, mysterious, 54
set of letters, linear, 181
set of properties, 126
set of rules, 126
sets of statefunctions, 124
severe short-distance singularities, 207
sex and divorce stories, 266
she, he or, 176
sheet, 255
shell, classical, 209
Shephard, Gordon M., 189
Shiekh, Anwar Y., xiii
Shih, Yanhua, 162, 217
Shimony, Abner, 179
short-distance singularities, severe, 207
short-lived, 261

shorthand, 225
should happen, 4
sight, 134
signals, acausal, 210
silence, vii
sillier and sillier, 263
silliness, 259
silly, 99, 184
Silverman, M. P., 201
similar circumstances, 178
similar perception, 178
simple group, noncompact, 13
simplified models, 9
simultaneously different properties, 227
single basis vector, amplitude of a, **90**
single case, generalize from a, 138
single irreducible representation of a compact semisimple group, 49
single particle, **88**
single particle, entanglement for a, 214
single photon, **86**
single Poincaré representation, 89
single representation, basis functions of, 46
single systems, 117
single-valued, 171
singularities, 30, 44
singularities, severe short-distance, 207
singularity, terminate on a, 225
slits, 202
sloppiness, 196
sloppy thought, 197
small distance, 10
small distances, **11**
small, atoms are, 188
smell the wood, 132
Smolin, Lee, 242
smooth geometry of space-time, 253
smothers, 167
social consensus, 4
social construct, 197
society, v
Society, American Physical, 162
society, deceiving, vi
sociologists of science, vi
sociology, v, 263
solace, emotional, 194
solid vacuum, 258
Solomon, Dan, xiii
solvable, 39
soothing, 110
soul, 186
souls, 136
sound, 4, 12
space, 139
space and time, distinctions between, 31
space, curved, 29
space, definition of, 141
space, dual, 138, 140
space, empty, 254
space, Euclidean, 29
space, fields in empty, 181
space, flat, 29
space, Hilbert, 63
space, isotropy of, 213
space, lost in, 240
space, pseudo-Euclidean, **38**
space, symmetry of, 43
spacelike measurements, 85, 222
spacelike-separated measurements, correlations between, 171
spacetime fluctuation, 241
spacetime fluctuations, 226
spacetime foam, 253
space-time, little pieces of, 253
space-time, rend, 253
space-time, smooth geometry of, 253
spacetime structure, 241
space-translation symmetry, 59
special positions, 46
special relativity, **41**
specific theories, **16**
speculation, xi
speculative, ix
speed limit, 110
speed of light, 234
speed of light, faster than the, 220
spell, **195**
spherical harmonics, 44, 204
spin, half-integer, 82
spin-$\frac{1}{2}$, 55, 82

spin-$\frac{1}{3}$, 47
spirits, 106
spontaneous decays, 259
spontaneously pop out of nothingness, 263
spookiness, vii
spooky, 197, 265
spooky action-at-a-distance, 231
spooky action-at-a-distance of classical physics, 210
spooky confusion-at-a-distance, 215
spots of ink, 12
spout nonsense, 266
spout nonsense, licensed to, 197
spray, 255
spray with bubbles, 255
spread throughout, 161
springiness, 258
springs, 144
Spruch, Larry, 256
square of the amplitude, 61
square root of -1, 56
SQUID, viii, 74
standards, 197
standards, minimal of, 223
standards of science, v
star, inside a, 10
star, statefunction of an electron in, 119
stars, electrons in, 18
stars form, 193
stars, interior of, 184
startled, 192
state, **145**
state, actual physical, **184**, 212
state, definite physical, 182
state, energy, **45**
state, entangled, **216**
state, excited, 42
state, macroscopic, 68
state, multi-particle, **88**
state of a system, 121
state of the system, **143**
state, pure, 72
state vector reduction, 76
state vector reduction, instantaneous, 78
statefunction, **126**, **143**, 144, 198
statefunction, classical, 73
statefunction, gravitational, **241**
statefunction, meaning of, 158
statefunction of a fundamental object, 48
statefunction of an electron in a star, 119
statefunction of pointers, 72
statefunction of the pointer, 69
statefunction, quantum mechanical, 1
statefunctions, absolute square of, 60
statefunctions are analytic functions, 158
statefunctions, class of, 126
statefunctions, complete set for, 45
statefunctions, entangled, 229
statefunctions, expressions for, 109
statefunctions of baryons, 206
statefunctions of classical objects, 122
statefunctions, sets of, 124
statements, beguiling, 188
statements, negative, 178
statevector, **143**
statistical mechanics, 233
statistically, 221
statistics, Bose-Einstein, 53
statistics of fermions and bosons, 100
Steane, Andrew, xiii
stellar interior, 131
Stenger, Victor J., xiii
Stephani, Hans, 140
steps, all, 265
stick together, 157
stimulate, 12
Stodolsky, Leo, xiii, 64
Stoler, David, 166, 179, 221
storage capacity, immense, 134
stories, sex and divorce, 266
strange, 265
strange concepts, xi
strange ideas, 79
strange phenomena, 243
strangeness of nature, 196
Streater, R. F., 55
string of words, contentless, 112

string of words, empty, 147
strings of letters, 179
strings of words, 147, 217, 252
strong interactions, 206
strongly opposed, v
structure, internal, 53
structure, logical, 151
Strunz, Walter T., 224
Stuckey, Mark, xiii
student, elementary physics, 223, 239
students, 154, 208, 259
stupendous number, 250
SU(2), 48
SU(3), 48
subconscious feeling, 179
subconsciously, 171
subgroup, inhomogeneous, 13
subjective, 19, 20, 149
subjective feel, 129
subjective feeling, 187
subjective feelings, 177
subjective images, 130
subjective implications, 187
subjects, crackpot, v, 266
subscripts, 239
substitute numbers in equations, 154
success, criteria for, 240
succor, emotional, 194
succor, psychological, 149
sudden change of our knowledge, 118
sum of states, incoherent, 75
sum of the angles of a triangle, 35
sum, semi-direct, 39
superconductivity, 16, 17
superficial, 153
superfluous, 79
superGod, 190
supernova, 18
supernovas, 16
superposition, 212
superposition, coherent, 75
superposition, decay of a, 245
superposition, incoherent, 73
superposition, macroscopic quantum, 244
superposition of distinct space-time geometries, 255
superposition of states, coherent, 70
superposition principle, 81, 244
superpositions, coherent, 73
supersaturated vapor, 71
superstition, 81, 216, 242, 255, 261
superstition, independent, 260
supplement science, 194
support, 181
Sussman, George, 249
swallowed, 181
swing, 224
Switzerland, 211
symbol, 104
symbols, 265
symbols as abbreviations, 78
symbols, logical, 151
symbols, mathematical, 154
symbols, rigorous manipulation of, 115
symmetric, circularly, 213
symmetrical, 45
symmetry, **40**
symmetry of space, 43
symmetry, broken, 213
symmetry, space-translation, 59
symmetry, time-translation, 59
synapse, 133
synaptic cytoskeleton, 133
syndrome, Tourette's, 267
system, measured, 243
system, state of a, 121
systemfunction, 145
systems, coordinate, 35
systems, individual, 109
systems, number, 56
systems, procedure for preparing, 118
systems, reference, 35
systems, rigorous, 104
systems, single, 117

table, 3, **132**, 101, 223
table, never see a, 132
tacks, 224
tantalizing word, 251
taste, 8, 121, 158
tautology, 165
taxpayers, viii
TCP theorem, 55, 143

teachers, 221
teachers, physics, 154
technique, fundamental, 138
technological, 144
teeming, 254
Tegmark, Max, 156
teleportation, quantum, 173, 229
telescopes, 4
television, 4
Teller, Paul, 164
temperature, 9, 10, 233
temperature of the vacuum, 258
tendency, unfortunate, 37
terminate on a singularity, 225
terminology, 245
terminology, misleading, 77
terms, cross, 70
terms, meaningless, 187
test, Rorschach, 181
theorem, Bell's, 221
theorem, CPT, 55, 143
theorem, Gleason's, 63
theorem, PCT, 55, 143
theorem, TCP, 55, 143
theoretical physicists, 229, 243, 262
theories, alternative, 194, 245
theories, competing, 16
theories, deterministic, 246
theories, foundation of physical, 151
theories meaningful, physical, 125
theories, phenomenological, 33
theories, specific, **16**
theory, **16**
theory, assumptions of the, **16**
theory corresponds to reality, 18
theory is consistent, **102**
theory is true, **19**
theory is useful, 18
theory of evolution, 15
theory of the atom, Bohr's, 78
theory, phenomenological, **20**
theory, physical, **11**
theory, scientific, 14
there, 185
thermal motion, 203
thermometer, 9
think, 135, 186, 188
think, ways that we, 192
thinking, God's, vi
thinking, leaping without, 77
Thomas Bradwardine, 252
Thorne, Kip S., 258, 259
thought, sloppy, 197
thought, unexamined habits of, 4
thoughtlessness, 196
tides, 185
time, 10, **58**
time and space, distinctions between, 31
time dependent, coefficients that are, 77
time expands, 163
time, flow of, 164
time, history of, 163
time, nature of, 162
time, Newton's absolute, 164
time reversal, 220
time scale of neural impulses, 236
time, zero, 76
time-dependent Schrödinger's equation, **136**, 137
time-independent Schrödinger's equation, **137**
time-translation operator, **58, 136**
time-translation symmetry, 59
Tipler, Frank J., 174
titles, self-important, 266
tongue of nature, mother, 172
tool, 197
tool, calculational, 233
tool of modern science, major, 170
tools, 19
top of a building, 4
topics, unpleasant, 248
total energy, 58
total internal reflection, 238
totality, 8
totter, 9
touch, 3, 8, 120
Tourette's syndrome, 267
Toyama, F. M., 247
track in a cloud-chamber, 203
tracks, 131
transformation, 34, **40**
transformation group, **40**
transformation group of Euclidean 3-space, 47

transformation group of geometry, 13
transformation group of the geometry, 1
transformation parameters, 107
transformational mechanics, 143
transformations, consistency under, 43
transformations, general coordinate, 95
transformed by, **38**
transistor, 17
transition from quantum to classical mechanics, 74
translation generator, 56
transmission of information, 173
treacherous, 178
treatises, learned, 191
tree, 17, 26
trees, 106, 189
triangle, sum of the angles of a, 35
trick, 266
Trimble, Virginia, 174
true, 8, **17**
true, theory is, **19**
tunnel from nothing, 237
tunnel, quantum weirdness allows electrons to, 238
tunneling, 236
two or more places, 224
types of nothingness, many, 182

ultimate meaning, **7**
unanswerable, 192
unbelievable, 265
uncertainty distance, minimum, 11
uncertainty principle, **108, 112,** 161, 253, 257, 261
uncertainty principle, Heisenberg, 227
uncertainty principle, violates the, 221
uncertainty, waves of quantum, 164
uncomfortable, 110, 149, 193
unconscious, 111
unconsciously, 187
uncorrelated, decays are, 106
underlying absurdity, 20
underlying assumptions, 20
underlying fabric, 254
underlying philosophy, 34
understand, **18**
understand nature, 17
unethical, vi
unexamined, 77
unexamined habits of thought, 4
unfathomable consciousness perceptions, 176
unfortunate name, 76
unfortunate tendency, 37
unglued, 157
unhappy, 110, 120, 148, 149, 159
unhelpful, **11**
unimaginable, x
unit cells, 16
unitary, 58, 68
units, arbitrary choice of, 235
universe, beginning of the, 192
universe, birth of the, 250
universe, consistent physical, 27
universe, design of the, 229, 243
universe, every particle in the, 229
universe is designed, 191
universe is governed by laws, 263
universe is purely mathematical, 130
universe, meaning of the, 190
universe, mysteries of the, 153
universe, newborn, 164
universe, nontrivial, 22
universe, probability for the, 111
universe, proper, 21
universe to exist, 125
universe, wavefunction of the, 111, 160, 237
universes, ensembles of, 112
universes, many, 112
universes, multitude of, 5, 174, 182
unknowable, 23
unlawful motion, 22
unlearn, 208
unpleasant reality, 194
unpleasant topics, 248
unreadable, 153
unsatisfying, emotionally, 148, 176
untruthful, vi
unwary, xi
useful description, 74
useful, 5, **17**

useful, theory is, 18
useless, 16, 17

Vaccaro, John, 214
vacuity, 191
vacuity of sentences and phrases, 14
vacuous, 43
vacuum, 183, 251
vacuum, dread idea of the, 253
vacuum field, fluctuating, 260
vacuum fluctuations, 259
vacuum has energy, 257
vacuum, melt the, 258
vacuum, nature abhors a, 261
vacuum of space, 264
vacuum, solid, 258
vacuum, temperature of the, 258
vacuum, zero-temperature, 257, 258
vague feeling, 187
vagueness, 7, 172
validation, 152
value, definite predetermined, 209
value, emotional, 194
value, heuristic, 20, 233
values, 195
values, actual physical, 226
van der Waals force, 256
van Dijk, W., 247
van Kampen, N. G., vi, 65, 70
vanish, 255
vanity, 228
vapor, supersaturated, 71
Varadarajan, V. S., 63
variable coupling constant, 34
variables, conjugate, 108
variables, hidden, 246
Varshalovich, D. A., 215
vary wildly, phases, 244
vast wealth, 264
Vatsya, Raj, xiii
vector potential, 97
vector potential, electromagnetic, 199
velocity, collapse of, 65
verbiage, meaningless, 167, 175, 186
vibrations in air, 12
Victoria-Monge, Carles, 231
view of nature, coherent, 7
view, orthodox, 170
view, philosophical, 179
violates the uncertainty principle, 221
violent fluctuations, 253, 263
virtual particle, **7**
virtual particles, 252, 260
virtual quanta, 83
vision, 26
visual cortex, 176
visualize, 204
visualized, 172
void, 252
void, emotional, 191
von Baeyer, Hans Christian, 165, 255
Voss, David, 254
vote, majority, 118, 238
vote, popular, 197
votes, 266

wall, ball hitting a, 65, 79
Wallace, Philip R., 146
Walther, Herbert, 118, 142, 249
want, 189
warm feeling, nice, 154
water, 165, 188, 204, 255
water, waves in, 84
wave, 201, **204**
wave equations, relativistic, 138
wave, classical light, 156
wave, electromagnetic, 204
wavefunction, 1, 143, 144, 198
wavefunction, collapse of the, 64, 160
wavefunction of the universe, 111, 160, 237
wavefunction, reduction of the, 175
waveguides, 144
wavelength, deBroglie, 233
wavepacket, 52, **91**
wave-particle duality, 97, 201, 205
waves, 144
waves in water, 84
waves of quantum uncertainty, 164
waves, what?, 144
way, magic, 209
ways of knowing, other, 194

ways that we think, 192
we are not alone, 246
wealth, vast, 264
wear, 19
web, complex, 262
web of concepts, 8, 10
web of interconnected observations, 131
week, 223
Weinberg, Steven, 238, 261
weird, v, vi, 80, 196, 265
weird, quantum mechanics is, 221
weirdness, vii
weirdness, classical, 238
weirdness of quantum mechanics, 223
weirdness, quantum, viii, 197
Weiss, Peter, 182, 257, 258
well of energy, 257
well, infinitely deep potential, 201
well-behaved, 171
well-hidden meanings, 111
Weng, Gezhi, 133
what waves?, 144
Wheeler, James T., xiii
Wheeler, John A., 258, 259
which-way, 217
Whitaker, Andrew, 170, 221
white, 224
white, European, male, 189
why probability must be conserved, 58
Wightman, A. S., 55
Wigner, E. P., 240
Wigner, Eugene, 27
Wilczek, Frank, 165, 167, 169
wild explanation, 156
wildest, xi
wildly, phases vary, 244
Wilford, John Noble, 163
wink in and out, 256
wires, 7
wiser, xi
wishful thinking, x
wishing, xi
wood, 132, 255
wood, smell the, 132
word, empty, 192
word, tantalizing, 251
word, wrong, 245
words are defined, 146
words, conglomerations of, 191
words, contentless string of, 112
words, empty string of, 147
words, esoteric, 153
words, strings of, 147, 217
workable astronomy, 4
works, everything, 189
world exists, **125**
world, external, xii, 18, 118
world, knowledge of the, 129
world, real, 177
world view, 229
worm, 26, 133, 134
worms, 226
worse than nonsense, 153
worshipped, 265
writer, 180
writers, science, 142
writing, poor, 197
wrong word, 245

Yam, Philip, 264
Yu, Rong, 162, 217
Yurke, Bernard, 166, 179, 221

Zanghi, Nino, 214
Zeh, H. D., 76, 199, 201
Zeh, H. Dieter, xiii
Zeilinger, Anton, 54, 173
zero point energy, 257
zero time, 76
zero-temperature vacuum, 257, 258
Zhou, Fei, 256
Zoller, Peter, 224
Zurek, Wojciech H., viii

AUTHORS GUILD BACKINPRINT.COM EDITIONS are fiction and nonfiction works that were originally brought to the reading public by established United States publishers but have fallen out of print. The economics of traditional publishing methods force tens of thousands of works out of print each year, eventually claiming many, if not most, award-winning and one-time best-selling titles. With improvements in print-on-demand technology, authors and their estates, in cooperation with the Authors Guild, are making some of these works available again to readers in quality paperback editions. Authors Guild Backinprint.com Editions may be found at nearly all online bookstores and are also available from traditional booksellers. For further information or to purchase any Backinprint.com title please visit www.backinprint.com.

Except as noted on their copyright pages, Authors Guild Backinprint.com Editions are presented in their original form. Some authors have chosen to revise or update their works with new information. The Authors Guild is not the editor or publisher of these works and is not responsible for any of the content of these editions.

THE AUTHORS GUILD is the nation's largest society of published book authors. Since 1912 it has been the leading writers' advocate for fair compensation, effective copyright protection, and free expression. Further information is available at www.authorsguild.org.

Please direct inquiries about the Authors Guild and Backinprint.com Editions to the Authors Guild offices in New York City, or e-mail staff@backinprint.com.

978-0-595-33690-6
0-595-33690-6

Printed in the United States
110077LV00004B/76/A

9 780595 336906